THEORY OF SUPERCONDUCTIVITY

THEORY OF SUPERCONDUCTIVITY

M. Crisan
Institute of Physics and Technology of Materials, Bucharest
and
Department of Physics, University of Cluj, Romania

World Scientific
Singapore • New Jersey • London • Hong Kong

Published by

World Scientific Publishing Co. Pte. Ltd.,
P O Box 128, Farrer Road, Singapore 9128
USA office: 687 Hartwell Street, Teaneck, NJ 07666
UK office: 73 Lynton Mead, Totteridge, London N20 8DH

Library of Congress Cataloging-in-Publication Data

Crisan, M. (Mircea)
Theory of superconductivity/M. Crisan.
p. cm.
ISBN 997150569X
ISBN 9971509970 (pbk)
1. Superconductivity. I. Title.
QC611.92.C75 1989
537.6'23--dc20 89-16480
CIP

Printed in Singapore by Utopia Press.

PREFACE

Superconductivity was discovered in 1911 by Kamerling Onnes and at that time, it was difficult to realise the importance of this phenomenon. In 1934, Gorter and Casimir proposed the first model — the two-fluid model — for the superconducting state and in 1935, F. London and H. London developed the electrodynamics of superconductors.

Ginzburg and Landau used in 1950 the Landau theory of phase transitions to treat the transition from the normal to the superconducting state. The method was successfully applied by Abrikosov in 1958 to predict the vortex structure in superconductors.

The first microscopic model was developed by Bardeen, Cooper and Schrieffer between 1956 and 1958 and had as its main point the occurrence of electron pairs due to the attractive electron-electron interaction mediated by virtual phonons. In 1958, Gor'kov showed that the model proposed by Bardeen, Cooper and Schrieffer can be treated within the framework of the Green function method. In the same period, the important contributions to the theory of superconductivity were made by Bogoliubov and Anderson. The role of the electron-phonon interaction in the occurrence of the superconducting state was elucidated by Eliashberg in 1960 and Scalapino, Wilkins and Schrieffer in 1961. In 1962, Josephson predicted additional tunnelling current when both sides of a junction are superconductors.

Starting from 1960, Matthias traced the occurrence of superconductivity in a large number of alloys and compounds containing magnetic impurities. During that period the main problem was the discovery of materials as well as microscopic mechanisms for the coexistence of superconductivity with magnetic order. The phenomenological ideas were transformed between 1961 and 1966 in more and more sophisticated models, using the many-body methods of Abrikosov, Gor'kov, Fulde, de Gennes and Maki to develop the theory of superconducting alloys. The experimental results obtained since 1973 by Maple and Fischer demonstrate clearly the coexistence between superconductivity and magnetic order in some ternary compounds with a special crystalline structure. In another class of rare-earth compounds, the narrow band of the "f"-electrons present superconducting pairing called "heavy fermion superconductivity". The theory of coexistence between superconductivity and magnetic order was developed beginning in 1978 by Matsubara, Maekawa, Machida, Tachiki, Keller, Fulde, Levin *et al.*

In 1986, Bendorz and Müller showed the possibility of obtaining the superconducting state in all oxides and in 1987 many groups from the USA, the Soviet Union, Japan, W. Germany, England and France announced the discovery of high temperature superconductivity.

Even before this discovery there was great interest in possible practical applications of superconductivity, but in future, the technology will be drastically affected by this discovery. For the time being, the theoretical models are elaborated by reconsidering the electron-phonon interaction, but till now there is no decisive answer concerning the mechanism which is responsible for high temperature superconductivity.

All these problems will be treated in this small book in a unified way to assure an equilibrium between the physical ideas and the mathematical formalism which is necessary in the treatment of this difficult subject.

A prerequisite to reading this book is some familiarity with solid state physics, quantum and statistical mechanics and the theory of many-body systems. I have not tried to incorporate such special problems as the theory of superconducting slabs, or the theory of nonstationary behaviour of superconductors because there are many books on these subjects. The other class of problems which has not been treated is the unsolved problems as to the critical behaviour of superconductors.

Last but not least, I would like to thank Dr. Ulf Lindström for his help.

I hope that this book will help solid-state theorists to approach the existing treatments of superconductivity and solid-state experimentalists to become acquainted with theory of superconductivity.

M. Crisan

CONTENTS

THEORY OF SUPERCONDUCTIVITY

I
PHENOMENOLOGICAL THEORY OF SUPERCONDUCTIVITY

1. Experimental Facts

Infinite conductivity: When any of a large class of crystalline or amorphous elements and compounds is cooled, the electrical resistivity disappears at a definite critical temperature T_c. In the first approximation, the transition is not accompanied by any change in structure or property of the crystal lattice and has been interpreted as an *electronic transition.*

If we assume the usual Ohm's law describing the superconducting state

$$\mathbf{j} = \sigma \mathbf{E}$$

and the Maxwell equation

$$\frac{\partial \mathbf{B}}{\partial t} = -\nabla \times \mathbf{E}$$

for $\rho = 0, \mathbf{E} = 0$ and $\mathbf{B}$ remains constant for such a medium.

Meissner effect: The infinite conductivity is one of the most important characteristics of a new state. However, the true nature of the superconducting state appears more clearly in an external magnetic field.

Let us consider a normal metal in a uniform magnetic field: when the sample is cooled and becomes superconducting, experiments performed by

Meissner and Ochenfeld demonstrated that all the magnetic flux was expelled to the exterior. This indicates that $\mathbf{B}$ = constant, and it is in fact zero. In a multiply connected sample, as a ring, the holes trap the magnetic flux.

Critical field: The Meissner effect occurs only for sufficiently low magnetic fields. For simplicity, we consider a long cylinder of a pure superconductor in a parallel applied field H. If the sample is superconducting at temperature T in zero field, there is a unique critical field $H_c(T)$ above which the sample becomes normal. This field is temperature dependent and the empirical equation which describes well this dependence is

$$H_c(T) = H_{c0}\left[1 - \left(\frac{T}{T_c}\right)^2\right] .$$

Persistent currents and flux quantization: A different case of magnetic behaviour is connected to the flux trapping in a superconductor ring.

Suppose a normal metallic ring is placed in a magnetic field perpendicular on its plane. When the temperature is lowered, the metal becomes superconducting and the flux is expelled. If the external field is removed, no flux passes through the superconducting metal and the trapped flux must remain constant. This flux is maintained by the circulating supercurrent in the ring itself. The flux trapped in sufficiently thick rings is quantized in units of $\Phi_0 = \pi/e$.

Specific heat: The superconducting materials also have distinctive thermal properties. In the superconducting state, the specific heat C_s initially exceeds the specific heat of the normal state C_n and

$$C_s(T) \propto \exp\left(-\frac{\Delta}{T}\right) .$$

This dependence indicates the existence of a gap in the energy spectrum separating the excited states from the ground states by the energy Δ.

Isotopic effect: The transition temperature varies with the ionic mass M,

$$T_c \propto M^{-\frac{1}{2}}$$

and that demonstrates the importance of the ionic lattice in superconductivity.

Normal tunnelling: The conduction electrons in a superconductor and a normal metal can be brought into thermal equilibrium with one another by placing the metals into such close contact that they are separated by a thin insulating layer, which the electrons can cross by quantum tunnelling. When both metals are in the normal state, application of a potential difference raises the chemical potential of one metal with respect to the other and further electrons tunnel through the insulating layer. However, when one of the metals becomes superconducting then no current is observed to flow until the potential V reaches the value $eV = \Delta$. The size of Δ is in good agreement with the value inferred from low-temperature specific heat measurements.

Frequency dependent electromagnetic behaviour: The response of a metal to electromagnetic field is determined by the frequency dependent conductivity, which depends on the available mechanisms for energy absorption by the conduction electrons at the given frequency. Because the electronic excitation spectrum in the superconducting state is characterized by an energy gap Δ, one would expect the AC conductivity to differ substantially from its normal state form at small frequencies compared with Δ, and to be the same in the superconducting and normal states at large frequencies compared with Δ. Except that near the critical temperature, Δ is in the range between microwave and infrared frequencies. In the superconducting state, the AC behaviour does not differ from that in the normal state at optical frequencies.

2. Gorter-Casimir Two-Fluid Model

The first attempts to apply thermodynamics to the superconducting phase have been made by Rutgers and Ehrenferst in 1933.

The discovery of the Meissner effect finally enabled Gorter and Casimir (1934) to develop a thermodynamic treatment of the transition from the normal to the superconducting state using the standard theory of the phase transition with two supplementary assumptions:

a) The system exhibiting superconductivity possesses an ordered or condensed state, the total energy of which is characterized by an order parameter. This order parameter is generally taken to vary from zero at $T = T_c$ to unity at $T = 0$ K, and thus it can be taken to indicate that fraction of the total system which finds itself in the superconducting state.

b) The entropy of the system is due to the disorder of noncondensed individual excited particles which behave as the particles in the normal state.

In particular, the two-fluid models make the conceptual assumption that in the superconducting state a fraction n_s of the conduction electrons are "superconducting" electrons condensed in an ordered phase, while the other fraction $n_n = n - n_s$ remains "normal". The free energy of the superconductor is given by

$$F_S = \frac{n'}{n} F_n + \frac{n_s}{n} F_s \tag{2.1}$$

where $n = n_n + n_s$, $n' = (n_n n)^{\frac{1}{2}}$, F_n is the free energy of the normal electrons and F_s is the free energy of the superconducting electrons.
Using the new variables

$$x = \frac{n_n}{n} \qquad 1 - x = \frac{n_s}{n} . \tag{2.2}$$

Equation (2.1) becomes

$$F_S = x^{\frac{1}{2}} F_n + (1-x) F_s . \tag{2.3}$$

The equilibrium state at a fixed temperature can be obtained from the condition

$$\left(\frac{\partial F_S}{\partial x}\right)_T = 0 \tag{2.4}$$

which gives

$$x = \frac{1}{4}\left(\frac{F_n}{F_s}\right)^2 \tag{2.5}$$

and the free energy (2.3) can be written as

$$F_S = F_s + \frac{1}{4}\frac{F_n^2}{F_s} . \tag{2.6}$$

In an external magnetic field H, the relation between F_n and F_s is

$$F_n - F_s = \frac{H_c^2}{2\mu_0} \tag{2.7}$$

and if $x = 1$, $F_S = F_n$, then all the electrons are in the normal state. At low temperatures

$$F_n = \frac{1}{2}\gamma T^2 \tag{2.8}$$

where γ is the Sommerfeld constant.

At $T = 0$, the electrons are in the superconducting state $H_c = H_{c0}$ and from (2.8), we get $F_n(T = 0) = 0$. The free energy of the superconducting electrons is

$$F_s(T = 0) = -\frac{H_{c0}^2}{2\mu_0} . \tag{2.9}$$

Using (2.6) and (2.9), the fraction x defined by (2.5) becomes

$$x = \frac{\mu_0^2 \gamma^2}{4H_{c0}^4} T^4 \tag{2.10}$$

and for $T = T_c$

$$x = \left(\frac{T}{T_c}\right)^4, \qquad \gamma = \frac{2H_{c0}^2}{\mu_0 T_0^2}. \tag{2.11}$$

The free energies for the metal in the normal and superconducting states are

$$F_n = -\frac{H_{c0}^2}{\mu_0}\left(\frac{T}{T_c}\right)^2, \qquad F_s = -\frac{H_{c0}^2}{2\mu_0}\left(1 + \frac{T^4}{T_c^4}\right) \tag{2.12}$$

and from Eq. (2.7), the critical field H_c can be calculated as

$$H_c(T) = H_{c0}\left(1 - \frac{T^2}{T_c^2}\right). \tag{2.13}$$

With these results, we can calculate the difference between the electronic specific heats in the superconducting and normal states and we obtain

$$\Delta C_e = C_{es} - C_{en} = \gamma T_c\left[3\left(\frac{T}{T_c}\right)^3 - \left(\frac{T}{T_c}\right)\right]. \tag{2.14}$$

If ΔC_e is finite, the temperature dependence of C_{es} can be approximated as

$$C_{es} \cong \gamma T^3. \tag{2.15}$$

The two-fluid model has a restricted applicability because the exponential was not obtained. The behaviour in the magnetic field as the Meissner effect cannot be explained in the framework of this model. Finally, the contribution at the normal phase of the $x^{\frac{1}{2}}$ fraction of electrons cannot be explained.

3. Electrodynamics of Superconductors

Perfect diamagnetism

Even in the absence of a microscopic explanation of the phenomenon of superconductivity, it is useful and reasonable to assume that the vanishing of the magnetic induction in a superconductor is due to induced surface currents. In an external magnetic field, the distribution of these currents is just to create an opposing interior field cancelling the applied one.

We can then give an image of the macroscopic superconductor in an external field:

a) in interior: $\mathbf{B} = \mathbf{H} = \mathbf{M} = 0$
b) at the surface: $\mathbf{j}_a \neq 0$ ($\mathbf{j}_a$ is the surface current density)
c) outside: $\mathbf{B} = \mathbf{H} + \mathbf{H}_a$ ($\mathbf{H}_a$ is the field due to the surface current).
Let us consider the surface of the superconductor. From the equation

$$\nabla \times \mathbf{B} = 4\pi \mathbf{j}_a \ ,$$

we get $\nabla \cdot \mathbf{B} = 0$ because in the bulk superconductor $\mathbf{j}_a = 0$. Taking the path 1-2-3-4 (Fig. 1) now we consider

$$\oint \mathbf{B} \cdot d\mathbf{l} = 4\pi I \ .$$

Using now

$$\oint \mathbf{B} \cdot d\mathbf{l} = \int_1^2 Bdl + \int_2^3 Bdl + \int_3^4 Bdl + \int_4^1 Bdl = Bl_{12} \ ,$$

we get

$$Hl_{12} = 4\pi j_a l_{12} \ ,$$

an equation which can be generalized as

$$\mathbf{j}_a = \frac{1}{4\pi} \mathbf{n} \times \mathbf{H}$$

where $\mathbf{n}$ is the unit vector normal to the surface.

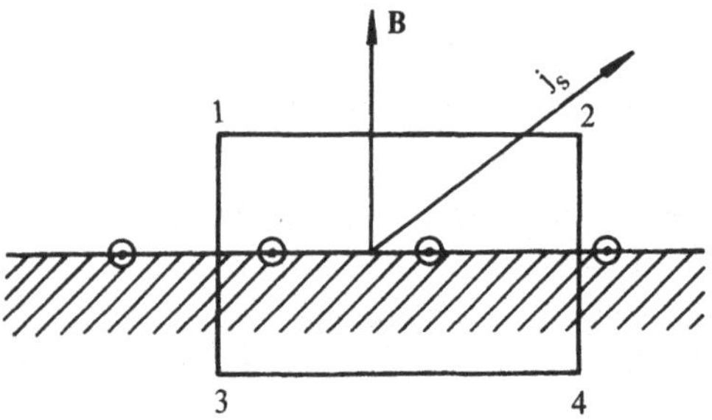

Fig. 1. The magnetic field $\mathbf{B}$ and the current $\mathbf{j}_s$ at the surface of a superconductor.

The intermediate state

The influence of the magnetic field on the superconductor is very sensitive to the geometry of the sample. If one considers an ellipsoidal superconducting sample in an applied magnetic field **H** (**H** is considered to have the direction of the major axis of the sample), the internal field $\mathbf{H}_i$ is

$$\mathbf{H}_i = \mathbf{H} - D\mathbf{M} \tag{3.1}$$

where D is the demagnetization factor and **M** the magnetization of the superconductor.

The magnetic induction is

$$\mathbf{B} = \mathbf{H}_i + 4\pi\mathbf{M}$$

which gives, for the magnetization,

$$\mathbf{M} = -\frac{1}{4\pi}\mathbf{H}_i \,. \tag{3.2}$$

From (3.1) and (3.2), we get

$$\mathbf{H}_i = \frac{4\pi}{4\pi - D}\mathbf{H} \tag{3.3}$$

which shows that $\mathbf{H}_i$ is dependent on the geometrical form of the superconducting sample and in the direction of the external field. The magnetic field in the exterior of the ellipsoidal sample $\mathbf{H}_e$ is different from the applied field, the field lines being deformed due to the magnetic properties and to the shape of the sample. Because $\nabla \cdot \mathbf{B} = 0$ and in the interior of the sample $\mathbf{B} = 0$, at the surface of the sample, the tangential components of the magnetic field must be continuous $H_i^t = H_e^t$ so that, projecting (3.3) on the tangent plane to the surface, we get

$$H_e^t = \frac{4\pi}{4\pi - D} H \sin\theta$$

where θ is the angle between **H** and the normal to the surface. For a spherical sample $D = 4\pi/3$ and $H_e^t = \frac{3}{2} H \sin\theta$, so that at the equator $(\theta = \pi/2)$, the effective field acting on the superconducting sphere has its maximum value $3/2\ H$ and its minimum one $H_e^t = 0$ is reached at the poles $(\theta = 0)$, see Fig. 2a.

If $H = H_c$ at the equator, the effective field overcomes H_c and penetrates the interior of the superconducting sphere to a certain depth on

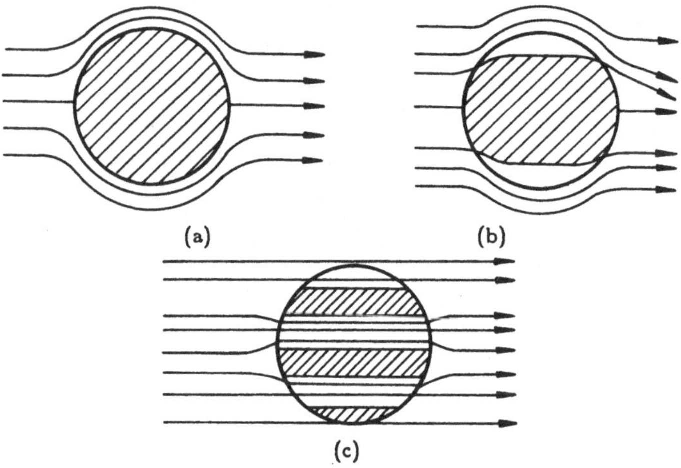

Figs. 2. The penetration of the magnetic field in a spherical sample.

which the sample becomes normal. The remaining superconducting region may be equivalent to an ellipsoid for which D is smaller and the effective field becomes smaller than the critical field. This leads to the dividing of the sample into normal and superconducting regions like in Fig. 2c. Using thermodynamics, one can show that this state, called *intermediate state*, is more favourable energetically than the one presented in Fig. 2b.

London equations

We consider a simple model in which the total number of electrons is: $n = n_s + n_n$, n_s being the number of superconducting electrons, n_n the number of normal electrons and $n_s(T_c) = 0$. The current due to the superconducting electrons can be easily calculated. Indeed, from the motion equation we have

$$n_s m \frac{d\mathbf{v}_s}{dt} = n_s e \mathbf{E}$$

where m is the electron mass, $\mathbf{v}_s$ the velocity associated with superconducting electrons. If we define

$$\mathbf{j}_s = e n_s \mathbf{v}_s \ ,$$

the electric field is

$$\mathbf{E} = \frac{d}{dt}(\Lambda \mathbf{j}_s)$$

with

$$\Lambda = \frac{m}{n_s e^2} \ .$$

If $\partial \mathbf{j}_s/\partial t = 0$, $\mathbf{E} = 0$. If we change the chemical potential of the superconducting electrons, $\mathbf{j}_s \neq 0$. This can be done by the superconducting electron slab.

Let us consider that in a superconductor in each point $\mathbf{r}$, we have, the magnetic field $\mathbf{H}(\mathbf{r})$. The density of the energy is:

$$W_c = n_s \frac{mv_s^2}{2} = \frac{mj_s^2}{2n_s e^2} .$$

Using now the equation

$$\nabla \times \mathbf{H} = 4\pi \mathbf{j}_s , \tag{3.4}$$

we get

$$W_c = \frac{\lambda^2}{8\pi}(\nabla \times H)^2$$

where

$$\lambda^2 = \frac{m}{4\pi n_s e^2} .$$

The total energy will be:

$$F_{sH} = F_{s0} + \frac{1}{8\pi}\int dV[H^2 + \lambda^2(\nabla \times H)^2]$$

and we look for the minimum value $\delta F_{sH} = 0$. We have,

$$\delta F_{sH} = \frac{1}{8\pi}\int dV[2\mathbf{H}\cdot\delta\mathbf{H} + 2\lambda^2(\nabla \times \delta\mathbf{H})\cdot(\nabla \times \mathbf{H})] .$$

Using

$$\mathbf{a}\cdot(\nabla \times \mathbf{b}) = \mathbf{b}\cdot(\nabla \times \mathbf{a}) - \nabla\cdot(\mathbf{a}\times\mathbf{b}) ,$$

we get

$$\int dV[H^2 + \lambda^2(\nabla \times (\nabla \times \mathbf{H}))]\cdot\delta\mathbf{H} - \int dV[\nabla\cdot((\nabla \times \mathbf{H})\times\delta\mathbf{H})] = 0 .$$

The second integral can be calculated using the Gauss theorem as

$$\int dV[\nabla\cdot((\nabla \times \mathbf{H})\times\delta\mathbf{H})] \to \oint d\mathbf{s}\cdot((\nabla \times \mathbf{H})\times\delta\mathbf{H})$$

and we get the general equation

$$\mathbf{H} + \lambda^2\nabla \times (\nabla \times \mathbf{H}) = 0 . \tag{3.5}$$

Using now $\mathbf{H} = \nabla \times \mathbf{A}$ and Eq. (3.4) we get

$$\mathbf{j}_s = -\frac{1}{4\pi\lambda^2}\mathbf{A}$$

if $\nabla \cdot \mathbf{A} = 0$ and $\mathbf{A} \cdot \mathbf{n} = 0$. This relation can be re-written as

$$\begin{aligned} \mathbf{j}_s &= -\frac{1}{\Lambda}\mathbf{A} \\ \Lambda &= 4\pi\lambda^2 \; . \end{aligned} \tag{3.6}$$

Let us consider Eq. (3.5) for the half space defined by $x > 0$. If the magnetic field is parallel to the axis Oz using

$$\nabla \times (\nabla \times \mathbf{H}) = -\nabla^2\mathbf{H} \quad \text{and} \quad \nabla \cdot \mathbf{H} = 0$$

we get

$$\frac{d^2H(x)}{dx^2} - \frac{1}{\lambda^2}H(x) = 0 \tag{3.7}$$

with $H_0 = H(0)$ and $H(\infty) = 0$, conditions which are consistent with the Meissner effect. The solution of this equation is

$$H(x) = H(0)\exp(-x/\lambda) \tag{3.8}$$

where $\lambda = m/(4\pi n_s e^2)$ is called the London penetration depth. A similar equation can be obtained for $\dot{\mathbf{H}} = \partial\mathbf{H}/\partial t$ which shows that for $x \gg \lambda$, $\dot{\mathbf{H}} \to 0$, then in a superconductor, the magnetic field cannot change in time from the value it had when the sample became perfectly conducting.

The London equations have been generalized by Pippard, considering that the wave functions of the superconducting electrons are not perfectly rigid. Thus the electrons which interact have a finite coherence length ξ_0 and the electromagnetic properties of the superconductor will be modified if $\mathbf{A}(\mathbf{r})$ varies rapidly at these distances.

In this case, the current at a certain point is determined by the value of the electromagnetic potential $\mathbf{A}(\mathbf{r})$ at the same point. Namely, the relation between the current and the electromagnetic field is a local one expressed by the equation

$$\mathbf{j}_s(\mathbf{r}) = -\frac{1}{\Lambda}\mathbf{A}(\mathbf{r}) \tag{3.9}$$

which is equivalent to the London equation (3.6) with certain conditions imposed on $\mathbf{A}(\mathbf{r})$.

For the superconductors in which the field (or $\mathbf{A}$) varies rapidly ($\xi_0 \gg \lambda$) the current at a certain point $\mathbf{r}$ is determined by the values of the electromagnetic field in the whole region surrounding that point. Now, in order to calculate the current, we have to perform the average of $\mathbf{A}(\mathbf{r})$ on this region. On the other hand, experiments have shown that the penetration depth λ is determined by the length of the mean free path l of the electrons in the normal metal. This parameter depends on the presence of the impurities and λ increases with the decreasing of l, namely, with the increasing of the concentration of impurities. The presence of impurities affects the superconducting state and Pippard proposed for the coherence length the equation

$$\frac{1}{\xi} = \frac{1}{\xi_0} + \frac{1}{l} \tag{3.10}$$

where ξ_0 is the coherence length in the pure superconductor. We have to note that the presence of non-magnetic impurities does not change the critical temperature. This assumption, known as the Anderson theorem, can be demonstrated in the microscopic theory. On these considerations Pippard uses, instead of the local Eq. (3.9), the new equation

$$\mathbf{j}_s = -\frac{3}{4\pi\xi_0\Lambda}\int d\mathbf{r}' \cdot \frac{\mathbf{R}(\mathbf{R}\cdot\mathbf{A}(\mathbf{r}))}{R^4} e^{-R/\xi} \tag{3.11}$$

where $\mathbf{R} = \mathbf{r} - \mathbf{r}'$ and the constant in front of the integral is chosen in such a way that we should obtain for $\lambda \gg \xi$ and $l \gg \xi_0$ (pure superconductor) the London equation (3.9). Indeed if $\lambda \gg \xi$, $\mathbf{A}$ varies slowly for $r \cong \xi$, where it may be considered constant and from (3.11), we get

$$\mathbf{j}_s(\mathbf{r}) = -\frac{1}{\Lambda}\frac{\xi}{\xi_0}\mathbf{A}(\mathbf{r}) \,. \tag{3.12}$$

This equation defines a penetration depth

$$\lambda_{\mathrm{p}} = \lambda\left(\frac{\xi}{\xi_0}\right)^{\frac{1}{2}} \tag{3.13}$$

and we can see that for $l \gg \xi_0$, $\xi = \xi_0$ and $\lambda_{\mathrm{p}} = \lambda$, Eq. (3.12) becomes identical to (3.9). If $l \ll \xi_0$ then $\xi = l$ and

$$\lambda_{\mathrm{p}} = \lambda\left(\frac{\xi_0}{l}\right)^{\frac{1}{2}} \,. \tag{3.14}$$

This theory defines two limits in the electrodynamics of superconductors, namely

$\xi \ll \lambda_{\mathrm{p}}$ London limit

and

$$\xi \gg \lambda_p \qquad \text{Pippard limit} .$$

Quantum effects versus classical treatment

If we consider that the superconducting state is a quantum state described by a wave function

$$\psi(\mathbf{r}) = (n_s/2)^{\frac{1}{2}} \exp(i\theta(r)) \tag{3.15}$$

and this wave function describes a particle of mass $2m$ and charge $2e$, the operator $\hat{p}$ will be defined as

$$\nabla\theta(\mathbf{r}) = 2m\mathbf{v}_s + 2e\mathbf{A} \tag{3.16}$$

and in the absence of a magnetic field, the density of the current is $n_s\mathbf{v}_s/2$, or using the following expression and (3.15)

$$\mathbf{j} = \frac{i}{4m}(\psi\nabla\psi^* - \psi^*\nabla\psi) ,$$

we get the result

$$\nabla\theta(\mathbf{r}) = 2m\mathbf{v}_s$$

and the superconducting current becomes

$$\mathbf{j}_s = n_s e\mathbf{v}_s = \frac{1}{\Lambda}\Big(\frac{\Phi_0}{2\pi}\nabla\theta(\mathbf{r}) - \mathbf{A}(\mathbf{r})\Big) .$$

If we consider now a cylinder which may become superconductor in an external magnetic field and if we take C contour from a surface at a distance $r \gg \lambda$, then $\mathbf{j}_s = 0$ and

$$\oint \mathbf{j}_s \cdot d\mathbf{l} = \frac{\Phi_0}{2\pi}\oint \nabla\theta \cdot d\mathbf{l} .$$

If we define

$$\Phi = \oint_c \mathbf{A} \cdot d\mathbf{l}$$

then we have

$$\Phi = \frac{\Phi_0}{2\pi}\oint_c \nabla\theta(r) \cdot d\mathbf{l} ,$$

where Φ is the total magnetic flux over the area enclosed by the contour C. We have to mention that $\theta(\mathbf{r})$ is a multiform function. However, we have to define $\psi(r)$ by (3.15) and if $\theta(\mathbf{r}) = 2\pi n$, and ψ is invariant (this is a gauge transform)

$$\oint_c \nabla\theta(\mathbf{r}) \cdot d\mathbf{l} = 2\pi n$$

and Φ becomes

$$\Phi = n\Phi_0$$

or $\Phi_0 = \pi/|e|$ in the unities system with $\hbar = c = 1$.

4. Ginzburg-Landau Theory of Superconductivity

General theory

We will consider a system of electrons in a superconducting state. The wave function Ψ of the electrons is defined by

$$\Psi(\mathbf{r}) = \left(\frac{n_s}{2}\right)^{\frac{1}{2}} \exp i\theta(\mathbf{r}) \tag{4.1}$$

and

$$|\Psi(\mathbf{r})|^2 = 1$$

appears as an *order parameter.*

In the absence of the magnetic field, the free energy density is written as

$$F_s = F_n + \alpha(T)|\Psi|^2 + \frac{\beta}{2}|\Psi|^4 \tag{4.2}$$

where F_n is the density of the free energy for the normal state, $\alpha(T)$ and $\beta(T)$ are temperature dependent coefficients. From the condition

$$\frac{\partial F_s}{\partial|\Psi|} = 0\ ,$$

we get

$$|\Psi|^2 = -\frac{\alpha(T)}{\beta(T)} \tag{4.3}$$

and

$$F_n - F_s = \frac{\alpha^2(T)}{2\beta(T)}$$

and in the presence of the magnetic field

$$F_n - F_s = \frac{H_c^2}{8\pi}\ . \tag{4.4}$$

The coefficient $\alpha(T)$ will be written as

$$\alpha(T) = a|T_c - T| \quad \text{and} \quad \beta(T) \cong \beta(T_c) \, .$$

Equation (4.4) defines the *thermodynamical critical field.*

For $T < T_c, \alpha < 0$ and for $T > T_c, \alpha > 0$, these conditions will assure the minimum of (4.2). However, for the superconductors, the magnetic field is important and we will consider the free energy density taking into consideration the magnetic field.

The Ginzburg-Landau equations

Near the critical temperature, the density of the free energy of a superconductor in a magnetic field is (see Ref. 5)

$$F_{sH} = F_n + \alpha(T)|\Psi|^2 + \frac{\beta}{2}|\Psi|^4 + \frac{1}{4m}|(-i\nabla - 2e\,\mathbf{A})\Psi|^2 \tag{4.5}$$

where $\mathbf{A}$ is the potential vector of the electromagnetic field. From (4.5), we write the free energy

$$\mathcal{F}_{sH} = \mathcal{F}_n + \int dV \Big[\alpha|\Psi|^2 + \frac{\beta}{2}|\Psi|^4 + \frac{1}{4m}|(-i\nabla - 2e\,\mathbf{A})\Psi|^2 + \frac{1}{8\pi}(\nabla \times \mathbf{A})^2 - \mathbf{H} \cdot \frac{\nabla \times \mathbf{A}}{4\pi}$$

which has to be minimum. If we consider $\Psi(\mathbf{r})$ and $\mathbf{A}(\mathbf{r})$ as independent variables, we have

$$\delta_{\psi^*} \mathcal{F}_{sH} = 0 \tag{4.6}$$

$$\delta_{\mathbf{A}} \mathcal{F}_{sH} = 0 \tag{4.7}$$

and from (4.6), we get

$$\alpha|\Psi| + \beta\psi|\Psi|^2 + \frac{1}{4m}(i\nabla + 2e\,\mathbf{A})\Psi = 0 \, , \tag{4.8}$$

$$(i\nabla\psi + 2e\,\mathbf{A}\psi) \cdot \mathbf{n} = 0 \tag{4.9}$$

where $\mathbf{n}$ is the unitary vector perpendicular to the superconducting surface. Let us calculate $\delta_{\mathbf{A}} \mathcal{F}_{sH}$. Performing the variation on $\mathbf{A}$, we get the expression for the superconducting current

$$\mathbf{j}_s = -\frac{i}{2m}(\Psi^* \nabla \Psi - \psi \nabla \Psi^*) - \frac{2e^2}{m}|\Psi|^2 \mathbf{A}(\mathbf{r}) \, . \tag{4.10}$$

If we change the variables as

$$\psi(\mathbf{r}) = \frac{\Psi(\mathbf{r})}{\Phi_0}, \qquad \Phi_0^2 = \frac{n_s}{2} = \frac{\alpha(T)}{\beta}$$
$$\xi^2(T) = \frac{1}{4m\alpha(T)}, \quad \lambda^2 = \frac{m}{4m_s e^2} = \frac{m\beta}{8\pi e^2 \alpha(T)} \tag{4.11}$$

then Eqs. (4.8) and (4.10) can be re-written as

$$\xi^2 \left(i\nabla\psi(\mathbf{r}) + \frac{2\pi}{\Phi_0}\mathbf{A}\right)^2 \psi(\mathbf{r}) - \psi(\mathbf{r}) + \psi(\mathbf{r})|\psi(\mathbf{r})|^2 = 0 \tag{4.12}$$

$$\nabla \times (\nabla \times \mathbf{A}(\mathbf{r})) = -\frac{i\Phi_0}{4\pi\lambda^2}(\psi^*\nabla\psi - \psi\nabla\psi^*) - \frac{|\psi|^2}{\lambda^2}\mathbf{A}(\mathbf{r})\ . \tag{4.13}$$

Using $\psi = |\psi|e^{i\theta(r)}$, Eq. (4.13) becomes

$$\nabla \times (\nabla \times \mathbf{A}(\mathbf{r})) = \frac{|\psi|^2}{\lambda^2}\left(\frac{\Phi_0}{8\pi}\nabla\theta(\mathbf{r}) - \mathbf{A}(\mathbf{r})\right)\ .$$

Here Eq. (4.9) is a boundary condition for (4.12). From (4.9), we can see that this condition has a simple physical meaning: the supercurrent passing through the surface of separation of a superconductor and an insulator is zero.

Using the microscopic theory, we may show that at the metal-normal superconductor contact, we have

$$\left(i\nabla + \frac{2\pi}{\Phi_0}\mathbf{A}\right)\mathbf{n}\psi = ia\psi$$

where a is a real constant.

Another important physical fact which can be observed from these equations is the *gauge invariance*.

Then we have

$$\mathbf{A} = \mathbf{A}' + \nabla\varphi(\mathbf{r})$$

where $\varphi(\mathbf{r})$ is a scalar function and we have

$$H = \nabla \times \mathbf{A} = \nabla \times \mathbf{A}'$$

because $\nabla \times \nabla\varphi(\mathbf{r}) = 0$.
If from ψ, we pass to ψ' defined by

$$\psi \equiv \psi' \exp\left[i\frac{2\pi}{\Phi_0}\varphi(\mathbf{r})\right]\ ,$$

we get from (4.13)

$$\nabla \times (\nabla \times \mathbf{A}') = \frac{|\psi'|^2}{\lambda^2}\left(\frac{\Phi_0}{2\pi}\nabla\theta' - A'\right) .$$

The characteristic length

Let us consider (4.12) for $\mathbf{A} = 0$, and we have

$$-\xi^2 \frac{d^2\psi}{dx^2} - \psi + \psi^3 = 0 \qquad (4.14)$$

for the one dimensional case. In the interior of the superconductor $|\psi| = 1$. At the surface, $\psi(x)$ will be written as

$$\psi(x) = 1 - f(x) .$$

From (4.14), we get

$$\xi^2 \frac{d^2 f(x)}{dx^2} - 2f(x) = 0 \qquad (4.15)$$

with the condition: $f(x) \to 0, x \to \infty$.
The solution of Eq. (4.15) is

$$f(x) = f(0) \exp\left[-\sqrt{2}\frac{x}{\xi}\right] \qquad (4.16)$$

where ξ is the characteristic length for the variation of $\psi(x)$.
From (4.11) and $\alpha(T) \sim |T_c - T|$, we get for $\lambda(T)$ and $\xi(T)$

$$\lambda(T) = \lambda(0)|T - T_c|^{\frac{1}{2}} , \quad \xi(T) = \xi(0)|T_c - T|^{\frac{1}{2}} . \qquad (4.17)$$

If we define now the *Ginzburg-Landau parameter* χ as

$$\chi(T) = \frac{\lambda(T)}{\xi(T)} , \qquad (4.18)$$

from (4.11), we get

$$\chi(T) = 2\sqrt{2}\, e\lambda^2 H_c(T) \qquad (4.19a)$$

or

$$\sqrt{2}\, H_c(T) = \frac{\phi_0}{2\pi\lambda(T)\xi(T)} \qquad (4.19b)$$

where $\phi_0 = \pi/|e|$.

As a function of this parameter, we have a possibility to define two classes of superconducting materials:

Type I — superconductors with $\chi < 1/\sqrt{2}$

Type II — superconductors with $\chi > 1/\sqrt{2}$.

If we consider now the variation of $\psi(x)$ in a magnetic field from (4.8), we get

$$\frac{1}{4m}(-i\nabla - 2e\,\mathbf{A})\psi = -\alpha\psi \tag{4.20}$$

where the nonlinear term has been neglected. In the xy-plane

$$-\alpha = \frac{2eH}{m}\left(n + \frac{1}{2}\right) + \frac{1}{2}mv_z^2$$

and the maximum value of H_{c2} which appears as a minimum eigenvalue of (4.20) is

$$-\alpha = \frac{eH}{m} \,.$$

This field denoted by H_{c2} is called the upper critical field and using $\alpha^2/2\beta = H_c^2/8\pi$ and (4.19) we get

$$H_{c2} = \chi(T)\sqrt{2}\,H_c(T) \,.$$

Superconductor-normal metal contact

If we consider the contact between a normal metal (N) and a superconducting one (S), the superconductivity can be induced in the normal metal by the extension of the $\psi(\mathbf{r})$, at the separation surface $\psi(\mathbf{r}) < 1$. Let us consider two metals, one in the normal state which may become superconductor at $T < T_{\rm cn}$ and a superconductor with the critical temperature $T_{\rm cs}$. The surface between two metals will be defined by the equation $x = 0$. The superconductor will be in the region $x > 0$ and the normal in the region $x < 0$. Equation (4.14) can be written as

$$-\xi^2\left(\frac{d\psi}{dx}\right)^2 - \psi^2 + \frac{1}{2}|\psi|^3 = C \tag{4.21}$$

where C is a constant. If we take for $C = -\frac{1}{2}$, from (4.21), we get

$$\psi(x) = \tanh\frac{x - x_0}{\xi\sqrt{2}} \tag{4.22}$$

where x_0 is a constant which can be determined from the behaviour of $\psi(x)$ at $x = 0$.

The boundary condition in this case will be written as

$$\frac{1}{\psi}\frac{d\psi}{dx} = \frac{1}{b} \tag{4.23}$$

where b has to be calculated from the microscopic theory. From (4.22) and (4.23), we get

$$-\sinh\frac{\sqrt{2}\,x_0}{\xi} = \frac{\sqrt{2}}{\xi}b$$

which is a relation between x_0 and b.

Let us consider the behaviour of the $\psi(x)$ in the normal region $x < 0$. Using also the Ginzburg-Landau approximation, the coefficient $\alpha_n(T) = |T - T_{\rm cn}|$ will be $\alpha < 0$ for $T < T_{\rm cn}$ and $\alpha > 0$ for $T > T_{\rm cn}$. In this case, Eq. (4.21) becomes

$$-\xi_{\rm n}'^2\frac{d^2\psi}{dx^2} + \psi + |\psi|^3 = 0\,. \tag{4.24}$$

In the normal state, $\psi \ll 1$ and (4.24) becomes

$$-\xi_{\rm n}^2\frac{d^2\psi}{dx^2} + \psi = 0 \tag{4.25}$$

an equation which has the solution

$$\psi(x) = \psi(0)\exp\left(-\frac{|x|}{\xi_{\rm n}}\right). \tag{4.26}$$

If $T_{\rm cn}$ and $T_{\rm cs}$ do not differ very much, we can consider ψ and $d\psi/dx$ continuous at the NS-border, and in (4.26) $b = \xi_{\rm n}$. In a pure normal metal, the mean path of the free electrons is greater than $\xi_{\rm n}(l \gg \xi_{\rm n})$ and the coherence length is

$$\xi_{\rm n} = \frac{v_{0\rm n}}{2\pi T}\,.$$

In a dirty normal metal, $l \ll \xi_{\rm n}$, the coherence length will be

$$\xi_{\rm n} = \left(\frac{v_0 l}{6\pi T}\right)^{\frac{1}{2}}.$$

The behaviour of ψ as function of x is given in Fig. 3.

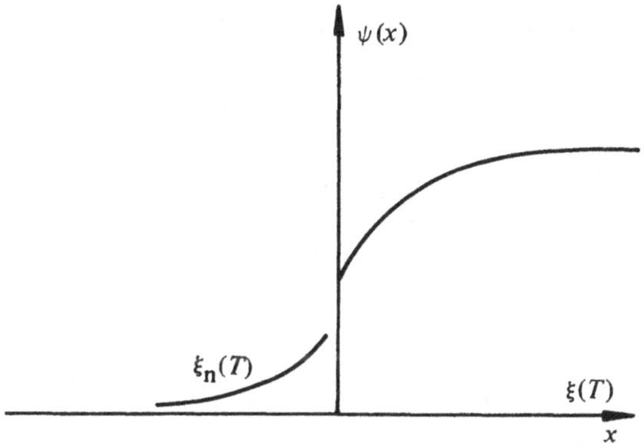

Fig. 3. The order parameter as function of x at the superconductor-normal contact.

Energy of the surface separation between normal and superconducting phases

Let us consider the NS contact and the surface of separation will be taken perpendicular on the x-axis and the magnetic field parallel with the z-axis.

The origin of the x-axis will be considered in the domain of the surface separation. Then we have

$$\mathbf{H} = (0, 0, H(x))$$
$$\mathbf{A} = (0, A(x), 0)$$
$$\psi = \psi(x) \, .$$

Then Eqs. (4.12) and (4.13) become

$$-\xi^2 \frac{d^2\psi}{dx^2} + \left(\frac{2\pi\xi}{\Phi_0}\right)^2 A^2\psi - \psi + |\psi|^3 = 0 \, ; \quad \frac{d^2A}{dx^2} = \frac{|\psi|^2}{\lambda^2} A(x) \, . \tag{4.27}$$

A first integral of the first equation is

$$\left[1 - \left(\frac{2\pi\xi_0}{\Phi_0}\right)^2\right]\psi^2 - \frac{1}{2}\psi^4 \left(\frac{2\pi\lambda}{\Phi_0}\right)\left(\frac{dA}{dx}\right)^2 + \xi^2\left(\frac{dA}{dx}\right)^2 = C \tag{4.28}$$

where C is a constant, which will be determined by the conditions

$$x \to \infty, \quad \psi(x) \to 1, \quad \frac{d\psi}{dx} \to 0, \quad A(x) \to 0 \, .$$

From these conditions and from (4.19b), we get $C = \frac{1}{2}$ and (4.28) becomes

$$\left[\left(\frac{2\pi\xi}{\Phi_0} A\right)^2 - 1\right]\psi^2 + \frac{1}{2}\psi^4 = \xi^2\left(\frac{d\psi}{dx}\right)^2 + \frac{H^2}{H_c^2} - \frac{1}{2} \, . \tag{4.29}$$

At the surface separation, the density for the Gibbs potential is

$$G_s = F_{s0} - \frac{HH_c}{8\pi} \tag{4.30}$$

and at long distance from the separation surface $G_s = F_{s0}$, where F_{s0} is the density of the free energy of a superconductor, and

$$F = F_{n0} + \frac{H_c^2}{8\pi} \,. \tag{4.31}$$

Thus for a normal metal, G_n is given by

$$\begin{aligned} G_n &= F - \frac{HH_c}{4\pi} = F_{n0} + \frac{H_c^2}{8\pi} - \frac{H_c^2}{4\pi} \\ &= F_{n0} - \frac{H_c^2}{8\pi} = F_{s0} \end{aligned} \tag{4.32}$$

where we used the conditions that in the normal state $H = H_c$ and

$$F_n - F_{s0} = \frac{H_c^2}{8\pi} \,.$$

Equation (4.32) is in fact the justification of the approximation which has been assumed.

In order to see what happened at the separation contact, we consider the geometry given in Fig. 4.

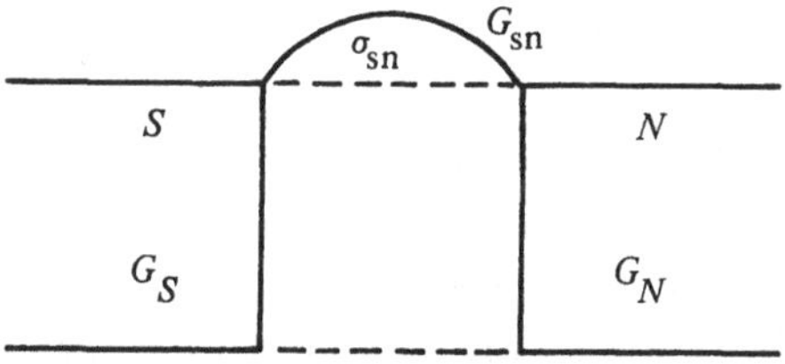

Fig. 4. Energy of the surface separation between normal and superconductor.

The surface energy will be written as

$$\sigma_{ns} = \int_{-\infty}^{\infty} dx (G_{sH} - G_n) \tag{4.33}$$

where

$$G_{sH} = F_{sH} - \frac{HH_c}{8\pi}$$

with

$$F_{sH} = F_n + \frac{H_c^2}{4\pi}\Big[-|\psi|^2 + \frac{1}{2}|\psi|^4 + \xi^2\Big|\Big(i\nabla + \frac{2\pi}{\Phi_0}\mathbf{A}\Big)\psi\Big|^2\Big] \tag{4.34}$$

and

$$G_n = F_n - \frac{H_c^2}{8\pi} .$$

Using (4.34), the surface energy (4.33) is

$$\sigma_{ns} = \int_{-\infty}^{\infty} dx\Big\{\frac{H_c^2}{8\pi}\Big[-|\psi|^2 + \frac{1}{2}|\psi|^4 + \xi^2\Big|\Big(i\nabla + \frac{2\pi}{\Phi_0}\mathbf{A}\Big)\psi\Big|^2\Big] + \frac{H^2}{8\pi} - \frac{HH_c}{4\pi} + \frac{H_c^2}{8\pi}\Big\}$$

and if $\mathbf{A} \equiv (0, A, 0)$, we get

$$\sigma_{ns} = \int_{-\infty}^{\infty} dx\Big\{\frac{H_c^2}{4\pi}\Big[-|\psi|^2 + \frac{1}{2}|\psi|^4 + \xi^2\Big(\frac{d\psi}{dx}\Big)^2 + \Big(\frac{2\pi\xi A}{\Phi_0}\Big)^2\psi^2\Big] + \frac{H^2}{8\pi} - \frac{HH_c}{4\pi} + \frac{H_c^2}{8\pi}\Big\}$$

which together with (4.29) gives

$$\sigma_{ns} = \frac{H_c^2}{2\pi}\int_{-\infty}^{\infty} dx\Big[\xi^2\Big(\frac{d\psi}{dx}\Big)^2 + \frac{H(H-H_c)}{2H_c^2}\Big] . \tag{4.35}$$

As $H_c \gg H$, the second term is always negative and in the London theory where the first term is zero we have $\sigma_{ns} < 0$. In the domain of separation $\xi(d\psi/dx) \sim 1$, $\xi^2(d^2\psi/dx^2) \sim 1$ and

$$\int_{-1}^{1} dx\xi^2\Big(\frac{dH}{dx}\Big)^2 \sim \xi .$$

In this domain, $H(H-H_c)/2H_c^2 \sim -1$ and this integral is of the order of $-\lambda$.

Let us consider now the two limits:

a) $\chi \ll 1, \lambda \ll \xi$. Then from (4.35) we get and an exact calculation gives

$$\sigma_{ns} \sim H_c^2\xi .$$

b) $\chi \gg 1$, $\lambda \gg \xi$. In this case, the most important contribution in (4.35) is given by $H(H-H_c)/2H_c^2$ and the energy becomes

$$\sigma_{ns} \sim -H_c^2\lambda$$

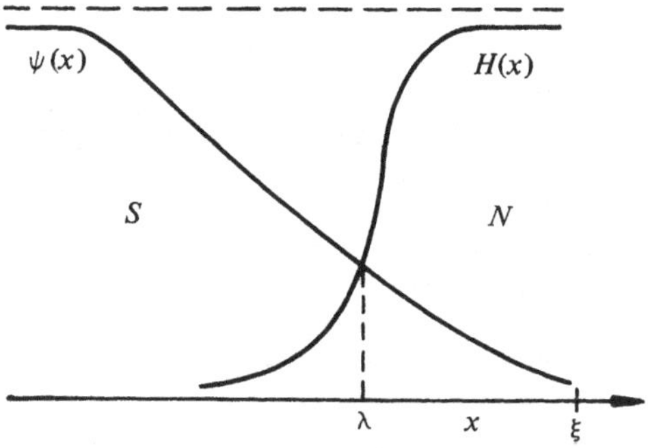

$\chi \ll 1$

(a)

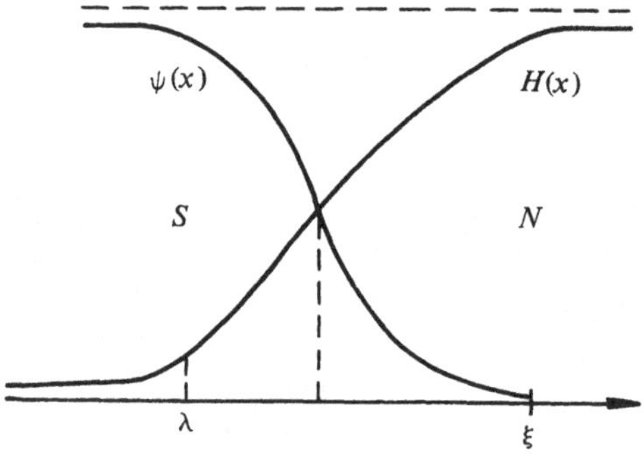

$\chi \gg 1$

(b)

Figs. 5. Spatial dependence of the order parameter $\psi(x)$ and the magnetic field $H(x)$ at the superconductor-normal interface.

and an exact calculation gives

$$\sigma_{\rm ns} = -\frac{H_{\rm c}^2}{8\pi}\lambda \ .$$

The physical aspects of these two results are shown in Figs. 5a and 5b.

In case a) $\psi(x)$ has an important variation at the distance ξ but the magnetic field changes at the distance λ. The energy of the magnetic field is $H_{\rm cn}^2/8$ and $\sigma_{\rm ns} > 0$.

In case b) $\psi(x) \sim 1$ on the domain λ and has a small variation at the distances of the order of ξ. The magnetic field presents a strong variation at the distance of the order of λ and $\sigma_{\rm ns} = -H_{\rm cn}^2/8\pi < 0$.

With these results, we can give a more accurate classification of the superconductors:

Type I	Type II
$\chi < \dfrac{1}{\sqrt{2}}$	$\chi > \dfrac{1}{\sqrt{2}}$
$\sigma_{ns} > 0$	$\sigma_{ns} < 0$.

5. Josephson Tunnelling

In 1962, Josephson[4] predicted a tunnelling current when both sides of a junction are superconductors. This current is due to the direct passage of coherent pairs from one side of the barrier to the other. The new effect called the Josephson effect can be understood using the wave function for the pair

$$\psi = \sqrt{n}\,\exp[i\theta(\mathbf{r})]\ . \tag{5.1}$$

To understand this, we simply consider two pieces of superconducting materials. The phase $\theta(\mathbf{r})$ of the pairs is conjugate to their total number, the absolute value of the phases in either of the pieces is arbitrary but their relative values are fixed. This is equivalent to the statement that all pairs have the same momentum. The common pair phase in one superconductor is independent of that in the other, since one can also change the number of pairs in one without affecting the other.

The Josephson current can be calculated in a simple way using the Schrödinger equation for the wave function (5.1) in case of two pieces of superconducting metals. These two equations are

$$i\frac{\partial\psi_1}{\partial t} = (E_1\psi_1 + B\psi_2) \tag{5.2a}$$

$$i\frac{\partial\psi_2}{\partial t} = (E_2\psi_2 + B\psi_1) \tag{5.2b}$$

where ψ_1 and ψ_2, and E_1, E_2 are the wave functions and the energies of the pairs in the superconductor (1) and respectively (2), B being the coupling energy due to the extension of the wave function through the barrier.

From (5.1) and (5.2), we get

$$\frac{\partial n_1}{\partial t} = 2B\sqrt{n_1 n_2}\sin(\theta_1 - \theta_2) = -\frac{\partial n_2}{\partial t} \tag{5.3}$$

$$n_1\frac{\partial\theta_1}{\partial t} - n_2\frac{\partial\theta_2}{\partial t} = (E_2 n_2 - E_1 n_1)\ . \tag{5.4}$$

Equation (5.3) expresses the conservation of the number of pairs in the total system. For $n_1 = n_2 = n$ we have

$$\frac{\partial n}{\partial t} = 2B\sin(\theta_2 - \theta_1) \tag{5.5}$$

$$\frac{\partial \theta_1}{\partial t} - \frac{\partial \theta_2}{\partial t} = (E_2 - E_1)\ . \tag{5.6}$$

The current through the junction defined as $J = 2e\frac{\partial n}{\partial t}\frac{V}{S}$ where V is the volume and S is the junction area, is

$$J = 4eB\frac{N}{S}\sin(\theta_1 - \theta_2) = J_0\sin(\theta_1 - \theta_2) \tag{5.7}$$

where J_0 can be calculated using the microscopic theory. The occurrence of the current J on a junction with zero bias voltage across the junction is called direct Josephson effect. If we apply on the junction a voltage V_0, then the difference $E_2 - E_1 = 2eV_0$ and from (5.6) we obtain

$$\theta_1 - \theta_2 = 2eV_0t + \theta_0 \tag{5.8}$$

where θ_0 is the initial phase. Using (5.8) in the equation for the current (5.7), we get

$$J = J_0\sin[(2eV_0)t + \theta_0] \equiv J_0\sin(\omega_0t + \theta_0)\ . \tag{5.9}$$

We have to note that under a constant voltage V_0, on the junction, the current becomes alternative with the frequency $\omega_0 = 2eV_0$. We can see that a Josephson junction can be considered the simplest type of oscillator.

If we apply a variable voltage besides the constant one, this also leads to some interesting phenomena. So, if the applied voltage on the junction has the form $V = V_0 + v\cos\omega t$, from (5.6) we obtain

$$\theta_1 - \theta_2 = \omega_0t + \theta_0 + \frac{2ev}{\omega}\sin\omega t \tag{5.10}$$

and from (5.7), we get

$$\begin{aligned} J &= J_0\sin\left[\omega_0t + \theta_0 + \frac{v}{\omega}\sin\omega t\right] \\ &= J_0\sum_{n=0}^{\infty} J_n\left(\frac{2eV}{\omega}\right)\sin[(n\omega + \omega_0)t + \theta_0] \end{aligned} \tag{5.11}$$

where $J_n(x)$ is the Bessel function of order n.

The Josephson effect is a fundamental quantum effect at macroscopic scale. Even more striking in this respect are the experiments which demonstrate the phase coherence. As a consequence, one can obtain a "superconducting interferometer" (there is a precise analogy with the optical interferometer), the phase and the current of the pairs are changed by an external magnetic field. Using the same model, the current in the case of the quantum interference has been calculated as

$$J(H) = J(0)\left|\frac{\sin \pi\Phi/\Phi_0}{\pi\Phi/\Phi_0}\right| \tag{5.12}$$

where $\Phi = LHd$ and $\Phi_0 = \pi/e$, L being the dimension of the junctions, and d the distance between the two junctions in a magnetic field H, parallel to Oz.

6. Influence of the Fluctuations

The phenomenological Ginzburg-Landau theory developed in Sec. 4 has as an important assumption that the wave function for the pair (the order parameter) depends only on the temperature. If we take

$$\psi \equiv \psi(\mathbf{r}, t) \tag{6.1}$$

the free energy functional (4.5) describes the fluctuations in the superconducting state due to the spatial and temporal variation of ψ.

In the simplest approximation, the time dependence of $\psi(\mathbf{r}, t)$ will be described by the Langevin equation

$$\begin{aligned}\frac{\partial\psi(\mathbf{r}, t)}{\partial t} &= -\gamma\frac{\delta G}{\delta\psi} + F(\mathbf{r}, t) \\ &= -\gamma\left[\alpha + \frac{1}{4m}(-i\nabla + 2e\,\mathbf{A})^2\psi(\mathbf{r}, t)\right] + F(\mathbf{r}, t)\end{aligned} \tag{6.2}$$

where γ^{-1} is the relaxation time and $F(\mathbf{r}, t)$ the random forces. In the absence of the magnetic field, the Gibbs potential will be approximated by

$$G_0 = \alpha\psi^2 \tag{6.3}$$

and the time evolution of the wave function which describes the fluctuations is

$$\frac{\partial\psi}{\partial t} = -\gamma\frac{\delta G}{\delta\psi} = -\gamma\alpha\psi\,. \tag{6.4}$$

In order to calculate the partition function Z of the fluctuation in an external magnetic field, we will use the simple form

$$G = \int d^3\mathbf{r} \left[\alpha|\psi|^2 - \frac{1}{4m}\psi^*(-i\nabla + 2e\,\mathbf{A})\psi \right] \tag{6.5}$$

and for the wave function

$$\psi(\mathbf{r}) = \sum_n \psi_n \Phi_n(\mathbf{r}) \tag{6.6}$$

where $\Phi_n(\mathbf{r})$ are the eigenfunctions of the equation

$$-\frac{1}{4m}(\nabla + 2ie\,\mathbf{A})^2\Phi_n = E_n\Phi_n \ . \tag{6.7}$$

The total Gibbs potential is

$$G = \sum_n |\psi_n|^2(\alpha + E_n) \ . \tag{6.8}$$

In order to calculate the partition function

$$Z = \sum_{\{\psi\}} \exp\left[-\frac{1}{T}G_{\mathrm{f}} \right] , \tag{6.9}$$

we perform the substitution

$$\psi_n = x_n + iy_n$$

and (6.9) becomes

$$Z = \prod_n \left[\int dx_n dy_n \exp -\frac{1}{T}(x_n^2 + y_n^2)(\alpha + E_n) \right] \ . \tag{6.10}$$

If we perform the integral in (6.10), the Gibbs potential

$$G_{\mathrm{f}} = -T\ln Z$$

will be given by

$$G_{\mathrm{f}} = -T\sum_n \ln\frac{\pi T}{(\alpha + E_n)} \ . \tag{6.11}$$

The thermodynamic quantities can be calculated if one defines the average values on the fluctuation states ψ. This average will be defined for a physical quantity X by the relation

$$\langle X\rangle = \frac{1}{Z}\sum_{\{\psi\}}\left\{\exp -\frac{1}{T}G_f(\psi)X(\psi)\right\} . \tag{6.12}$$

In this way, we get

$$\begin{aligned}\langle\psi_n\rangle &= 0\\ \langle\psi_n^*\psi_{n'}\rangle &= \delta_{n,n'}T/(\alpha(T)+E_n)\end{aligned} \tag{6.13}$$

and we will consider $E_{n+1} \geq E_n \geq 0$.
For $T > T_c, \alpha(T) > 0$ and $\alpha(T) + E_n > 0$. At T_c, the coefficient $\alpha(T)$ changes sign and there is a temperature T_c which satisfies

$$\alpha(T_c^*) + E_0 = 0 , \tag{6.14}$$

and at this temperature the Gibbs potential and the correlation function of the order parameter are singular. The most important contribution in these divergences is given by $\alpha(T), \beta(T)$ being approximately constant and $\beta(T) \simeq \beta(T_c)$.

From (6.12), the contribution to the divergencies can be obtained as

$$G_d = T\sum_n \ln[\alpha(T) + E_n] . \tag{6.15}$$

Using for the wave function $\psi(\mathbf{r})$ a plane wave of the form $V^{-\frac{1}{2}}\exp(i\mathbf{k}\cdot\mathbf{r})$ and taking for E_n the kinetic energy $E(\mathbf{k}) = k^2/2m$, the correlation function (6.13) is

$$\langle\psi_{\mathbf{k}}^*\psi_{\mathbf{k}'}\rangle = \delta_{\mathbf{k},\mathbf{k}'}T/[\alpha(T) + k^2/2m] \tag{6.16}$$

which diverges for $T = T_c$. The Gibbs potential (6.15) becomes

$$G_d = T_c\sum_{\mathbf{k}} \ln[\alpha(T) + k^2/4m] \tag{6.17}$$

and the specific heat due to the fluctuations can be calculated from (6.17) as

$$C = T_c^2\frac{1}{(2\pi)^3}\int d^3\mathbf{k}\frac{(d\alpha(T)/dT)^2}{[\alpha(T)+k^2/4m]^2} . \tag{6.18}$$

If we perform the integral in (6.18)

$$C_{d=3} = \frac{1}{\xi_0^3}\frac{1}{\sqrt{\tau}} \tag{6.19}$$

where $\tau = |T - T_c|/T_c$ and the coefficient $\alpha(T)$ has been expressed using

$$\xi(T) = \xi_0|4m\alpha(T)|^{-\frac{1}{2}} = \xi_0\tau^{-\frac{1}{2}} \,. \tag{6.20}$$

The divergences in the thermodynamic quantities are essentially determined by the dimension of the system. In a two-dimensional system, we get

$$C_{d=2} = \frac{\tau^{-1}}{4\pi a(\xi_0)^2} \tag{6.21}$$

where a is a constant. For one-dimensional systems, the divergence is even stronger, and in a similar way we get for the specific heat

$$C_{d=1} = \frac{1}{4b\xi_0}\tau^{-\frac{3}{2}} \tag{6.22}$$

where b is a constant.

If we analyse the behaviour of the fluctuations in an electric field one can calculate, following the standard methods, the contribution of these fluctuations on the conductivity. In the linear response approximation, the conductivity $\sigma(\omega)$ has been obtained as

$$\sigma(\omega) = \frac{e^2}{m^2}\frac{1}{V}\sum_{\mathbf{q}}\langle\psi_{\mathbf{q}}^*\psi_{\mathbf{q}}\rangle\left[\frac{q_x^2}{\alpha + q^2/4m - i\gamma} - \frac{q_x^2}{\alpha + q^2/4m + i\gamma}\right] \tag{6.23}$$

which will be considered in the static limit $\omega \to 0$ and we get

$$\sigma(0) = \frac{e^2}{2m\gamma}\frac{1}{V}\sum_{\mathbf{q}}\frac{q_x^2}{(\alpha(T) + q^2/4m)^3} \,. \tag{6.24}$$

The summation over $\mathbf{q}$ can be transformed to an integral which is again divergent in different dimensions as it follows

$$\begin{aligned} \sigma_{d=3} &= e^2(32\xi_0)^{-1}\tau^{-\frac{1}{2}} \\ \sigma_{d=2} &= e^2(16a)^{-1}\tau^{-1} \\ \sigma_{d=3} &= e^2(16b/(\pi\xi_0))\tau^{-\frac{3}{2}} \,. \end{aligned} \tag{6.25}$$

All these results give a qualitative description of the critical behaviour of the superconductors. At the calculation of the conductivity in the microscopic theory one also has to consider the Maki diagrams.

On the other hand, until now the Renormalization Group theory fails the general expectation in solving this problem.

7. Type-II Superconductors

General properties

We have obtained an important result in the framework of the Ginzburg-Landau theory, namely for type-II superconductors, the surface energy $\sigma_{ns} < 0$, in the presence of the magnetic field. This suggests that in any conditions in these materials, the superconducting and normal states may coexist. The induction $\mathbf{B}(\mathbf{H})$ and magnetization $\mathbf{M}(\mathbf{H})$ as function of $\mathbf{H}$ are shown in Figs. 6 and 7, and the phase diagram H-T shows the critical fields which will appear in theory of superconductivity.

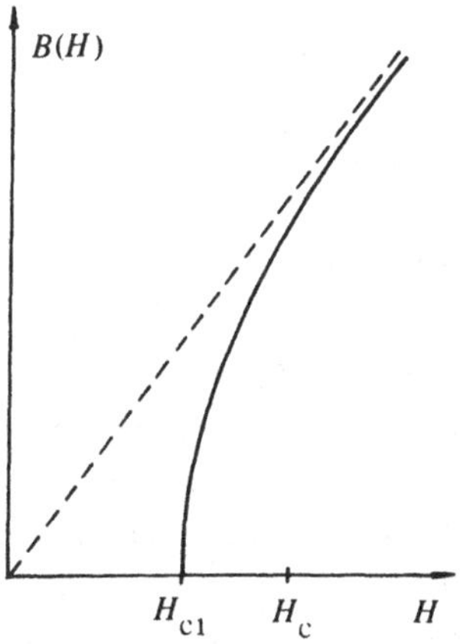

Fig. 6. The induction B as function of the magnetic field.

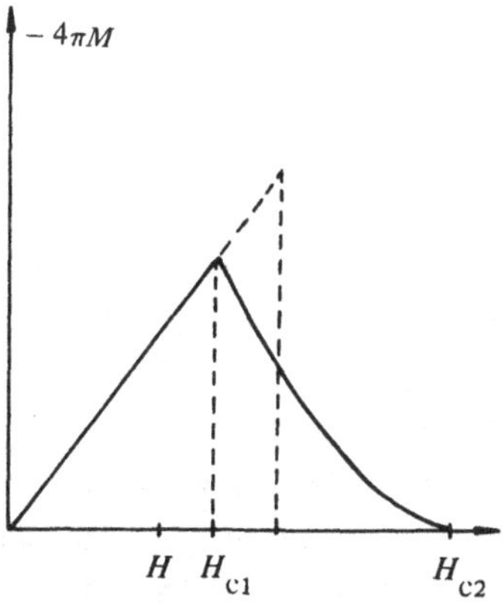

Fig. 7. Magnetization M as function of the magnetic field H.

The dependence of B and M on the magnetic field shows that $B = 0$ for $H < H_{c1}$. This critical field is called lower critical field and below the transition line $H_{c1}(T)$, the superconductor presents a complete Meissner effect with $\mathbf{B} = 0$. For $H_{c1} < H < H_{c2}$, $\mathbf{B} \neq 0$ the state is still superconducting and at $H = H_{c2}(T)$, the superconductivity can be destroyed. Some materials have a superconducting state for $H_{c2} < H < H_{c3}$ called surface superconductivity. The mixed state is specific for the type-II superconductors and the field penetration gives rise to the vortex-state. As indicated in Fig. 7, the area under the $M(H)$ curve is identical to the area obtained by assuming perfect diamagnetism up to H_c with discontinuous disappearance of M at H_c. The continuous disappearance of the magnetization at H_{c2} establishes that the transition from the normal phase in the mixed state is a second order phase transition. The vortex-state is usually called mixed-state or intermediate-state, and the occurrence of such a state may be explained using the fact that the magnetic flux is quantized.

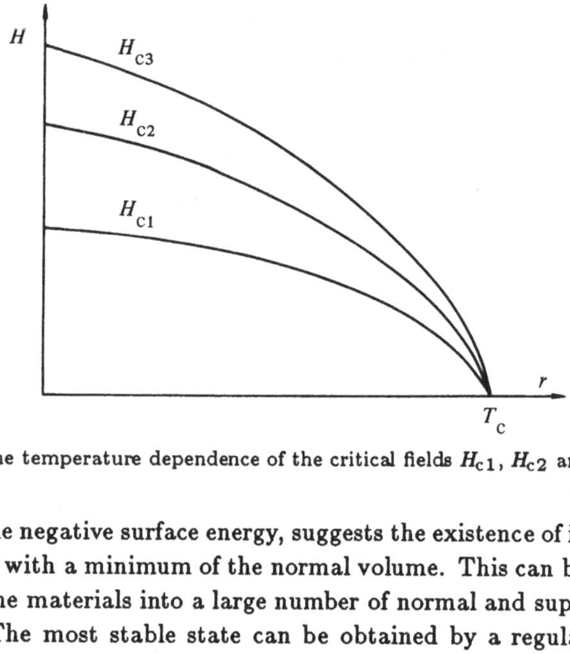

Fig. 8. The temperature dependence of the critical fields H_{c1}, H_{c2} and H_{c3}.

Indeed, the negative surface energy, suggests the existence of interface as is compatible with a minimum of the normal volume. This can be obtained by dividing the materials into a large number of normal and superconducting sheets. The most stable state can be obtained by a regular array of normal filaments of negligible thickness which are parallel to the external field and surrounded by superconducting phase.

At the normal filaments, the superconducting order parameter vanishes and then increases linearly with the distance. It reaches its maximum value at $r \sim \xi$. The magnetic field has a maximum value at the normal filaments and decreases with the distance at $r \sim \lambda \gg \xi$. We can imagine the mixed state as in the superconducting material there have been performed a number of filamentary holes, regularly spaced, parallel to the external field and each of them containing magnetic flux. The magnetic flux associated with each normal hole is quantized in units of Φ_0 and cannot penetrate into the superconducting materials because the superconducting currents circulate in planes perpendicular to the filament. (See Fig. 9.)

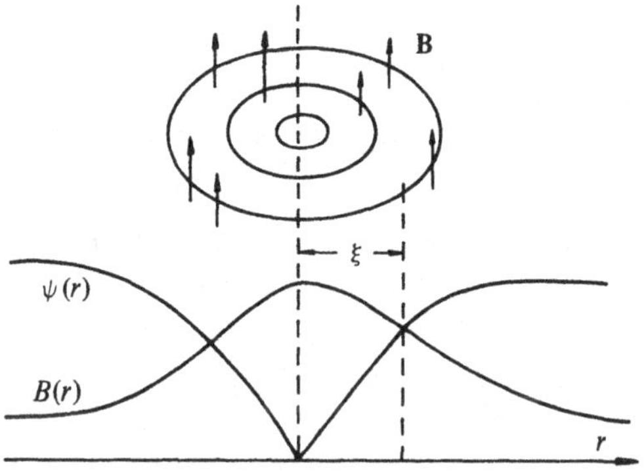

Fig. 9. The vortex line in the mixed state.

The region over which $\psi(\mathbf{r})$ varies from zero to its maximum value $\psi(\infty)$ is called core of the vortex, which contains low-lying energy levels of single electrons rather like a normal metal. Near H_{c2}, the overlap between the vortex lines modifies the core state and it is better to consider the material as being in a gapless state.

Properties of the isolated vortices

Let us consider a type-II superconductor with $\chi \gg 1$. At the distance $r \gg \xi$ we will consider $|\psi(\mathbf{r})|^2 = 1$. The Ginzburg-Landau result

$$\nabla \times (\nabla \times \mathbf{A}) = \frac{|\psi|^2}{\lambda^2}\Big(\frac{\Phi_0}{2\pi}\nabla\theta(\mathbf{r}) - \mathbf{A}(\mathbf{r})\Big)$$

becomes

$$\nabla \times (\nabla \times \mathbf{A}) = \frac{1}{\lambda^2}\Big(\frac{\Phi_0}{2\pi}\nabla\theta(\mathbf{r}) - \mathbf{A}(\mathbf{r})\Big) \qquad (7.1)$$

and using the definition $\nabla \times \mathbf{A} = \mathbf{H}$, we get

$$\lambda^2 \nabla \times \mathbf{H} = \frac{\Phi_0}{2\pi}\nabla \times \nabla\theta(\mathbf{r})$$

which can be re-written as

$$H^2 + \lambda^2 \nabla \times (\nabla \times \mathbf{H}) = \frac{\Phi_0}{2\pi}\nabla \times \nabla\theta(\mathbf{r}) \ . \qquad (7.2)$$

In every point of the vortex, except for the center, $\nabla\boldsymbol{\theta}(\mathbf{r}) = 0$, while in the center of the vortex $|\nabla\theta(\mathbf{r})| \to \infty$. In order to study the behaviour of the $\nabla \times \nabla\theta(\mathbf{r})$ in the center of the vortex, we consider the integral of this expression on a circle with the radius in the center of the vortex:

$$\int \nabla \times \nabla\theta(\mathbf{r}) \cdot d\mathbf{S}$$

which will be transformed using the Stokes theorem as

$$\oint_S \nabla \times \nabla\theta(\mathbf{r}) \cdot d\mathbf{S} = \oint_l \nabla\theta(\mathbf{r}) \cdot d\mathbf{l}$$

where $d\mathbf{l}$ is on the circle and with a radius which is not too great. If the phase $\theta(\mathbf{r})$ changes by 2π at one rotation we have

$$\oint \nabla \times \nabla\theta(\mathbf{r}) \cdot d\mathbf{S} = 2\pi\ .$$

Thus $\nabla \times \nabla\theta(\mathbf{r})$ is a function different from zero except for the center of the vortex, where it is singular, and in fact infinite. Then the equation for the magnetic field is

$$\mathbf{H} + \lambda^2 \nabla \times (\nabla \times \mathbf{H}) = \Phi_0 \delta(\mathbf{r}) \mathbf{e}_v\ , \tag{7.3}$$

$\mathbf{e}_v$ being a unit vector along the axis.
From the condition $H(\infty) = 0$, the solution of this equation is

$$H(r) = \frac{\Phi_0}{2\pi\lambda^2} K_0(r/\lambda) \tag{7.4}$$

where $K_0(z)$ is the MacDonald function.

The asymptotic behaviour of $K_0(z)$ is

$$K_0(z) = \begin{cases} \ln \frac{1}{z}\ ; & z \ll 1 \\ \frac{\exp - z}{\sqrt{z}}\ ; & z \gg 1 \end{cases} \tag{7.5}$$

and for $r \sim \xi$,

$$H(0) \cong \frac{\Phi_0}{2\pi} \ln \chi\ . \tag{7.6}$$

In the domain of the vortex region, from the Ginzburg-Landau theory, we obtain

$$H(0) = \frac{\Phi_0}{2\pi} (\ln \chi - 0.18)$$

where the last term is given by the core contribution. Let us calculate the first critical field H_{c1}. This is the minimum magnetic field at which the vortices appear in a type-II superconductor. In order to calculate H_{c1}, we have to calculate the Gibbs free energy of a superconductor with vortices and one without vortices. We will consider $\chi \gg 1$ ($\lambda \gg \xi$), the London case where in the Gibbs free energy there are no corrections of the type $\nabla\psi$ and for the Gibbs free energy we can use

$$E_v = \frac{1}{8\pi}\int dV[H^2 + \lambda^2(\nabla \times \mathbf{H})^2] \, . \tag{7.7}$$

Equation (7.7) can be considered as the sum of the kinetic energy and the energy of the magnetic field of a vortex. Using now

$$(\nabla \times \mathbf{H})^2 = \mathbf{H} \cdot (\nabla \times (\nabla \times \mathbf{H})) - \nabla \cdot ((\nabla \times \mathbf{H}) \times \mathbf{H}) \, ,$$

from (7.7), we get

$$E_v = \frac{1}{8\pi}\int dV \mathbf{H} \cdot (\mathbf{H} + \lambda^2 \nabla \times (\nabla \times \mathbf{H})) - \lambda^2 \int dV \nabla \times ((\nabla \times \mathbf{H}) \times \mathbf{H}) \, . \tag{7.8}$$

We consider a special geometry in which $(\nabla \times \mathbf{H}) \times \mathbf{H} = 0$ and because $\mathbf{H} \to 0$ as $r \to \infty$, the second integral vanishes. From (7.6), we have

$$E_v = (\Phi_0/4\pi\lambda)^2 \ln\chi = \frac{1}{8\pi}\int dV \mathbf{H} \cdot (\mathbf{H} + \lambda^2 \nabla \times (\nabla \times \mathbf{H})) \, . \tag{7.9}$$

Let us calculate the minimum magnetic field which can be applied so that vortices should appear in a superconductor. The Gibbs free energy for the unit length is

$$\mathcal{G} = E_v - \int dV \frac{\mathbf{B} \cdot \mathbf{H}_0}{4\pi} \tag{7.10}$$

where $\mathbf{H}_0$ is the external magnetic field, and E_v is the Gibbs free energy of the unity length. The density of the Gibbs free energy is

$$G = F - \frac{\mathbf{B} \cdot \mathbf{H}_0}{4\pi} \tag{7.11}$$

where F is the free energy for $H_0 = 0$.

For one vortex, (7.11) may be written as

$$\mathcal{G} = E_v - \frac{\Phi_0 H_0}{4\pi} \tag{7.12}$$

and for a sufficiently small external field H_0, we have $\mathcal{G} > 0$ and the state with vortices is unstable. But we can calculate a critical field where $\mathcal{G}$ becomes negative and the vortex state becomes energetically favourable. This is called the first critical field and is given by

$$H_{c1} = \frac{4\pi}{\Phi_0} E_v$$

or

$$H_{c1} = \frac{\Phi_0}{4\pi\lambda} \ln \chi \ . \tag{7.13}$$

Let us consider the interaction between vortices. For simplicity, we will take two parallel vortices, with the coordinates of the centers $\mathbf{r}_1$ and $\mathbf{r}_2$. The energy of the system can be written as

$$E = \frac{1}{8\pi} \int dV [H^2 + \lambda^2 (\nabla \times \mathbf{H})^2] \tag{7.14}$$

where H can be obtained as a solution of the equation

$$\mathbf{H} + \lambda^2 \nabla \times (\nabla \times \mathbf{H}) = \Phi_0 [\delta(\mathbf{r} - \mathbf{r}_1) + \delta(\mathbf{r} - \mathbf{r}_2)] \mathbf{e}_v \ . \tag{7.15}$$

The energy (7.14) can be written as

$$E = \frac{\Phi_0}{8\pi} (H(\mathbf{r}_1) + H(\mathbf{r}_2)) \tag{7.16}$$

where $\mathbf{H}(\mathbf{r})$ is the field in the center of the first vortex. This field is the sum between the field of vortex 1 and the field resolution from the action of vortex 2 which will be denoted by $H_{1,2}(\mathbf{x})$ where $\mathbf{x} = \mathbf{r}_1 - \mathbf{r}_2$. In a similar way, we may calculate $\mathbf{H}(\mathbf{r}_2)$. From (7.16), we get

$$E = 2E_v + 2(\Phi_0/8\pi) H_{1,2}(\mathbf{r})$$

and if we denote

$$U = \frac{\Phi_0}{4\pi} H_{1,2}(\mathbf{x})$$

we get

$$f = -\frac{\partial U(x)}{\partial x} = -\frac{\Phi_0}{8\pi} \frac{dH_{1,2}(x)}{dx} \ .$$

If we apply for these two vortices the Maxwell equation

$$\frac{dH_{1,2}}{dx} = 4\pi j_{1,2}(x)$$

where $j_{1,2}$ is the density of the current due to the first vortex at the point of the second vortex, we have

$$|f| = j_{1,2} \Phi_0 \ .$$

This equation is a special form of a more general equation

$$\mathbf{f}_1 = \mathbf{j} \times \mathbf{\Phi}_0 \ , \qquad \mathbf{\Phi}_0 = \Phi_0 \mathbf{e}_v \ .$$

The upper critical field

In the mixed phase of a type-II superconductor, the vortices form a lattice so that the energy has the minimum value. If this lattice is in an external magnetic field and this field gradually increases, the lattice period increases and when it becomes of the order of the coherence length ξ the superconducting phase is destroyed. The transition from the mixed state to the normal phase is a second order phase transition and it appears at $H = H_{c2}$ (if the surface effects are neglected) which is called the upper critical field. This critical field is proportional to H_c and from the Ginzburg-Landau classification of superconductors

$$H_{c2} = \sqrt{2}\,\chi H_c$$

or using the relation $\sqrt{2}\,H_c = \Phi_0(2\pi\lambda\xi)$ obtained from the Ginzburg-Landau theory, the upper critical field H_{c2} becomes

$$H_{c2} = \frac{\Phi_0}{2\pi\xi^2}\,. \tag{7.17}$$

We have to mention that the transition from the mixed state to the normal state is a second order phase transition in an external magnetic field and $H_{c2}(T)$ appears in the H_c-T plane as a transition line. There is no contradiction with the general theory of the phase transitions from thermodynamics because the magnetic field is not a conjugate field to the order parameter of the superconducting state as it is for the magnetization, the order parameter from the magnetism.

The upper critical field will be calculated using the microscopic theory and the main point of this calculation is that on the transition line $H_{c2}(T)$ the order parameter is very small.

The magnetic moment of a type-II superconductor

Let us calculate the magnetic moment of a type-II superconductor in the mixed state when $H \gg H_{c1}$. A simple case is to consider a cylinder with $Oz \| H$ and the coherence length $\xi = \xi(x)$ a monotonic function. We will also take λ as independent of x. From these assumptions, we can see that H_{c2} will depend on x. In the volume of the sample, the current is

$$j = \frac{dM(x)}{dx}$$

and $B = H_0 + 4\pi M$ where $M = M(x)$.

Thus the current j is

$$j = \frac{dM(x)}{dx} .$$

The Lorentz force will act on each vortex written as

$$f_L = j\Phi_0 = \Phi_0 \frac{dM}{dx} \tag{7.18}$$

but if the vortices are in equilibrium, the forces between them will be in a concrete configuration which will assure the equilibrium. The energy of vortex E_v will also be a function of x and a force $-\nabla E_v$ will act on a vortex where

$$E_v = \left(\frac{\Phi_0}{4\pi\lambda}\right)^2 \ln\chi = \left(\frac{\Phi_0}{4\pi\lambda}\right)^2 \ln\frac{\lambda}{\xi} . \tag{7.19}$$

If we calculate now $f = -dE_v/dx$, and using the Lorentz force we get the equilibrium if

$$\Phi_0 \frac{dM}{dx} = \left(\frac{\Phi_0}{4\pi\lambda}\right)^2 \frac{1}{\xi}\frac{d\xi}{dx} \tag{7.20}$$

with the solution

$$M = \frac{\Phi_0}{16\pi^2\lambda^2} \ln\frac{\xi}{\lambda} \tag{7.21}$$

where l is a constant with dimensions of length.

In (7.21) we still have $l = l(H)$, because $\xi(x)$ increases with x and H_{c2} decreases, it is clear that there is a point x_0 where $H_{c2}(x_0) = H_0$, then $M(x_0) = 0$ and $l = \xi(x_0)$.

On the other hand, using the relations

$$H_{c2} = \chi\sqrt{2}\,H_c\,, \quad \sqrt{2}\,H_c = \frac{\Phi_0}{2}\pi\xi\,,$$

we get

$$\Phi_0 = 2\pi\xi^2 H_{c2}$$

and in x_0, we have

$$2\pi l^2 H_0 = \Phi_0$$

and from these two equations, we get

$$\frac{\xi}{l} = [H_0/H_{c2}]^{\frac{1}{2}} .$$

The magnetization (7.21) can be written as

$$M = -\frac{\Phi_0}{16\pi^2\lambda^2} \ln\left(\frac{H_{c2}}{H_0}\right)^2 = -\frac{\Phi_0}{32\pi\lambda^2} \ln\frac{H_{c2}}{H_0} \tag{7.22}$$

and the induction B can be calculated now as

$$B = H_0 - \frac{\Phi_0}{8\pi\lambda^2}\ln\frac{H_{c2}}{H_0} \ . \tag{7.23}$$

If we now write $H_{c2}/H_0 = (H_{c2} - H_0/H_0) + 1$, Eqs. (7.22) becomes

$$M = -\frac{\Phi_0}{32\pi\lambda^2}\ln\left[1 - \frac{H_{c2} - H_0}{H_0}\right] \cong -\frac{\Phi_0}{32\pi^2\lambda^2}\frac{H_{c2} - H}{H_0} \tag{7.24}$$

or

$$-4\pi M = \frac{H_{c2} - H}{4\chi^2} \ . \tag{7.25}$$

This expression can be compared with

$$-4\pi M = \frac{H_{c2} - H}{1.16(2\chi^2 - 1)} \tag{7.26}$$

obtained by Abrikosov[2] using a more sophisticated calculation of the magnetization from the Ginzburg-Landau theory.

Surface superconductivity

The above results refer to an infinite superconductor where the influence of the sample surfaces can be neglected. A sample surface represents an inhomogeneity in the material, and during the reduction of an external magnetic field we expect the occurrence of superconductivity at the surface to set in at a higher field than in the interior of a bulk sample. Since, this inhomogeneity arising from the surface of a specimen is rather well defined, a quantitative treatment of this problem is possible using the Ginzburg-Landau theory.

We assume that the superconductor occupies the half-space $x > 0$ and that its surface coincides with the y-z plane at $x = 0$. The surface is assumed to be electrically insulating. An external magnetic field is applied parallel to the surface along the z-direction. We start with the linearized Ginzburg-Landau equation (4.20). Setting $A_x = A_z = 0$ and $A_y = Hx$, we have

$$-\frac{1}{4m}\frac{\partial^2\psi}{\partial x^2} + \frac{1}{4m}[-i\nabla_y - 2eHx]\psi = -\alpha\psi \ . \tag{7.27}$$

We are looking for a solution of the form

$$\psi = f(x)\exp(iky) \ . \tag{7.28}$$

The function $f(x)$ then satisfies the equation

$$-\frac{1}{4m}\frac{d^2 f(x)}{dx^2} + \frac{1}{4m}[k - 2eHx]^2 f(x) = -\alpha f(x) . \tag{7.29}$$

At the insulating surface, we have the boundary condition

$$\left.\frac{d\psi}{dx}\right|_{x=0} = \left.\frac{df}{dx}\right|_{x=0} = 0 . \tag{7.30}$$

Equation (7.29) is analogous to the Schrödinger equation for a harmonic oscillator of frequency $\omega = eH/m$ with the equilibrium position given by $x_0 = k/2mH$.

The boundary condition (7.30) becomes important for an evaluation of the eigenvalue of Eq. (7.27). If the position $x_0 \gg \xi_0(T)$, the wave function φ will be localized near x_0, and its magnitude will be close to zero at the surface. The boundary condition (7.30) is then automatically satisfied. The solution of Eq. (7.29) has the form

$$f(x) = \exp\left[-\frac{1}{2}\left(\frac{x - x_0}{\xi(T)}\right)^2\right] \tag{7.31}$$

with the eigenvalue $|\alpha| = eH_{c2}/2m$ and the upper critical field equal to H_{c2}. The same result has been obtained for $x_0 = 0$. In these two cases, the potential energy from (7.29) has the form

$$V(x) = \frac{(2eH)^2}{4m}(x - x_0)^2 . \tag{7.32}$$

In order to obtain the lowest eigenvalue which corresponds to the higher field, when the position x_0 is placed approximately at a distance ξ away from the surface, we extend $V(x)$ beyond the point $x = 0$ by its mirror image.

The lowest eigenfunction of such a potential is an even function of x, satisfying the condition (7.30). Obviously, the lowest eigenvalue of this potential is lower than for the potential (7.32). The exact calculation yields for the lowest eigenvalue

$$|\alpha| = 0.59\frac{e}{2m}H_c \tag{7.33}$$

and for the new critical field in the presence of a surface, we find

$$H_{c3} = \frac{1}{0.59}H_{c2} , \tag{7.34}$$

i.e., a value larger than H_{c2} as we have expected. Experimentally, the phenomenon of surface superconductivity in a magnetic field in the range $H_{c2} < H < H_{c3}$ oriented parallel to the sample surface has often been observed. However, surface superconductivity can easily be suppressed by plating the surface with a normal metal. The suppression of surface superconductivity arises, of course, from the pair-breaking process in the normal metal. The temperature dependence of the critical fields H_c, H_{c1}, H_{c2} and H_{c3} are given in Fig. 8.

Abrikosov vortex state

In 1957, Abrikosov[5] showed that the type-II superconductors characterized by the vortex state can be treated using the Ginzburg-Landau theory. Abrikosov's result essentially introduced a new phenomenon which is being referred to, since, as vortex state and is typical for type-II superconductors.

We consider a cylinder of type-II superconductor placed in an external magnetic field H oriented along the z-direction and parallel to the axis of the cylinder. The diameter of the cylinder is small enough, so that the demagnetization can be neglected. During the decreasing of the magnetic field, the occurrence of the superconducting state in the sample begins when H becomes H_{c2}. At the beginning, the order parameter $|\psi|$ is small and the linearized Ginzburg-Landau equation can be used for obtaining the quantity ψ. If the field is decreased appreciably below H_{c2}, the order parameter $|\psi|$ becomes larger, and the complete, nonlinear Ginzburg-Landau equation must be used.

If the magnetic field H is only slightly less than H_{c2}, the solution ψ_L of the complete Ginzburg-Landau equation must have a strong similarity to a certain solution ψ_L of the linearized equation. The solution ψ_L satisfies the equation

$$\frac{1}{4m}(i\nabla + 2e\,\mathbf{A})^2\psi_L = -\alpha\psi_L \tag{7.35}$$

with

$$\nabla \times \mathbf{A} = (0, 0, H_{c2}) \; . \tag{7.36}$$

Equation (7.35) has many degenerate eigenvalues corresponding to solutions describing the occurrence of the superconducting state in different regions of the sample.

As in the case of Eq. (7.27), solutions of (7.35) have the form

$$\psi_k = \exp(iky)\exp\left[-\frac{1}{2}\left(\frac{x - x_0}{\xi(T)}\right)^2\right] \tag{7.37}$$

with

$$x_0 = \frac{k}{2eH_{c2}} \tag{7.38}$$

where k is an arbitrary parameter. The solutions describe a gaussian band of superconductivity of width $\xi(T)$ extending perpendicular to the x-axis at the location $x = x_0(k)$. A general solution ψ_L must be a linear combination of ψ_L. We are interested in a solution which is periodic both in the directions x and y. Periodicity in the y-direction is achieved by setting

$$k = k_n = nq \tag{7.39}$$

yielding the period

$$\Delta y = \frac{2\pi}{q} \, . \tag{7.40}$$

The general solution will have the form

$$\psi_L = \sum_n C_n \exp[inqy] \exp\left[-\frac{1}{2}\left(\frac{x - x_n}{\xi(T)}\right)^2\right] \tag{7.41}$$

with

$$x_n = \frac{nq}{2eH_{c2}} \, . \tag{7.42}$$

The function ψ_L defined by (7.41) is periodic in the y-direction. Periodicity in the x-direction can also be established if the coefficients C_n are periodic functions of n, so that $C_{n+\nu} = C_n$, where ν is integer. The particular choice of ν determines the type of periodic lattice structure. For $\nu = 1$, we get from the calculations a square lattice and for $\nu = 2$ a triangular lattice. Abrikosov obtained a square lattice, but later Kleiner *et al.*[6], de Gennes and Matricon[7] showed that a triangular lattice has a lower energy throughout the mixed state. Furthermore, a triangular lattice is microscopically stable against small vibrational deviations.[6] De Gennes and Matricon[7] suggested the possibility of investigating the nature of the vortex line structure by slow neutron diffraction. The results[6−7] indicate unambiguously the formation of the triangular lattice.

From (7.42), we note that the periodicity in the x-direction is

$$\Delta x = \frac{2\pi}{\Delta y} \frac{1}{2eH_{c2}} \tag{7.43}$$

yielding

$$\Delta x \Delta y H_{c2} = \Phi \, , \tag{7.44}$$

i.e., each unit cell of the periodic lattice contains one flux quantum.

From (7.41), we get some general conclusions independent of the choices for C_n and q. Using this function, we can calculate the current

$$J_{Lx} = -\frac{e}{2m}\frac{\partial}{\partial y}|\psi_L|^2 \tag{7.45a}$$

$$J_{Ly} = \frac{e}{2m}\frac{\partial}{\partial x}|\psi_L|^2 \tag{7.45b}$$

equations which indicate that the lines of constant $|\psi_L|^2$ coincide with the lines of constant local field H_s and with stream lines of the current J_L. Contour diagrams of the quantity for the triangular vortex lattice are shown in Fig. 10.

Fig. 10. The vortices lattice in the mixed state.

If we consider now the case of the complete Ginzburg-Landau equation which is nonlinear, the normalization of the function ψ_L becomes important. The normalization will determine the strength of the supercurrent and therefore the magnetic induction $\mathbf{B}$ and the free-energy F. We assume that the free energy F remains stationary, if ψ_L is replaced by the function $(1+\varepsilon)\psi_L$ where ε is a small constant. To first order in ε, the variation in the free energy is

$$\delta F = 2\varepsilon \int d\mathbf{r} \left[\alpha(T)|\psi_L|^2 + \beta|\psi_L|^4 + \frac{1}{4m}|(i\nabla + 2e\mathbf{A})\psi_L|^2\right] . \tag{7.46}$$

Writing integrals such as $\int d\mathbf{r}|\psi_L|^2$ in the form $\Omega\overline{|\psi_L|^2}$, where Ω denotes the macroscopic volume, the condition $\delta F = 0$ yields

$$\alpha\overline{|\psi_L|^2} + \beta\overline{|\psi_L|^4} + \frac{1}{4m}\overline{|(i\nabla + 2e\,\mathbf{A})\psi_L|^2} = 0 . \tag{7.47}$$

We consider now the potential vector $\mathbf{A}$ as

$$\mathbf{A} = \mathbf{A}_0 + \mathbf{A}_1 \tag{7.48}$$

where $\mathbf{A}_0$ is the vector potential for $H = H_{c2}$. The correction $\mathbf{A}_1$ arises from the fact that the applied magnetic field is smaller than H_{c2} and that the supercurrents also contribute to the field. The function ψ_L has to satisfy (7.35) and we will only keep the linear terms in $\mathbf{A}_1$ and from (7.47) we obtain

$$\beta\overline{|\psi_L|^4} - \overline{\mathbf{A}_1 J_L} = 0 \tag{7.49}$$

with

$$\mathbf{J}_L = \frac{2e}{4mi}(\psi_L^* \nabla\psi_L - \psi_L(\nabla\psi_L^*)) \tag{7.50}$$

which is in fact the current associated with the unperturbed solution.

Integrating the second term in (7.49) by parts and setting $\nabla \times \mathbf{A}_1 = \mathbf{H}_1$ and $\nabla \times \mathbf{H}_0 = 4\pi\mathbf{J}_L$ we obtain

$$\beta\overline{|\psi_L|^4} - \frac{1}{4\pi}\overline{\mathbf{H}_1 \cdot \mathbf{H}_0} = 0\ . \tag{7.51}$$

Noting that $\mathbf{H}_1$ and $\mathbf{H}_s$ are parallel to the z-direction, we can write

$$\mathbf{H}_1(\mathbf{r}) = \mathbf{H} - \mathbf{H}_{c2} + \mathbf{H}_s(\mathbf{r})\ . \tag{7.52}$$

Inserting (7.52) into (7.51) and using the magnetic field H_s associated to the supercurrents

$$\mathbf{H}_s = -\frac{2\pi e}{m}|\psi_L|^2 \tag{7.53}$$

we have

$$\beta\overline{|\psi|^4} + \overline{\frac{e}{m}|\psi|^2(\mathbf{H} - \mathbf{H}_{c2} - \frac{2e\pi}{m}|\psi_L|^2)} = 0 \tag{7.54}$$

or setting again $|\psi_L| = \psi_0 f$ and using the definition for k we finally obtain

$$\overline{f^4}\left(1 - \frac{1}{2k^2}\right) - \overline{f^2}\left(1 - \frac{H}{H_{c2}}\right) = 0\ . \tag{7.55}$$

Equation (7.55) is a rather general result, which is independent of detailed behaviour of ψ_L. For a particular lattice type, one can calculate

$$\beta = \frac{\overline{f^4}}{(\overline{f^2})^2}\ , \tag{7.56}$$

which is independent of the normalization of ψ. Using (7.53), the magnetic induction is

$$B = H + H_s = H - \frac{2e}{m}\overline{|\psi_L|^2} \, . \tag{7.57}$$

From (7.55) and (7.56), we find

$$B = H - \frac{H_{c2} - H}{(2\chi^2 - 1)\beta} \tag{7.58}$$

which gives for the magnetization M

$$M = \frac{B - H}{4\pi} = -\frac{H_{c2} - H}{4\pi\beta(2\chi^2 - 1)} \, . \tag{7.59}$$

Using the identity for the Gibbs free-energy density g

$$(\partial g/\partial M)_T = -M \, ,$$

we can calculate g by integrating down from the normal state at H_{c2} where $g_s(H_{c2}) = g_n(H_{c2})$. In this way, we obtain

$$g_s(H) = g_n(H_{c2}) - \frac{(H_{c2} - H)^2}{8\pi(2\chi^2 - 1)} \, . \tag{7.60}$$

Equation (7.60) indicates that the configurations with the smallest value of β is thermodynamically most stable. Equation (7.60) applies to the regime $H < H_{c2}$ and $\chi > 1/\sqrt{2}$. It will be rather useful to discuss the validity of the Abrikosov theory. This theory of the vortex state is based on the phenomenological concepts developed in the Ginzburg-Landau theory. However, as we will show, Gor'kov demonstrated that for superconductors with arbitrary electron mean free path the Ginzburg-Landau theory follows from the microscopic theory for the temperature region near T_c where the order parameter, which can be identified with $\psi(\mathbf{r})$, is small.

Using the microscopic theory in the dirty-limit and a small order parameter, Maki[8] and de Gennes[8] showed that this theory remains valid near H_{c2} at all temperatures with the introduction of temperature-dependent Ginzburg-Landau parameters

$$k_1(T) = \frac{H_{c2}(T)}{\sqrt{2}\, H_c(T)}$$

and

$$-4\pi\left(\frac{dM}{dH}\right)_{H_{c2}} = \frac{1}{\beta(2k_2^2(T) - 1)} \, .$$

Another direction for the generalization of the theory was to obtain theoretical results valid for all temperatures, magnetic fields and mean free paths. This approach is known as the Eilenberger equation who transformed the Gor'kov equations into a set of transport-like equations. As Usadel showed for the case of dirty limit, the Eilenberger equations can be transformed into even simpler diffusion-like equations.[10]

In the next chapter, we shall see the approximations in which using quantum-mechanics microscopic description, we obtain phenomenological results.

References

1. D. Schoenberg, *Superconductivity* (Cambridge University Press, 1952).
2. A. B. Pippard, *Proc. Roy. Soc.* **A216** (1953) 547.
3. V. L. Ginzburg and L. D. Landau, *Zh. Exp. Teor. Fiz.* **20** (1950) 1064.
4. B. D. Josephson, *Phys. Rev. Lett.* **1** (1962) 251.
5. A. A. Abrikosov, *Zh. Exp. Teor. Fiz.* **32** (1957) 1442; [*Sov. Phys. JETP* **5** (1957) 1174].
6. W. H. Kleiner, L. M. Roth and S. H. Autler, *Phys. Rev.* **133A** (1964) 1226.
7. P. G. de Gennes and J. Matricon, *Rev. Mod. Phys.* **36** (1964) 45.
8. K. Maki, *Prog. Theor. Phys.* **29** (1963) 603.
9. P. G. de Gennes, *Phys. Kondens Materie* **3** (1964) 79.
10. R. D. Parks, *Superconductivity* (Marcel Deker, New York, 1968).

II
MICROSCOPIC THEORY OF SUPERCONDUCTIVITY

The explanation of superconductivity based on the quantum mechanics ideas appeared only in 1958 and was given by Bardeen, Cooper and Schrieffer. The main point of this theory is the occurrence of the electron pairs responsible for superconductivity which are given by the electron-electron attraction. This interaction is mediated by virtual phonons and has to be greater than the Coulomb repulsion.

The methods from the many-body theory, namely the Green function method, have been successfully applied by Gor'kov in the theory of superconductivity. This method is in fact equivalent to the self-consistent field method given by Bogoliubov. In the framework of the microscopic theory and using the Green function method in the linear response approximation, one can study the interaction between the magnetic field and the superconducting state. Near the critical temperature, the microscopic theory gives the Ginzburg-Landau results. The inhomogeneous state can be studied using a similar method, as that given by Gor'kov developed by Ushadel and Eilenberger. The influence of the real non-virtual phonons on the superconducting state has been pointed out by Eliashberg, Scalapino, Schrieffer, Wilkins, Anderson and Morel. They considered the electron-phonon interactions as well as the Coulomb interaction and calculated the critical temperature in the perturbation theory. These results known as strong coupling theory of superconductivity, have been extended for different problems as the theory of dilute alloys, coexistence between magnetic order and superconductivity. The influence of the spin fluctuations (paramagnons) has been considered using this theory by Berk and Schrieffer.

Finally, we have to mention that the triplet pairing considered at the beginning to be energetically unfavourable seems to be possible in the new class of superconductors known as heavy fermion superconductors.

8. The Cooper Instability of the Fermi Gas

In 1956, Cooper suggested a simple idea whereby the instability of the Fermi gas under attractive electron interactions could be understood in terms of the bound state between a pair of fermionic particles interacting with each other but not with the other electrons from the Fermi gas.

The main problem which has to be solved is the origin of the electron-electron attraction.

The experimental data showed that the gap Δ is of the order $\Delta \cong T_c = 10^{-3}$ eV/electron, which is a small quantitative effect to produce a large qualitative effect. Thus the Coulomb correlation energy is too big, and can therefore be neglected. This reason for neglecting the Coulomb energy in the occurrence of the superconductivity seems odd, but has a deep meaning. Indeed, if some effect is associated with an energy too large, then we know that the effect is not the cause of superconductivity. On the other hand, the isotopic effect gives a reasonable idea that the electron-phonon interaction can be important for the superconductivity.

Frohlich[1] showed that the electron-phonon interaction can mediate an attractive electron-electron attraction (see Appendix 1) and Cooper[2] treated the pair formation using quantum mechanics.

At the very beginning, we are not interested in the origin of the attraction between the electrons, the electron-phonon interaction being a possible interesting explanation but in a more accurate treatment, the Coulomb interaction has to be considered.

The simple problem of the pairs of electrons will be treated supposing that the total momentum of the pair is zero and we can consider the interaction between electrons in the center-of-mass frame. The Schrödinger equations for the wave function of a pair is

$$\frac{p^2}{2m}\psi(\mathbf{p}) + \int d^3\mathbf{p}' g(\mathbf{p}-\mathbf{p}')\psi(\mathbf{p}') = E\psi(\mathbf{p}) \tag{8.1}$$

where we considered only the states below the Fermi surface and so, Eq. (8.1) becomes

$$\varepsilon(\mathbf{p})\psi(\mathbf{p}) + gN(0)\int_{\varepsilon_0}^{\varepsilon_0+D} d\varepsilon'\psi(\mathbf{p}') = E\psi(\mathbf{p}) \tag{8.2}$$

and writing the integral as

$$A = \int_{\varepsilon_0}^{\varepsilon_0+D} d\varepsilon'\psi(\mathbf{p}')$$

we can solve (8.2), and the wave function ψ is

$$\psi(\mathbf{p}) = \frac{-gN(0)A}{\varepsilon(\mathbf{p}') - E}$$

and with this result (8.2) becomes

$$1 = -gN(0)\int_{\varepsilon_0}^{\varepsilon_0+D} \frac{d\varepsilon'}{\varepsilon' - E} \tag{8.3}$$

which gives

$$\ln\frac{\varepsilon_0 - E + D}{\varepsilon_0 - E} = -\frac{1}{gN(0)} \; . \tag{8.4}$$

This equation has a solution if $g < 0$ as it is obtained by the electron-electron interaction mediated by the phonons (see Appendix 1). The binding energy of the bound state is

$$W = \varepsilon_0 - E \cong D\exp(-1/N(0)g) \; . \tag{8.5}$$

From this relation, we can see that the energy of a Cooper pair is lower than the energy of the free electrons which is ε_0. We say that the increasing electron system is unstable to small attraction. The parameter D is of the order of Debye energy ω_{D} and the lifetime of the phonons which mediate the interaction is $1/\tau \simeq \omega_{\mathrm{D}}$, where ω_{D} is the Debye frequency. As for the superconductors, ω_{D} is large, $1/\tau$ is very small, and these phonons with a very short lifetime are called virtual phonons.

The Cooper pair idea has been used by Bardeen, Cooper and Schrieffer[3] in order to describe the superconducting state. Indeed using the product function

$$\psi_{\mathrm{BCS}} = \prod_{p<p_0}(u_p + v_p c^\dagger_{\mathbf{p}\downarrow} c^\dagger_{-\mathbf{p}\uparrow})|0\rangle$$

where u_p and v_p are subject to the constrain

$$u_p^2 + v_p^2 = 1 ,$$

they obtained the energy of the superconducting state as

$$E^2(\mathbf{p}) = \varepsilon^2(\mathbf{p}) + \Delta^2$$

where $\varepsilon(\mathbf{p})$ is the kinetic energy of the electrons and Δ is the gap which is proportional to the mean value $\langle c_{\mathbf{p}\uparrow} c_{-\mathbf{p}\downarrow}\rangle$. After their paper was published, the method of the Green function has been generalized by Gor'kov[5] in order to describe the superconducting state. In the same period when Bardeen, Cooper and Schrieffer worked out their theory, Bogoliubov[4] developed a method which gives exactly the same results.

9. Self-Consistent Field Method — Gor'kov Equations

We start with the usual Hamiltonian

$$\mathcal{H} = \mathcal{H}_0 + \mathcal{H}_i$$

where

$$\mathcal{H}_0 = \sum_\alpha \int d^3\mathbf{r}\psi_\alpha^\dagger(\mathbf{r})\varepsilon\psi_\alpha(\mathbf{r})$$
$$\mathcal{H}_i = g\int d^3\mathbf{r}\psi_\uparrow^\dagger(\mathbf{r})\psi_\downarrow^\dagger(\mathbf{r})\psi_\downarrow(\mathbf{r})\psi_\downarrow(\mathbf{r}) .$$

In the self consistent method, this Hamiltonian will be written as

$$\mathcal{H}_{\text{eff}} = \sum_\alpha \int d^3\mathbf{r}\psi_\alpha^\dagger(\mathbf{r})\varepsilon\psi_\alpha(\mathbf{r}) + \int d^3r[\Delta\psi_\uparrow^\dagger(\mathbf{r})\psi_\downarrow^\dagger(\mathbf{r}) + h.c.] \tag{9.1}$$

where $\varepsilon = \frac{p^2}{2m} - \mu$ is the kinetic energy, $\Delta(\mathbf{r})$ is the self-consistent field defined as

$$\Delta(\mathbf{r}) = g\langle\psi_\downarrow(\mathbf{r})\psi_\uparrow(\mathbf{r})\rangle \tag{9.2}$$

and $\psi_\alpha^\dagger, \psi_\alpha$ are the operators describing the electrons.

In order to describe the behaviour of superconductors at $T \neq 0$, we will develop the method of anomalous Gor'kov[5] Green function using the operators $\psi_\alpha(\mathbf{r}, \tau)$ and $\psi_\alpha^\dagger(\mathbf{r}, \tau)$ where $\tau = it$.

We define

$$G(\mathbf{r}_1,\mathbf{r}_2;\tau_1,\tau_2) = -\langle T_\tau \psi_\uparrow(\mathbf{r}_1,\tau_1)\psi^\dagger(\mathbf{r}_2,\tau_2)\rangle \tag{9.3}$$

where the operator has been written as τ-dependent using

$$A(\tau) = \exp(\tau\mathcal{H})A(0)\exp(-\tau\mathcal{H}) \ .$$

Using now the Hamiltonian (9.1), we get

$$[\psi_\uparrow(\mathbf{r},\tau),\mathcal{H}_{\text{eff}}]_- = \varepsilon\psi_\uparrow(\mathbf{r}_1,\tau_1) + \Delta(\mathbf{r}_1)\psi_\uparrow^\dagger(\mathbf{r}_1,\tau_1)$$

and the equation of motion for (9.3) becomes

$$\left[\frac{\partial}{\partial\tau_1} + \varepsilon\right] G(\mathbf{r}_1,\mathbf{r}_2;\tau_1,\tau_2) - \Delta(\mathbf{r}_1)\langle T_\tau \psi_\downarrow^\dagger(\mathbf{r}_1,\tau_1)\psi_\uparrow^\dagger(\mathbf{r}_2,\tau_2)\rangle$$
$$= \delta(\tau_1-\tau_2)\delta(\mathbf{r}_1-\mathbf{r}_2) \ .$$

If we introduce a new Green function

$$F^\dagger(\mathbf{r}_1,\mathbf{r}_2;\tau_1,\tau_2) = \langle T_\tau \psi_\downarrow^\dagger(\mathbf{r}_1,\tau_1)\psi_\uparrow^\dagger(\mathbf{r}_2,\tau_2)\rangle \ ,$$

we get

$$\left[\frac{\partial}{\partial\tau_1} + \varepsilon\right] G(\mathbf{r}_1,\mathbf{r}_2;\tau_1,\tau_2) - \Delta(\mathbf{r}_1)F^\dagger(\mathbf{r}_1,\mathbf{r}_2;\tau_1,\tau_2) = \delta(\tau_1-\tau_2)\delta(\mathbf{r}_1-\mathbf{r}_2) \ . \tag{9.4}$$

In the same way, we get the equation for $F^\dagger(\mathbf{r}_1,\mathbf{r}_2;\tau_1,\tau_2)$ as

$$\left[\frac{\partial}{\partial\tau_1} - \varepsilon\right] F^\dagger(\mathbf{r}_1,\mathbf{r}_2;\tau_1,\tau_2) - \Delta(\mathbf{r}_1)G(\mathbf{r}_1,\mathbf{r}_2;\tau_1,\tau_2) = 0$$

where

$$\Delta(\mathbf{r}) = |g|F(\mathbf{r},\mathbf{r};\tau,\tau) \tag{9.5}$$

which is in fact equivalent to (9.2).

In the superconducting state which is in equilibrium, the Green functions depend on the difference $\tau - \tau'$. Then we can write the transforms

$$G(\mathbf{r},\mathbf{r}';\tau) = T\sum_n G(\omega;\mathbf{r},\mathbf{r}')e^{-\omega\tau} \ , \qquad F(\mathbf{r},\mathbf{r}';\tau) = \sum_n F(\omega;\mathbf{r},\mathbf{r}')e^{-\omega\tau}$$

and from (9.4)–(9.5), we get

$$[i\omega - \hat{\tau}_3\varepsilon - \frac{1}{2}(\Delta(\mathbf{r}_1)\hat{\tau}_+ + \Delta^\dagger(\mathbf{r}_1)\hat{\tau}_-)]\widehat{G}(\omega;\mathbf{r}_1,\mathbf{r}_2) = \delta(\mathbf{r}_1-\mathbf{r}_2) \tag{9.6}$$

where

$$\widehat{G}(\omega;\mathbf{r}_1,\mathbf{r}_2) = \begin{bmatrix} G(\omega;\mathbf{r}_1,\mathbf{r}_2) & -F^\dagger(\omega;\mathbf{r}_1,\mathbf{r}_2) \\ -F(\omega;\mathbf{r}_1,\mathbf{r}_2) & G(\omega;\mathbf{r}_1,\mathbf{r}_2) \end{bmatrix}$$

and $\widehat{\tau}_3$, $\widehat{\tau}_\pm = \widehat{\tau}_1 \pm i\widehat{\tau}_2$ are matrices in the pseudospin space. The free energy can be obtained using the Pauli theorem as

$$\Omega_s = \Omega_0 + \int_0^1 dg \int d\mathbf{r} \left\{ \Delta(\mathbf{r})\pi T \sum_n F(\omega;\mathbf{r},\mathbf{r}) \right.$$
$$\left. + \Delta^\dagger(\mathbf{r})\pi T \sum_n F^\dagger(\omega;\mathbf{r},\mathbf{r}) \right\} .$$

From this equation, we obtain the general formula

$$\Omega_s - \Omega_0 = \frac{1}{|g|}\int d\mathbf{r}|\Delta(\mathbf{r})|^2 - \int_0^1 dg \int d\mathbf{r} \left\{ \Delta(\mathbf{r})\pi T \sum_\omega F(\omega;\mathbf{r},\mathbf{r}',g) + h.c. \right\} . \tag{9.7}$$

Let us consider the case of the spatial homogeneous superconductor. In this case, we write the operators $\psi_\alpha(\mathbf{r})$ and $\psi^\dagger_\alpha(\mathbf{r}')$ as

$$\psi_\alpha(\mathbf{r}) = \frac{1}{\sqrt{V}}\sum_{\mathbf{p},\alpha} c_{\mathbf{p},\alpha} e^{i\mathbf{p}\cdot\mathbf{r}} , \quad \psi^\dagger_\alpha(\mathbf{r}) = \frac{1}{\sqrt{V}}\sum_{\mathbf{p},\alpha} c^\dagger_{\mathbf{p},\alpha} e^{-i\mathbf{p}\cdot\mathbf{r}}$$

and the Hamiltonian (9.1) becomes

$$\mathcal{H}_{\rm BCS} = \sum_{\mathbf{p},\alpha} \varepsilon_\alpha(\mathbf{p}) c^\dagger_{\mathbf{p}\alpha} a_{\mathbf{p}\alpha} + \sum_{\mathbf{p}} (\Delta(p) c^\dagger_{\mathbf{p}\uparrow} c^\dagger_{-\mathbf{p}\downarrow} + h.c.) . \tag{9.8}$$

Equation (9.6) becomes

$$\begin{aligned} (i\omega - \varepsilon(\mathbf{p}))G(\omega;\mathbf{p}) + \Delta(\mathbf{p})F^\dagger(\omega;\mathbf{p}) &= 1 \\ (i\omega + \varepsilon(\mathbf{p}))F^\dagger(\omega;\mathbf{p}) + \Delta^\dagger(\mathbf{p})G(\omega;\mathbf{p}) &= 0 \end{aligned} \tag{9.9}$$

which have the solutions

$$G(\omega,\mathbf{p}) = \frac{i\omega + \varepsilon(\mathbf{p})}{(i\omega)^2 - \varepsilon^2(\mathbf{p}) - \Delta^2} ; \quad F^\dagger(\omega,\mathbf{p}) = \frac{\Delta^\dagger(\mathbf{p})}{(i\omega)^2 - \varepsilon^2(\mathbf{p}) - \Delta^2} \tag{9.10}$$

which may also be written as

$$\begin{aligned} G(\omega,\mathbf{p}) &= \frac{1+\varepsilon(\mathbf{p})/E(\mathbf{p})}{i\omega - E(\mathbf{p})} + \frac{1-\varepsilon(\mathbf{p})/E(\mathbf{p})}{i\omega + E(\mathbf{p})} ; \\ F^\dagger(\omega,\mathbf{p}) &= \frac{\Delta^\dagger}{2E(\mathbf{p})}\left(\frac{1}{i\omega + E} - \frac{1}{i\omega - E}\right) \end{aligned} \tag{9.11}$$

where $\omega = \pi(n+\frac{1}{2})T$.

From (9.10), we see that the poles of the Green functions G and F at $T=0$ are

$$\omega(\mathbf{p}) = \pm E(\mathbf{p}) = \pm\sqrt{\varepsilon^2(\mathbf{p})+\Delta^2}$$

which satisfies

$$\lim_{\mathbf{p}\to 0}\omega(\mathbf{p}) = \Delta$$

which shows the existence of a "gap" in the energy of the electrons forming the pairs. The dispersion law of the energy $E(\mathbf{p})$ is shown in Fig. 11.

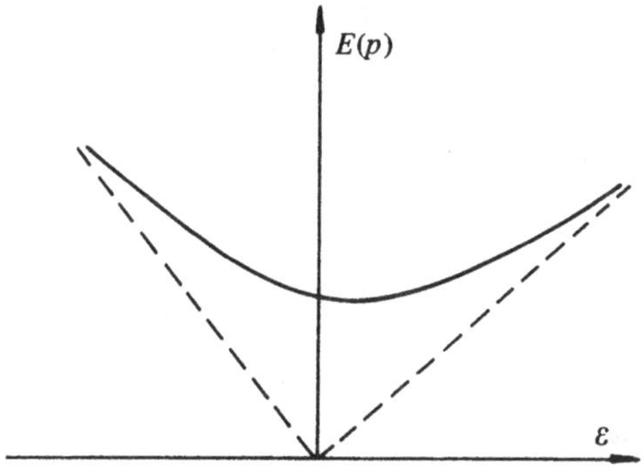

Fig. 11. The dependence of the energy E on the kinetic energy ε of the normal electrons.

For $T \neq 0$, the gap is temperature dependent and is in fact the order parameter of the superconducting state.

For the homogeneous case, (9.5) may be written using $F^\dagger$ from (9.10) as

$$\Delta^\dagger = gT\pi\sum_\omega\int\frac{d^3\mathbf{p}}{(2\pi)^3}\frac{\Delta^\dagger}{(i\omega)^2-\varepsilon^2(\mathbf{p})-\Delta^2} \tag{9.12}$$

which can be transformed as

$$g\int\frac{d^3\mathbf{p}}{(2\pi)^3}\frac{1}{\sqrt{\varepsilon^2(\mathbf{p})+\Delta^2}}\tanh\frac{1}{2T}\sqrt{\varepsilon^2(\mathbf{p})+\Delta^2} = 1\,. \tag{9.13}$$

In the limit $T\to 0$, from (9.13) we get

$$\Delta(0) = 2\omega_D\exp\left[-\frac{1}{N(0)g}\right] \tag{9.14}$$

with the approximations

$$\sinh x \simeq \frac{1}{x}$$

and where we considered a constant density of states $N(0)$ at the Fermi surface.

Let us consider $\Delta \to 0$. In this case, (9.13) becomes

$$\int_0^{\omega_D/2T_0} dx \, \frac{\tanh x}{x} = \frac{1}{N(0)g} \tag{9.15}$$

which can be re-written as

$$\ln \frac{\omega_D}{2T_c} \tanh \frac{\omega_D}{2T_c} - \int_0^{\omega_D} dx \, \frac{\ln x}{\cosh^2 x} = \frac{1}{N(0)g}$$

and using

$$\int_0^{\infty} dx \, \frac{\ln x}{\cosh^2 x} = \ln \frac{4\gamma}{\pi} \,, \quad \gamma = \exp C \,, \quad C = 0.057 \,,$$

we get

$$T_c = 2\omega_D \gamma/\pi \exp \left[\frac{-1}{N(0)g} \right] \,.$$

In order to obtain the temperature dependence of the gap, we write (9.13) as

$$\frac{1}{N(0)g} = \int_0^{\omega_D} \frac{d\varepsilon}{\sqrt{\varepsilon^2 + \Delta^2}} - 2 \int_0^{\omega_D} \frac{d\varepsilon}{\sqrt{\varepsilon^2 + \Delta^2}} f(\beta\sqrt{\varepsilon^2 + \Delta^2})$$

where $\beta = 1/T$ and $f(x)$ is the Fermi function. Using the same equation for $T = 0$, we get

$$\ln \frac{\Delta(T)}{\Delta(0)} = -2 \int_0^{\infty} \frac{d\varepsilon}{\sqrt{\varepsilon^2 + \Delta^2}} f(\beta\sqrt{\varepsilon^2 + \Delta^2})$$

an equation which can be written as

$$\ln \frac{\Delta(0)}{\Delta(T)} = 2 \sum_{n=1}^{\infty} (-1)^n K_0(n\Delta(T)/T) \tag{9.16}$$

where $K_0(x)$ is the MacDonald function.
In the low temperature region, we approximate $K_0(x) \simeq \sqrt{\pi/2x}\, e^{-x}$ and for $n = 1$

$$\ln \frac{\Delta(0)}{\Delta(T)} = \sqrt{\frac{2\pi}{\beta\Delta(T)}} \exp[-\beta\Delta(T)] \,.$$

If $\Delta(T)$ is in the low temperatures region, we may expand the logarithm as

$$\frac{\Delta(0) - \Delta(T)}{\Delta(0)} \cong \ln \frac{\Delta(0)}{\Delta(T)}$$

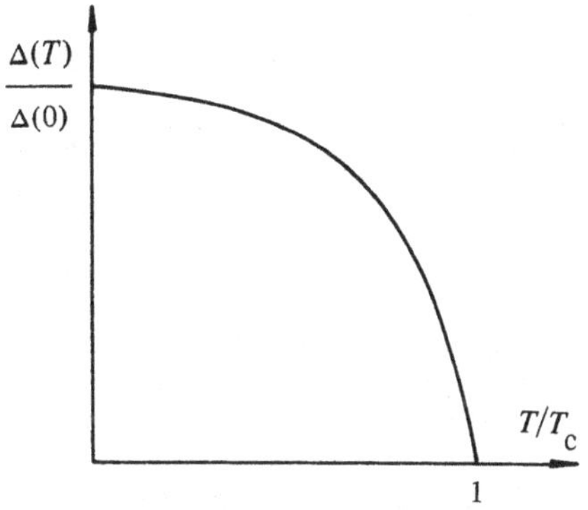

Fig. 12. The temperature dependence of the gap $\Delta(T)$.

and

$$\Delta(T) \cong \Delta(0) - [2\pi\beta^{-1}\Delta(0)]^{\frac{1}{2}} \exp[-\beta\Delta(0)] \,, \tag{9.17}$$

which is the equation for $\Delta(T)$ in the low temperatures domain. Let us discuss now the behaviour of Δ in the domain of high temperatures. Equation (9.12) can be re-written as

$$\frac{1}{N(0)g} = 2\beta\pi T \sum_{\omega} \int_0^{\omega_D} \frac{d\varepsilon}{\omega^2 + \varepsilon^2 + \Delta^2} \,.$$

In this domain where $\Delta \to 0$, we get

$$\frac{1}{N(0)g} = \ln\frac{2\gamma\omega_D}{\pi T} + 2\Delta^2\beta_c^{-1}\pi T_c \sum_{\omega} \int \frac{d\varepsilon}{(\omega^2 + \varepsilon^2)^2}$$

and using (9.14) we transform the above equation as

$$\ln\frac{T_c}{T} = \beta_c^{-1}\pi T_c \sum_{\omega} \frac{1}{\omega^3} = \frac{7\varsigma(3)\Delta^2}{8\pi^2 T_c}$$

which can be approximated as

$$\frac{T_c - T}{T_c} = \frac{7\varsigma(3)}{8\pi^2 T_c}$$

which gives for $\Delta(T)$ near T_c

$$\Delta(T) \cong 3.06 T_c [1 - T/T_c]^{\frac{1}{2}}$$

where $\varsigma(n)$ is the Riemann function.

The temperature dependence of the order parameter $\Delta(T)$ is given in Fig. 12.

The free energy for a spatial homogeneous superconductor can be calculated from (9.7) and (9.10), and we obtain the expression

$$\Omega = \frac{V}{|g|}\Delta^2 - 2T\sum_{\mathbf{p}} \ln\cosh\frac{E(\mathbf{p})}{2T} + \sum_{p} \varepsilon(\mathbf{p}) \tag{9.18}$$

or using the equation for the gap (9.13), the free energy becomes

$$\Omega = \sum_{\mathbf{p}} \left[\frac{\Delta^2}{2E(\mathbf{p})}\tanh\frac{E(\mathbf{p})}{2T} - 2T\ln\cosh\frac{E(\mathbf{p})}{2T}\right] . \tag{9.19}$$

If we define $\Omega_n = \Omega(\Delta = 0)$, the free energy of the normal state, we get

$$\frac{\Omega - \Omega_n}{V} = \int \frac{d^3\mathbf{p}}{(2\pi)^3}\left[\tanh\frac{E(\mathbf{p})}{2T} - E(\mathbf{p}) - 2T\ln\frac{1-\exp(-E(\mathbf{p})/T)}{1+\exp(-E(\mathbf{p})/T)}\right] \tag{9.20}$$

which gives at $T = 0$, the result

$$\frac{E_s - E_n}{V} = -\frac{1}{2}N(0)\Delta^2 . \tag{9.21}$$

Near the critical temperature $T \to T_c$ and $\Delta \to 0$, the free energy can be expanded in series up to Δ^4 and the free energy becomes

$$\begin{aligned}\Omega = \Omega_n - V\Bigg\{&\frac{\Delta^2}{2}\int\frac{d\varepsilon}{\varepsilon}\tanh\frac{\varepsilon}{2T} - \frac{\Delta^2}{|g|} + \frac{T_c - T}{4T_c^2}\int\frac{d^3\mathbf{p}}{(2\pi)^3}\cosh^{-2}\frac{\varepsilon(\mathbf{p})}{2T}\\ &+\frac{\Delta^4}{8}\int\frac{d^3\mathbf{p}}{(2\pi)^3}\frac{1}{\varepsilon^2(\mathbf{p})}\left[\frac{1}{2T_c}\cosh^{-2}\frac{\varepsilon(\mathbf{p})}{2T_c} - \frac{1}{\varepsilon(\mathbf{p})}\tanh\frac{\varepsilon(\mathbf{p})}{2T_c}\right]\Bigg\}\end{aligned} \tag{9.22}$$

or if we neglect the term of the fourth order, (9.22) gives

$$\Omega = \Omega_n - VN(0)\Delta^2\left[1 - \frac{T}{T_c} - \frac{7\zeta(3)}{16\pi^2T_c^2}\Delta^2\right] . \tag{9.23}$$

With these results, we can calculate the entropy

$$S = -\frac{1}{V}\frac{\partial\Omega}{\partial T} \tag{9.24}$$

and the specific heat

$$C_s = T\frac{\partial S}{\partial T} \tag{9.25}$$

of the superconducting state.

From the free energy (9.20), we get for the entropy

$$S = 2\int \frac{d^3\mathbf{p}}{(2\pi)^3}\left[\ln(1+\exp(-E(\mathbf{p})/T)) + \frac{E(\mathbf{p})}{T}\left(1+\frac{E(\mathbf{p})}{T}\right)^{-1}\right] . \quad (9.26)$$

This integral can be performed for the case of the constant density of states and we get

$$S = \frac{4N(0)}{T}\int_0^\infty \frac{(2\varepsilon^2+\Delta^2)d\varepsilon}{\sqrt{\varepsilon^2+\Delta^2}\,(1+\exp(E(\mathbf{p})/T))} \quad (9.27)$$

which can be expressed as

$$S = \frac{4N(0)\Delta^2}{T}\sum_{n=0}^{\infty}(-1)^n K_2\left(\frac{n\Delta}{T}\right) \quad (9.28)$$

where $K_2(x)$ is the second order MacDonald function.
In the low temperatures domain ($T \ll \Delta(0)$), the entropy given by (9.28) can be approximated as

$$S = 2N(0)\left(\frac{2\pi\Delta^3(0)}{T}\right)^{\frac{1}{2}} \exp\left(-\frac{\Delta(0)}{T}\right) . \quad (9.29)$$

Near the critical temperature, the entropy can be approximated by

$$S = S_{\mathrm{n}} - \frac{N(0)\Delta^2}{T_{\mathrm{c}}}$$

where $S_{\mathrm{n}} = \frac{2}{3}N(0)T$.
From (9.28) we calculate, using (9.25), the behaviour of the specific heat at low temperatures as

$$C_{\mathrm{s}} = 2N(0)\left(\frac{2\pi\Delta^3(0)}{T}\right)\frac{\Delta(0)}{T}\exp\left(-\frac{\Delta(0)}{T}\right) \quad (9.30)$$

and taking $C_{\mathrm{s}}(T_{\mathrm{c}}) = C_{\mathrm{n}}$ we get

$$\frac{C_{\mathrm{s}}}{C_{\mathrm{n}}} = \frac{3}{\gamma}\sqrt{\frac{2}{\pi}}\left(\frac{\Delta(0)}{T}\right)^{\frac{3}{2}} \exp\left(-\frac{\Delta(0)}{T}\right) . \quad (9.31)$$

The behaviour of $C_{\mathrm{s}}(T)$ near the critical temperature can be characterized by

$$\frac{C_{\mathrm{s}} - C_{\mathrm{n}}}{C_{\mathrm{n}}} = \frac{12}{7\zeta(3)} \simeq 1.43 .$$

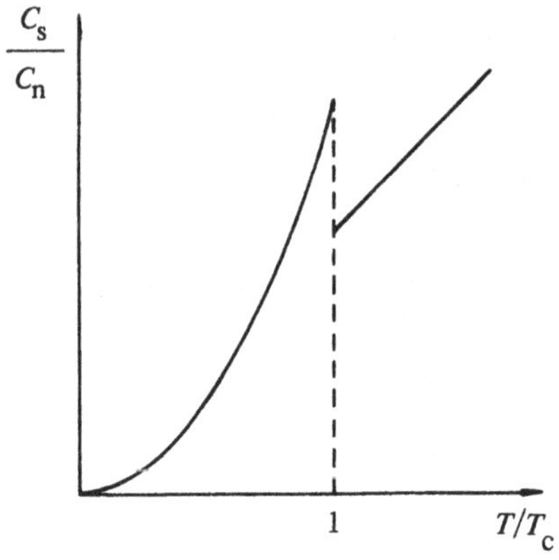

Fig. 13. The temperature dependence of the specific heat of a superconductor.

The temperature dependence of the specific heat is shown in Fig. 13.

Now we can calculate, using the Green functions, the density of states $N(\omega)$ using the general relation

$$N(\omega) = \operatorname{Im} \operatorname{Tr} G(\omega + i\gamma)\,, \qquad \gamma \to 0\,.$$

For metal in the normal state ($\Delta = 0$)

$$\begin{aligned} G(\omega, \mathbf{p}) &= \frac{1}{\omega - \varepsilon(\mathbf{p}) + i\gamma} \\ \operatorname{Im}(\omega - \varepsilon(\mathbf{p}) + i\gamma)^{-1} &= -\pi\delta(\omega - \varepsilon(\mathbf{p})) \\ N(\omega) &= \int \frac{d^3\mathbf{p}}{(2\pi)^3}\delta(\omega - \varepsilon) = \frac{1}{2\pi^2}\left(p^2\frac{dp}{d\varepsilon}\right)_{\varepsilon=0}\,. \end{aligned}$$

We will use the same method to calculate the density of states of a superconductor. In this case, the Green function is given by (9.11) and using a similar method we get

$$N_s(\omega) = \frac{1}{2}N(0)\int_{-\infty}^{\infty} d\varepsilon[\delta(\omega - E(\mathbf{p})) + \delta(\omega + E(\mathbf{p}))]$$

which gives

$$N_s(\omega) = \begin{cases} N(0)\frac{\omega}{\sqrt{\omega^2 - \Delta^2}}\,; & |\omega| > \Delta \\ 0 & \end{cases}\,.$$

The behaviour of the density of state given by (9.32) is shown in Fig. 14. This quantity can be measured with a great accuracy by tunnelling experiments.

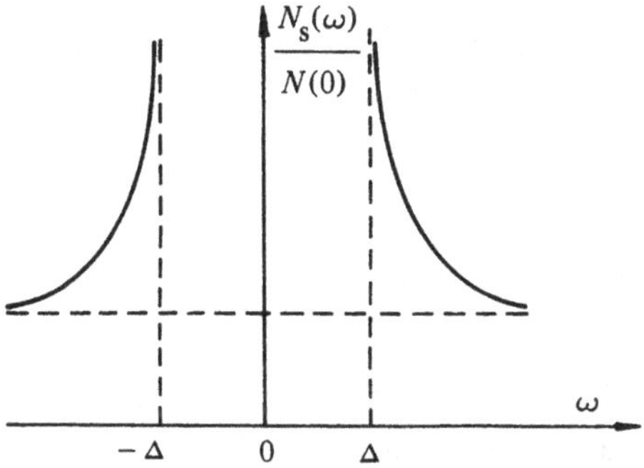

Fig. 14. The energy dependence of the density of states of a superconductor.

10. Linear Response to a Magnetic Field

The magnetic field may act on a superconductor in different ways and this action is dependent on the intensity of the field. In this chapter, we consider all the fields weaker than the critical field which destroys superconductivity. Using the Gor'kov equation, we will calculate the transverse current $\mathbf{j}(\mathbf{r},t)$ induced by an applied magnetic field, specified by the transverse vector $\mathbf{A}(\mathbf{r},t)$.

In the next calculations, we will use the London gauge

$$\nabla\cdot\mathbf{A}(\mathbf{r},t)=0\,,\qquad \mathbf{q}\cdot\mathbf{A}(\mathbf{q};\omega)=0\,. \tag{10.1}$$

The vector field $\mathbf{A}(\mathbf{q},\omega)$ can always be separated into its longitudinal and transverse parts

$$A^{\mathrm{l}}(\mathbf{q},\omega)=\mathbf{q}[\mathbf{q}\cdot\mathbf{A}(\mathbf{q};\omega)]\,,\quad A^{\mathrm{t}}(\mathbf{q};\omega)=A(\mathbf{q},\omega)-A^{\mathrm{l}}(\mathbf{q},\omega)\,. \tag{10.2}$$

In presenting this problem we will follow the method from Ref. 6 which appears to be most useful for the next application and on the other hand, makes transparent the method in the framework of the microscopic theory.

For the superconducting state, we use the Hamiltonian (9.1) in the presence of the electromagnetic field

$$\mathcal{H}_{\mathrm{A}}=\int d^3\mathbf{r}[\psi^\dagger_\alpha(\mathbf{r})\nabla\psi_\alpha(\mathbf{r})-(\nabla\psi^\dagger_\alpha(\mathbf{r}))\psi_\alpha(\mathbf{r})]\cdot\mathbf{A}(\mathbf{r})$$
$$+\frac{e^2}{2m}|\mathbf{A}(\mathbf{r})|^2\psi^\dagger_\alpha(\mathbf{r})\psi_\alpha(\mathbf{r})\,. \tag{10.3}$$

For any operator $\widehat{O}(\mathbf{r},t)$, the average $\langle\widehat{O}(\mathbf{r},t)\rangle_{\mathrm{A}}$ in the presence of the magnetic field described by a vector potential $\mathbf{A}$ can be expressed as

$$\langle\widehat{O}(\mathbf{r},t)\rangle_{\mathrm{A}}=\langle\widehat{O}_{\mathrm{H}}(\mathbf{r},t)\rangle-\int_{-\infty}^{t}dt'\langle[\widehat{O}_{\mathrm{H}}(\mathbf{r},t'),\mathcal{H}_{\mathrm{A}}]_-\rangle\,. \tag{10.4}$$

We have to mention that using (10.4) we will calculate the total current $\mathbf{j}$ in the presence of $\mathbf{A}$. This current has the form

$$\mathbf{j} = -\frac{1}{2} e[\psi_\alpha^\dagger \mathbf{v}_s \psi_\alpha + \mathbf{v}_s \psi_\alpha^\dagger \psi_\alpha]$$

where $\mathbf{v}_s$ is given by

$$\mathbf{v}_s = \frac{1}{m}(-i\nabla + e\,\mathbf{A})$$

and $\mathbf{j}$ becomes

$$\mathbf{j} = -\frac{e}{2mi}[\psi_\alpha^\dagger \nabla \psi_\alpha - (\nabla \psi_\alpha^\dagger)\psi_\alpha] \,. \tag{10.5}$$

Using now

$$\langle \mathbf{j}(\mathbf{r},t)\rangle = \langle \mathbf{j}_{\widetilde{\mathcal{H}}}(\mathbf{r},t)\rangle - i\int_{-\infty}^{t} dt' \langle [j(\mathbf{r},t'), \mathcal{H}_{\mathrm{A}}]\rangle \tag{10.6}$$

where $\widetilde{\mathcal{H}} = \mathcal{H}_{\mathrm{A}} - \mu \widehat{N}$ and

$$\widehat{O}_{\widetilde{\mathcal{H}}} = \exp[-i\widetilde{\mathcal{H}}t]\,\widehat{O}(\mathbf{r})\,\exp[i\widetilde{\mathcal{H}}t] \,. \tag{10.7}$$

In the linear response approach we will expand (10.6) to the first order in $\mathbf{A}$, and since the expectation value of $\mathbf{j}$ vanishes in zero field we find

$$\begin{aligned} \mathbf{j}(\mathbf{r},t) &= \langle \mathbf{j}(\mathbf{r},t)\rangle_{\mathrm{A}} \\ &= -\frac{e^2\mathbf{A}(\mathbf{r},t)}{m}\langle \psi_\alpha^\dagger(\mathbf{r},t)\psi_\alpha(\mathbf{r},t)\rangle \\ &\quad + i\int_{-\infty}^{\infty} dt' \int d^3\mathbf{r}' \langle [\mathbf{j}^0(\mathbf{r},t), \mathbf{j}(\mathbf{r},t)\cdot \mathbf{A}(\mathbf{r},t)]\rangle \end{aligned} \tag{10.8}$$

where $\psi(\mathbf{r},t), \mathbf{j}(\mathbf{r},t)$ are defined using (10.7) and

$$\mathbf{j}^0(\mathbf{r},t) = -\frac{e}{2mi}[\psi_\alpha^\dagger(\mathbf{r},t)\nabla\psi_\alpha(\mathbf{r},t) - (\nabla\psi_\alpha^\dagger(\mathbf{r},t))\psi_\alpha(\mathbf{r},t)] \tag{10.9}$$

is the current operator for $\mathbf{A} = 0$.

In the linear response approximation, the relation between $\mathbf{j}$ and $\mathbf{A}$ has the form

$$\mathbf{j}_k(\mathbf{r},t) = -\frac{1}{4\pi}\int d^3\mathbf{r}' K_{k,l}(\mathbf{r},t;\mathbf{r}',t')\mathbf{A}_l(\mathbf{r}',t') \,. \tag{10.10}$$

If we compare (10.8) and (10.10), the kernel may be written as

$$K_{k,l}(\mathbf{r},t;\mathbf{r}',t') = \frac{4\pi e^2}{m}\delta(\mathbf{r}-\mathbf{r}')\delta_{k,l}\langle \widehat{n}(\mathbf{r},t)\rangle + 4\pi P_{k,l}(\mathbf{r},t;\mathbf{r}',t') \tag{10.11}$$

where $P_{k,l}(\mathbf{r},t;\mathbf{r}',t')$ is the retarded propagator

$$P_{k,l}(\mathbf{r},t;\mathbf{r}',t') = -i\theta(t-t')\langle[\mathbf{j}_k^0(\mathbf{r};t),\mathbf{j}_l^0(\mathbf{r}',t')]\rangle \,. \tag{10.12}$$

The retarded function (8.11) is inconvenient for a perturbation analysis, and instead we introduce a temperature function

$$\mathcal{P}_{k,l} = -\langle T_\tau[\mathbf{j}_\kappa^0(\mathbf{r},\tau),\mathbf{j}_l^0(\mathbf{r}',\tau')]\rangle$$

where j^0 is given by (8.8).

Using now for $\mathcal{P}_{k,l}(l,l')$, the representation

$$\mathcal{P}_{k,l}(1,1') = -\frac{e}{2mi}\Big\{(\nabla_1-\nabla_2)_k\cdot(\nabla_{1'}-\nabla_{2'})_l \\ \times\langle T_\tau[\psi_\alpha^\dagger(2)\psi_\alpha(1)\psi_\beta^\dagger(2')\psi_\beta(1')]\rangle_{2'=1',1=1'}\Big\}$$

we will evaluate the product using the Gor'kov factorization. In the simple case of uniform and time-independent system where $\mathcal{P}_{k,l}(\mathbf{r},t;\mathbf{r}',t') = \mathcal{P}_{k,l}(\mathbf{r}-\mathbf{r}';t-t')$ we introduce the Fourier transform and

$$\mathcal{P}_{k,l}(\mathbf{r}-\mathbf{r}';\tau-\tau') = \frac{1}{(2\pi)^3}\int d^3q\, e^{i\mathbf{q}\cdot(\mathbf{r}-\mathbf{r}')}\pi T\sum_\nu e^{-i\nu(\tau-\tau')}\mathcal{P}_{k,l}(\mathbf{q};\nu)$$

and the Fourier transform becomes

$$\mathcal{P}_{k,l}(q,\nu) = 2\left(\frac{e}{m}\right)^2\int\frac{d^3\mathbf{p}}{(2\pi)^3}\pi T\sum_{\omega_1}\left(p+\frac{q}{2}\right)_k\left(p+\frac{q}{2}\right)_l \\ \times[G(\omega_1+\nu;\mathbf{p}+\mathbf{q})G(\mathbf{p},\omega_1)+F(\omega_1+\nu;\mathbf{p}+\mathbf{q})F^\dagger(\omega_1,\mathbf{p})]\,. \tag{10.13}$$

The expression (10.13) can be calculated using the Green functions (9.10) and after some algebra we get for the Fourier transform of the kernel (10.11) the equation

$$K_{k,l}(\omega,\mathbf{q}) = \frac{4\pi ne^2}{m}\delta_{k,l} + \frac{4\pi e}{m}\int\frac{d^3\mathbf{p}}{(2\pi)^3}p_kp_l\Big\{I_+(\varepsilon_+,\varepsilon_-;\Delta_+,\Delta_-) \\ \times[f(E_+)-f(E_-)]\left(\frac{1}{\omega+i\gamma-E_-+E_+}-\frac{1}{\omega+i\gamma+E_--E_+}\right) \\ +I_-(\varepsilon_+,\varepsilon_-;\Delta_+,\Delta_-)[1-f(E_+)-f(E_-)] \\ \times\left(\frac{1}{\omega+i\gamma-E_--E_+}-\frac{1}{\omega+i\gamma+E_++E_-}\right)\Big\} \tag{10.14}$$

where

$$E_{\pm} = E(p \pm q/2) \ , \qquad \varepsilon_{\pm} = \varepsilon(p \pm q/2)$$
$$I_{\pm}(\varepsilon_+, \varepsilon_-; \Delta_+, \Delta_-) = \frac{1}{2}\left(1 \pm \frac{\varepsilon_+\varepsilon_- + \Delta^2}{E_+E_-}\right) . \tag{10.15}$$

This kernel describes the general response of a uniform superconductor at temperature T to a weak magnetic field described by the transverse vector potential $\mathbf{A}$. The static limit $K(\omega = 0, \mathbf{q})$ will be necessary in order to explain the Meissner effect. From (10.10) and (10.15) considered in the static limit, we get

$$\begin{aligned} \mathbf{j}(\mathbf{q}) = &-\frac{ne^2}{m}\mathbf{A}(\mathbf{q}) - \frac{e}{m}\int \frac{d^3\mathbf{p}}{(2\pi)^3}\mathbf{p}\cdot\mathbf{p}[\mathbf{p}\cdot\mathbf{A}(\mathbf{p})] \\ &\times \left\{ I_+(\varepsilon_+, \varepsilon_-; \Delta_+, \Delta_-)\frac{f(E_+) - f(E_-)}{E_+ - E_-} \right. \\ &\left. - I_-(\varepsilon_+, \varepsilon_-; \Delta_+, \Delta_-)\frac{1 - f(E_+) - f(E_-)}{E_+ + E_-} \right\} . \end{aligned} \tag{10.16}$$

In the limit of small q, we get from (10.16)

$$\mathbf{j}(\mathbf{q}) = -\mathbf{A}(q)\frac{n_0e^2}{m} \ , \qquad n_0 = n - \frac{1}{32\pi^2 m}\int_0^\infty dpp^4\left(\frac{\partial f(E_p)}{\partial E_p}\right) . \tag{10.17}$$

From (10.17), we can see that $\lim K(q) \neq 0$ and then at $T \neq 0$ there is Meissner effect in the BCS superconductor.

Now, we can calculate the current $j(\mathbf{r})$ in the Pippard (nonlocal) limit.

Indeed, if we consider the general case of an electromagnetic field, the effects of this field on the superconducting state are sensitive to the wave length of the field. If this wave length is comparable to the penetration depth $\lambda(T)$ then the previous limit $q\xi_0 \ll 1$ describes the London superconductor and the limit $q(\xi_0) \gg 1$ will describe the pure Pippard superconductor. In addition, we shall consider $q \ll p_0$ in the next calculations because the penetration depth is always much larger than the interatomic spacing.

In order to study the Pippard limit we write (10.15) in spherical coordinates with $\mathbf{q}$ as polar axis, and the kernel K becomes

$$\begin{aligned} K(|\mathbf{q}|, \omega) \equiv K(q) = &\frac{4\pi e^2}{m} + \frac{4\pi e^2}{m^2}\int_0^\infty \frac{d\mathbf{p}d\omega}{(2\pi)^4}p^4\int_0^\pi d\theta \sin^3\theta \\ &\times \left[I_+(\varepsilon_+, \varepsilon_-; \Delta_+, \Delta_-)\frac{f(E_+) - f(E_-)}{E_+ - E_-} \right. \\ &\left. - I_-(\varepsilon_+, \varepsilon_-; \Delta_+, \Delta_-)\frac{1 - f(E_+) - f(E_-)}{E_+ - E_-} \right] \end{aligned} \tag{10.18}$$

which can be re-written as

$$K(q) = \frac{4\pi e^2}{m}\left\{1 - \frac{3}{16}\int_{-\infty}^{\infty} d\varepsilon \int_{-1}^{1} dz(1-z^2) \times \left[I_+(\varepsilon_+,\varepsilon_-;\Delta_+,\Delta_-)\frac{f(E_+)-f(E_-)}{E_+-E_-} - I_-(\varepsilon_+,\varepsilon_-;\Delta_+,\Delta_-)\frac{1-f(E_+)-f(E_-)}{E_+-E_-}\right]\right\} \tag{10.19}$$

where $\varepsilon_\pm \cong \varepsilon \pm \frac{1}{2}v_0 qz$, and terms of order q^2/p_0^2 have been neglected. If in (10.19) we consider $\Delta = 0$, the kernel $K_n(q)$ gives zero, for $q \ll p_0$ which means that the normal state is non-magnetic except for the Pauli paramagnetism and the Landau diamagnetism which have been neglected in this approximation.

The integral (10.19) has been calculated by Khalatnikov *et al.*[6]

$$K(q) = \frac{3\pi^2 ne^2}{mq\xi_0}\frac{\Delta(T)}{\Delta(0)}\tanh\frac{\Delta(T)}{2T}\,.$$

The phenomenological theory of Pippard can be obtained from these results. Indeed (10.16) will be calculated in the coordinate space at $T = 0$. Using the substitutions $\mathbf{p}+\mathbf{q}/2 = \mathbf{k}, \mathbf{p}-\mathbf{q}/2 = \mathbf{l}$, the Fourier transform of

$$\mathbf{j}(\mathbf{r}) = -\frac{ne^2}{m}\mathbf{A}(\mathbf{r}) - \frac{e}{4m}\int d^3\mathbf{r}'\int\frac{d^3\mathbf{k}\,d^3\mathbf{l}}{(2\pi)^6}(\mathbf{k}+\mathbf{l})[(\mathbf{k}+\mathbf{l})\cdot\mathbf{A}(\mathbf{r}') \times \exp[i(\mathbf{k}-\mathbf{l})\cdot(\mathbf{r}-\mathbf{r}')]F(\varepsilon_k,\varepsilon_l)] \tag{10.20}$$

where

$$F(\varepsilon,\varepsilon') = \frac{\varepsilon\varepsilon' + \Delta^2 - EE'}{EE'(E+E')}\,. \tag{10.21}$$

The second term from (10.20)

$$S_{i,j}(\mathbf{r}) = \frac{1}{(2\pi)^6}\int d^3\mathbf{k}\int d^3\mathbf{l}(\mathbf{k}+\mathbf{l})_i\cdot(\mathbf{k}+\mathbf{l})_j\exp[i(\mathbf{k}-\mathbf{k}')\cdot\mathbf{r}]F(\varepsilon_k,\varepsilon_l) \tag{10.22}$$

will be written in a more general form

$$S_{i,j}(\mathbf{r},\mathbf{r}') = \frac{1}{(2\pi)^6}\int d^3\mathbf{k}\int d^3\mathbf{l}(\mathbf{k}+\mathbf{l})_i(\mathbf{k}+\mathbf{l})_j \times \exp[i(\mathbf{k}-\mathbf{l})\cdot\mathbf{r} + i(\mathbf{k}+\mathbf{l})\cdot\mathbf{r}']F(\varepsilon_k,\varepsilon_l) \tag{10.23}$$

which after some transformations becomes

$$S_{i,j}(\mathbf{r},\mathbf{r}') = -\frac{4}{(2\pi)^4}\int_0^\infty k^2 dk \int_0^\infty l^2 dl F(\varepsilon_k,\varepsilon_l) \times \nabla_i'\nabla_l' j_0(\mathbf{k}\cdot(\mathbf{r}+\mathbf{r}'))j_0(\mathbf{l}\cdot(\mathbf{r}+\mathbf{r}')) \tag{10.24}$$

where $j_0(z) = \sin z/z$.

The function $F(\varepsilon_k,\varepsilon_l)$ has a peak near $k \cong l \cong p_0$ while the relevant distances are of the order $r \cong 1/p_0$. Then, because the trigonometric functions oscillate rapidly, the variation of the denominator of $j_0(z)$ can be neglected and (10.24) can be written as

$$S_{i,j}(\mathbf{r}) = S_{i,j}(\mathbf{x},0) = 8p_0^2\frac{x_i x_j}{x^4}\int_0^\infty \frac{kdk}{(2\pi)^2}\int_0^\infty \frac{ldl}{(2\pi)^2}F(\varepsilon_k,\varepsilon_l)\cos[(\mathbf{k}-\mathbf{l})\cdot\mathbf{r}]$$

or as an integral over energies

$$S_{i,j} = \frac{x_i x_j}{x^4}\frac{m^2p_0^2}{(2\pi)^4}\int_{-\infty}^\infty d\varepsilon d\varepsilon' F(\varepsilon,\varepsilon')\cos\left[(\varepsilon-\varepsilon')\frac{r}{v_0}\right]. \tag{10.25}$$

If we change the variables in (10.25) as

$$\varepsilon = \Delta_0\sinh(x+x'), \qquad \varepsilon' = \Delta_0\sinh(x-x')$$

we get

$$S_{i,j}(r) = -\frac{2\Delta_0 r_i r_j}{r^4}\frac{m^2p_0^2}{(2\pi)^4}\int_{-\infty}^\infty dx dx' \times \cos\left[\frac{2\Delta_0 r}{v_0}\sinh x'\cosh x\right]\frac{\sinh^2 x'}{\cosh x\cosh x'}$$

and using the substitution $z = \sinh x'$, we perform the integral over x, and $S_{i,j}(\mathbf{r})$ becomes

$$S_{i,j}(\mathbf{r}) = -\frac{\Delta_0 m^2 p_0^2}{\pi^4}\frac{r_i r_j}{r^4}\int_{-\infty}^\infty \frac{dx}{\cosh x}\left[2\delta\left(\frac{2\Delta_0 x\cosh x}{v_0}\right) - \pi\exp\left(-\frac{2\Delta_0 r\cos x}{v_0}\right)\right]. \tag{10.26}$$

Equation (10.26) can be written as

$$S_{i,j}(r) = \left(\frac{mp_0}{\pi^2}\right)^2 \frac{r_i r_j}{r^4}[-2\pi v_0 \delta(x) + \pi^2 \Delta_0 J(r)]$$

where

$$\begin{aligned} J(r) &= \frac{2}{\pi}\int_0^\infty dx \operatorname{sech} \exp\left[-\frac{2\Delta_0 r}{v_0}\cosh x\right] \\ &= \frac{2}{\pi}\int_{2\Delta_0/v_0}^\infty d\varsigma K_0(\varsigma) \end{aligned} \tag{10.27}$$

and K_0 is the MacDonald function.
From (10.20) and (10.27), we get

$$\mathbf{j}(\mathbf{R}) = -\frac{3ne^2\pi\Delta_0}{4\pi m v_0}\int d^3\mathbf{r}' \frac{\mathbf{R}[\mathbf{R}\cdot\mathbf{A}(\mathbf{r}')]}{R^4} \tag{10.28}$$

where $\mathbf{R} = \mathbf{r} - \mathbf{r}'$.

The expression $\mathbf{j}(\mathbf{R})$ is given by (10.28) which is well approximated by

$$J(R) \cong J(0)\exp[-R/\xi_0]$$

where $\xi_0 = v_0/\pi\Delta_0$. Then we get in the BCS theory similar results with the Pippard phenomenological theory. The above calculation can be performed at $T \neq 0$, $\Delta \neq 0$ and for an alloy with finite mean free path. In all cases, the results may be cast in the Pippard form, justifying the phenomenological theory.

11. Microscopic Derivation of the Ginzburg-Landau Equations

The Ginzburg-Landau equations used in different particular cases were purely phenomenological with the various constants fixed by the experiment. However, the Ginzburg-Landau equations (see Sec. 4) can be obtained from the Gor'kov equations. This calculation is important because it determines the phenomenological constants in terms of microscopic parameters. For normal metal, equations for the Green function are

$$\begin{aligned} &\left[i\omega - \frac{1}{2m}(\nabla + ie\,\mathbf{A})^2 + \mu\right]G^0(\mathbf{r},\mathbf{r}',\omega) = \delta(\mathbf{r}-\mathbf{r}') \\ &\left[i\omega + \frac{1}{2m}(\nabla - ie\,\mathbf{A})^2 + \mu\right]G^0(\mathbf{r},\mathbf{r}',\omega) = \delta(\mathbf{r}-\mathbf{r}') \,. \end{aligned} \tag{11.1}$$

Using these functions, the Gor'kov equations (9.6) can be written as

$$G(\mathbf{r},\mathbf{r}';\omega) = G^0(\mathbf{r},\mathbf{r}';\omega) - \int d^3y\, G^0(\mathbf{y},\mathbf{r};\omega)\Delta(\mathbf{y})F^\dagger(\mathbf{y},\mathbf{r};\omega) \tag{11.2}$$

$$F^\dagger(\mathbf{r},\mathbf{r}';\omega) = \int d^3y\, G^0(\mathbf{y},\mathbf{r};-\omega)G^0(y,\mathbf{r};\omega)\;. \tag{11.3}$$

If we consider that Δ is small, from (11.2) and (11.3), we get

$$g^{-1}\Delta^\dagger(\mathbf{r}) = \int d^3\mathbf{y}Q(\mathbf{r},\mathbf{y})\Delta^\dagger(\mathbf{y}) + \int d^3\mathbf{y}\, d^3\mathbf{z}\, d^3\mathbf{w}R(\mathbf{r},\mathbf{y},\mathbf{z};\mathbf{w})\Delta^\dagger(\mathbf{z})\Delta^\dagger(\mathbf{w}) \tag{11.4}$$

where

$$Q(T) = \pi T\sum_\omega G^0(\mathbf{y},\mathbf{r},-\omega)G^0(\mathbf{y},\mathbf{r},\omega) \tag{11.5}$$

$$R(\mathbf{r},\mathbf{y},\mathbf{z},\mathbf{w}) = \pi T\sum_\omega G^0(\mathbf{y},\mathbf{r};-\omega)G^0(\mathbf{y},\mathbf{z};\omega)G^0(\mathbf{w},\mathbf{z};-\omega)G^0(\mathbf{w},\mathbf{z};\omega)\;. \tag{11.6}$$

We see that the assumption of the small gap $|\Delta|$ leads only to a nonlinear integral equation. We will assume now that $|T_c - T| \leq T_c$ since $\Delta^\dagger$ and $\mathbf{A}$ vary slowly with respect to the range of kernels Q and R.

In order to calculate Q and R, we evaluate $G^0(\mathbf{r},\omega)$ in the absence of a magnetic field. This function can be written as

$$G^0(\mathbf{r},\omega) = \frac{1}{(2\pi)^3}\int d^3\mathbf{p}\,\frac{\exp(i\,\mathbf{p}\cdot\mathbf{r})}{i\omega - \varepsilon(\mathbf{p})} \tag{11.7}$$

which will be evaluated for $p_0 r \gg 1$ and $|\omega| \ll \mu$. Then the dominant contribution arises from the vicinity of the Fermi surface and we find

$$\begin{aligned} G^0(\mathbf{r},\omega) &\cong N(0)\int \frac{d\varepsilon}{i\omega-\varepsilon} j_0(pr)\\ &= \frac{N(0)}{2ip_0 r}\int \frac{d\varepsilon}{i\omega-\varepsilon}\exp\left[\left(ip_0 + \frac{\varepsilon}{v_0}\right)r - \exp\left(ip_0 + \frac{\varepsilon}{v_0}\right)r\right]\\ &= -\frac{\pi N(0)}{p_0 r}\exp\left[ip_0 r\,\mathrm{sign}\,\omega - \frac{r|\omega|}{v_0}\right]\;. \end{aligned} \tag{11.8}$$

The magnetic field will be considered in $G^0(\mathbf{r},\mathbf{r}';\omega)$ by a phase factor $\theta(\mathbf{r},\mathbf{r}')$ and

$$G(\mathbf{r},\mathbf{r}';\omega) \cong \exp[i\theta(\mathbf{r},\mathbf{r}')]G^0(\mathbf{r}-\mathbf{r}';\omega) \tag{11.9}$$

with the properties

$$(\nabla + ie\,\mathbf{A})^2 G = \exp(i\theta(\mathbf{r}))\{\nabla^2 G^0(\mathbf{r}) + 2i(\nabla\theta(\mathbf{r}) + e\,\mathbf{A}(\mathbf{r}))\cdot\nabla G^0(\mathbf{r})$$
$$+ [i\nabla^2\theta(\mathbf{r}) + ie\nabla\cdot A(\mathbf{r}) - (\nabla\theta(\mathbf{r}) + e\,\mathbf{A}(\mathbf{r}))^2]G^0(\mathbf{r})\}$$

and $\theta(\mathbf{r},\mathbf{r}') = 0$.

Taking terms which contain the first power of $\mathbf{A}$, we get

$$[\nabla\theta(\mathbf{r},\mathbf{r}') + e\,\mathbf{A}]\cdot(\mathbf{r}-\mathbf{r}') = 0 \tag{11.10}$$

an equation which determines the phase $\theta(\mathbf{r},\mathbf{r}')$. Let us calculate the kernel $Q(\mathbf{r},\mathbf{r}')$ from (11.5), and using (11.9) we get

$$Q(\mathbf{r},\mathbf{r}') = \pi T\sum_\omega \exp[2i\theta(\mathbf{r},\mathbf{r}')]G^0(\mathbf{y},-\mathbf{r};-\omega)G^0(\mathbf{y},-\mathbf{r};\omega)\,. \tag{11.11}$$

The kernel $Q(\mathbf{r},\mathbf{r}')$ will be evaluated from (11.11) and using (11.8) we obtain

$$Q(\mathbf{r},\mathbf{r}') = \left(\frac{\pi N(0)}{p_0 r}\right)^2 \pi T\sum_\omega \exp\left[-\frac{2r|\omega|}{v_0}\right]$$
$$= \left(\frac{\pi N(0)}{p_0 r}\right)^2 \left[\frac{\sin 2\pi r}{v_0}\right]^{-1}\,. \tag{11.12}$$

Near T_c, from (10.10), the phase $\theta(\mathbf{r},\mathbf{r}')$ is

$$\theta(\mathbf{r},\mathbf{r}') = -\frac{e}{2}[\mathbf{A}(\mathbf{r}) + \mathbf{A}(\mathbf{r}')]\cdot(\mathbf{r}-\mathbf{r}') \tag{11.13}$$

and in the same temperature regions $H \simeq H_c$, (11.13) can be expanded in powers of θ. If we define now $\mathbf{z} = \mathbf{y} - \mathbf{r}$ in the second term of (11.4), we have

$$\int d^3\mathbf{y}Q(\mathbf{r},\mathbf{y})\Delta^\dagger(\mathbf{y}) = \int d^3\mathbf{z}Q^0(\mathbf{z})\{-ie[\mathbf{A}(\mathbf{r}) + \mathbf{A}(\mathbf{r}')]\cdot\mathbf{z}\Delta^\dagger(\mathbf{r}+\mathbf{z})\}$$
$$= \int d^3\mathbf{z}Q^0(\mathbf{z})[1 - ie(\mathbf{A}(\mathbf{r}) + \mathbf{A}(\mathbf{r}+\mathbf{z}))\cdot\mathbf{z}]$$
$$-\frac{1}{2}e^2\{[\mathbf{A}(\mathbf{r}) + \mathbf{A}(\mathbf{r}+\mathbf{z})]\cdot\mathbf{z} + \dots\}\Delta^\dagger(\mathbf{r}+\mathbf{z})\,. \tag{11.14}$$

The short range Q^0 requires for z the condition $z \leq \xi_0$ and the remaining function may be expanded in a Taylor series about $z = 0$, and retaining the leading correction term we find

$$\int d^3\mathbf{y}Q(\mathbf{r},\mathbf{y})\Delta^\dagger(\mathbf{y}) \cong \Delta^\dagger(\mathbf{r})\int d^3\mathbf{z}Q^0(\mathbf{z})$$
$$+\frac{1}{6}[(\nabla - 2ie\,\mathbf{A}(\mathbf{r}))]^2\Delta(\mathbf{r})\int d^3\mathbf{z}\,\mathbf{z}^2Q^0(\mathbf{z})\,. \tag{11.15}$$

In (11.15), the integral $\int d^3\mathbf{z}\, Q^0(\mathbf{z})$ diverges logarithmically at the origin, and we will overcome this difficulty given by the short range potential from BCS by considering a cut off in the momentum space at $|\varepsilon(\mathbf{p})| \ll \omega_D$ and

$$\int d^3\mathbf{z} Q(\mathbf{z}) = \frac{1}{(2\pi)^3} \int d^3\mathbf{p} \pi T \sum_{\omega} (\omega^2 + \varepsilon^2(\mathbf{p}))^{-1}$$
$$= N(0) \ln \frac{T_c}{T} + \frac{1}{|g|} . \tag{11.16}$$

The next integral from (11.15) is

$$\int d^3\mathbf{z}\, z^2 Q^0(z) \cong \frac{1}{4} N(0) \frac{v_0^2}{(\pi T)^2} \int_0^\infty dy \, \frac{y^2}{\sinh y}$$
$$= \frac{7\varsigma(3)}{8} N(0) \left(\frac{v_0}{\pi T_c}\right)^2 . \tag{11.17}$$

In (11.4), we still have a nonlinear term which will be evaluated in the lowest order by setting $G^0 = G^0$ and taking $\Delta^\dagger(y)\Delta(z)\Delta^\dagger(w)$ as $\Delta^\dagger(\mathbf{r})|\Delta(\mathbf{r})|^2$. Using the result

$$\int d^3\mathbf{y}\, d^3\mathbf{z}\, d^3\mathbf{w}\, R(\mathbf{r}, \mathbf{y}, \mathbf{z}, \mathbf{w}) = -N(0) \frac{7\varsigma(3)}{8} \frac{1}{\pi T_c} ,$$

we get from (11.14), (11.15), (11.16) and (11.18), the equation

$$\frac{1}{2m} [\nabla - 2ei\, \mathbf{A}(\mathbf{r})]^2 \Delta^\dagger(\mathbf{r}) + \frac{6\pi^2 T_c^2}{7\varsigma(3)\varepsilon_0} \left[\frac{T - T_c}{T_c} \Delta^\dagger(\mathbf{r}) - \frac{7\varsigma(3)\Delta^\dagger(\mathbf{r})}{8(\pi T_c)^2} |\Delta(r)|^2 \right] = 0. \tag{11.18}$$

If we introduce the notation

$$\Psi(\mathbf{r}) = \frac{7\varsigma(3) m}{8(\pi T_c)} \Delta(\mathbf{r}) \tag{11.19}$$

we get from (11.18)

$$\frac{1}{4m} [-i\nabla + 2e\, \mathbf{A}(\mathbf{r})]^2 \Psi(\mathbf{r}) + \frac{6\pi^2 T_c}{7\varsigma(3)\varepsilon_0} \left(\frac{T - T_c}{T_c} \right) \Psi(\mathbf{r}) + \frac{1}{m} |\Psi|^2 \Psi(\mathbf{r}) = 0 , \tag{11.20}$$

which is identical to the phenomenological Ginzburg-Landau equation.

12. Quasiclassical Approximation

The Gor'kov equations can be generalized for pairs with velocity $\mathbf{v}_s$ different from zero. If we use Eq. (3.16) (taking $\mathbf{v}_s \equiv \mathbf{v}_s/2m$)

$$\mathbf{v}_s(\mathbf{r}) = \nabla\theta(\mathbf{r}) - \frac{e}{m}\mathbf{A}(\mathbf{r}) \,, \tag{12.1}$$

the Gor'kov equations become

$$\begin{aligned} &\left[i\omega - \frac{1}{2m}(\mathbf{p}_1 + m\mathbf{v}_s(\mathbf{r}_1))^2 + \mu\right]\widetilde{G}(\mathbf{r}_1, \mathbf{r}_2; \omega) + |\Delta(\mathbf{r}_1)|\widetilde{F}(\mathbf{r}_1, \mathbf{r}_2; \omega) \\ &\qquad = \delta(\mathbf{r}_1 - \mathbf{r}_2) \\ &\left[i\omega + \frac{1}{2m}(\mathbf{p}_1 - m\mathbf{v}_s(\mathbf{r}_1))^2 - \mu\right]\widetilde{F}(\mathbf{r}_1, \mathbf{r}_2; \omega) + |\Delta(\mathbf{r}_1)|\widetilde{G}(\mathbf{r}_1, \mathbf{r}_2; \omega) = 0 \end{aligned} \tag{12.2}$$

where

$$\begin{aligned} G(\mathbf{r}_1, \mathbf{r}_2, ;\omega) &= \exp[im(\theta(\mathbf{r}_1) - \theta(\mathbf{r}_2))]\widetilde{G}(\mathbf{r}_1, \mathbf{r}_2; \omega) \\ F(\mathbf{r}_1, \mathbf{r}_2, ;\omega) &= \exp[im(\theta(\mathbf{r}_1) - \theta(\mathbf{r}_2))]\widetilde{F}(\mathbf{r}_1, \mathbf{r}_2; \omega) \,. \end{aligned} \tag{12.3}$$

The Green functions from (12.2) describe states with $\mathbf{j}_0 \neq 0$. From (12.1), one obtains for the (superfluid) velocity of the pairs

$$\nabla \times \mathbf{v}_s = -\frac{e}{m}\mathbf{H} \,. \tag{12.4}$$

The physical properties of the superconducting state with $\mathbf{v}_s \neq 0$ can be studied if one gives a method to solve these equations. We will start with the local approximation in which we will change the variables

$$(\mathbf{r}_1, \mathbf{r}_2) \to \mathbf{r} = \mathbf{r}_1 - \mathbf{r}_2 \,, \quad R = \frac{1}{2}(\mathbf{r}_1 + \mathbf{r}_2) \tag{12.5}$$

$$\begin{aligned} \mathbf{p}_1 &= -i\left(\frac{1}{2}\nabla_R + \nabla_r\right) = \mathbf{p} - \frac{i}{2}\nabla_\mathbf{R} \\ \mathbf{r}_1 &= \mathbf{R} + \frac{i}{2}\nabla_\mathbf{p} \,. \end{aligned} \tag{12.6}$$

This new representation is called the Wigner representation. In addition, we shall neglect the spatial gradients of fields and order parameter. The general formula for the Fourier transform in this case is

$$f(\mathbf{R}, \mathbf{r}) = \int \frac{d^3\mathbf{p}}{(2\pi)^3} f(\mathbf{R}, \mathbf{p}) e^{i\mathbf{p}\cdot\mathbf{r}} \tag{12.7}$$

and the Green functions $G(\mathbf{R},\mathbf{p};\omega)$ and $F(\mathbf{R},\mathbf{p};\omega)$ satisfy the equations

$$\begin{aligned}&\left\{i\omega-\frac{1}{2m}\left[\mathbf{p}-\frac{i}{2}\nabla_R+m\mathbf{v}_s(\mathbf{R}+\frac{i}{2}\nabla_s)\right]^2+\mu\right\}G(\mathbf{R},\mathbf{p};\omega)\\&\quad+|\Delta(\mathbf{R}+\frac{i}{2}\nabla_\mathbf{p})|F(\mathbf{R},\mathbf{p};\omega)=1\\&\left\{i\omega+\frac{1}{2m}\left[\mathbf{p}-\frac{i}{2}\nabla_R-m\mathbf{v}_s(\mathbf{R}+\frac{i}{2}\nabla_p)\right]^2-\mu\right\}F(\mathbf{R},\mathbf{p};\omega)\\&\quad+|\Delta(\mathbf{R}+\frac{i}{2}\nabla_\mathbf{p})|G(\mathbf{R},\mathbf{p};\omega)=0\,.\end{aligned}\tag{12.8}$$

The order parameter and the current are defined as

$$|\Delta(\mathbf{R})|=|g|\pi T\sum_\omega\int\frac{d^3\mathbf{p}}{(2\pi)^3}F(\mathbf{R},\mathbf{p};\omega)\tag{12.9}$$

$$J(\mathbf{R})=\frac{2e}{\pi}\pi T\sum_\omega\int\frac{d^3\mathbf{p}}{(2\pi)^3}(\mathbf{p}+m\mathbf{v}_s)G(\mathbf{R},\mathbf{p};\omega)\,.\tag{12.10}$$

In the zero order approximation to the gradients. Eqs. (12.8) become

$$\begin{aligned}&\left[i\omega-\frac{1}{2m}(\mathbf{p}+m\mathbf{v}_s)^2+\mu\right]G(\mathbf{R},\mathbf{p};\omega)+|\Delta(\mathbf{R})|F(\mathbf{R},\mathbf{p};\omega)=1\\&\left[i\omega+\frac{1}{2m}(\mathbf{p}-\mathbf{v}_s m)^2-\mu\right]F(\mathbf{R},\mathbf{p};\omega)+|\Delta(\mathbf{R})|G(\mathbf{R},\mathbf{p};\omega)=0\end{aligned}\tag{12.11}$$

with solutions

$$\begin{aligned}G(\mathbf{R},\mathbf{p};\omega)&=\frac{i\omega-\mathbf{p}\cdot\mathbf{v}_s+\varepsilon}{[(i\omega-\mathbf{p}\cdot\mathbf{v}_s-\varepsilon)(i\omega-\mathbf{p}\cdot\mathbf{v}_s+\varepsilon)]-|\Delta(\mathbf{p})|^2}\\F(\mathbf{R},\mathbf{p};\omega)&=\frac{|\Delta|}{[(i\omega-\mathbf{p}\cdot\mathbf{v}_s-\varepsilon)(i\omega-\mathbf{p}\cdot\mathbf{v}_s+\varepsilon)]-|\Delta(\mathbf{p})|^2}\end{aligned}\tag{12.12}$$

where $\varepsilon=p^2/2m-\mu$ and $\mu'=\mu-mv_s^2/2$.
From Eq. (12.9) and using the expression (12.12) for F, the equation for the order parameter becomes

$$1=\frac{|g|}{2}\int\frac{d^3\mathbf{p}}{(2\pi)^3}\left\{1-f\left(\frac{E(\mathbf{p})+\mathbf{p}\cdot\mathbf{v}_s}{T}\right)-f\left(\frac{E(\mathbf{p})-\mathbf{p}\cdot\mathbf{v}_s}{T}\right)\right\}\,.\tag{12.13}$$

For $p \simeq p_0$ (p_0 the Fermi momentum), one may perform the integration over angles in (12.13) and

$$\frac{1}{gN(0)} = \int_0^{\omega_D} \frac{d\varepsilon}{E(p)} \left\{1 - \frac{T}{p_0 v_s} \ln \frac{1-\exp[(p_0 v_s - E(\mathbf{p}))/T]}{1+\exp[(p_0 v_s - E(\mathbf{p}))/T]}\right\} \qquad (12.14)$$

where $E(\mathbf{p}) = \varepsilon^2(\mathbf{p}) + |\Delta|^2$ and $gN(0) > 0$. This equation gives the order parameter Δ as a function of the temperature T and the superfluid velocity v_s. In the case $T = 0$ from (12.14), we get

$$\Delta(0) = 2\omega_D \exp\left[-\frac{1}{gN(0)}\right] \equiv \Delta_0(0) \qquad (12.15)$$

if $v_s < \Delta(0)/p_0$, where $\Delta_0(0)$ is the gap for $v_s = 0$. For $v_s > \Delta(0)/p_0$, one obtains

$$\ln \frac{p_0 v_s}{\Delta_0(0)} = \left[1 - \left(\frac{\Delta(0)}{p_0 v_s}\right)^2\right]^{\frac{1}{2}} - \ln\left[1 + \sqrt{1 - \left(\frac{\Delta(0)}{p_0 v_s}\right)^2}\right] . \qquad (12.16)$$

From these equation one can see that $\Delta(0)$ becomes zero for a critical velocity

$$v_s^c = \frac{e}{2}\frac{\Delta_0(0)}{p_0} . \qquad (12.17)$$

This velocity is called critical because the current $j_s(v_s^c) = 0$. If $v_s \neq 0$, the order parameter is identical to a gap in the energy of excitations. The gap in the energy of excitations is $\Delta_0(0) - p_0 v_s$ if $v_s < \Delta_0(0)/p_0$ and zero for $v_s > \Delta(0)/p_0$. In the interval $\Delta_0(0)/p_0$ to v_s^c we have gapless superconductivity. Using the local approximation, we can give a new set of equations to describe the superconducting state. Experimental results showed that $T_c/E_0 \ll 1$ for the superconductors. On the other hand, (12.16) gives $v_s^c \sim T_c/p_0$ which is equivalent to $mv_s^c/p_0 \sim T_c/E_0 \ll 1$ and we can conclude that $v_s \ll v_0$ (v_0 is the Fermi velocity) which means that the electrons in the superconducting state are strongly degenerated, but the Cooper pairs can be described quasiclassically. This observation gives us the possibility to reconsider Eqs. (12.8) using the approximations

$$\frac{1}{2m}\left(\mathbf{p} - \frac{i}{2}\nabla_R + mv_s\right)^2 - \mu \cong \varepsilon - \frac{i}{v_0}\mathbf{n}\cdot\nabla_R + p_0\mathbf{n}\cdot\mathbf{v}_s \qquad (12.18)$$

where $\mathbf{p} = \mathbf{n}(p_0 + \varepsilon/v_0)$, $\varepsilon \sim T_c$ and the operator acting on a function f satisfies

$$\nabla_R f = f/\xi_0 \, .$$

In the approximation (12.18) we neglect terms of the order

$$\varepsilon^2/2mv_s \sim T_c\,, \quad mv_s^2 \sim T_c\,, \quad \nabla_R^2 \sim \frac{1}{\xi_0^2}\,, \quad \frac{\varepsilon}{v_0} \sim T_c\,. \tag{12.19}$$

Now we have to consider $v_s(\mathbf{R} + \frac{i}{2}\nabla_p)$ and $\Delta(\mathbf{R} + \frac{i}{2}\nabla_p)$. Using the approximations

$$\begin{aligned} &\frac{\partial}{\partial p} = \frac{\partial\varepsilon}{\partial p}\frac{\partial}{\partial\varepsilon} \cong v_0\frac{\partial}{\partial\varepsilon_0}\,, \quad \frac{1}{p}\frac{\partial}{\partial\theta} = \frac{1}{p_0 + \varepsilon/v_0}\frac{\partial}{\partial\theta}\,, \\ &\frac{1}{p\sin\theta}\frac{\partial}{\partial\varphi} \simeq \frac{\partial/\partial\varphi}{p_0\sin\theta}\,, \end{aligned} \tag{12.20}$$

Eqs. (12.8) can be written in the quasiclassical approximation as

$$\begin{aligned} &\left[i\omega - \varepsilon + \frac{i}{2}v_0\mathbf{n}\nabla_{\mathbf{R}} - p_0\mathbf{n}\cdot\mathbf{v}_s\left(\mathbf{R} + \frac{i}{2}\mathbf{n}v_0\frac{d}{d\varepsilon}\right)\right]G(\mathbf{R},\mathbf{n},\varepsilon;\omega) \\ &\quad + \Delta\left(\mathbf{R} + \frac{i}{2}v_0\mathbf{n}\frac{d}{d\varepsilon}\right)F(\mathbf{R},\mathbf{n},\varepsilon;\omega) = 1 \\ &\left[i\omega + \varepsilon - \frac{i}{2}v_0\mathbf{n}\cdot\nabla_{\mathbf{R}} - p_0\mathbf{n}\cdot\mathbf{v}_s\left(\mathbf{R} + \frac{i}{2}\mathbf{n}v_0\frac{d}{d\varepsilon}\right)\right]F(\mathbf{R},\mathbf{n},\varepsilon;\omega) \\ &\quad + \Delta\left(\mathbf{R} + \frac{i}{2}\mathbf{n}v_0\frac{d}{d\varepsilon}\right)G(\mathbf{R},\mathbf{n},\varepsilon;\omega) = 0\,. \end{aligned} \tag{12.21}$$

Using now the Fourier transform

$$f(t) = \int \frac{d\varepsilon}{2\pi} f(\varepsilon)\exp(i\varepsilon t)$$

Eqs. (12.21) become

$$\begin{aligned} &\left[i\omega + i\frac{d}{dt} + \frac{i}{2}v_0\mathbf{n}\cdot\nabla_R - p_0\mathbf{n}\cdot v_s\left(\mathbf{R} + \frac{1}{2}\mathbf{n}v_0t\right)\right]G(\mathbf{R},\mathbf{n},t;\omega) \\ &\quad + \left|\Delta\left(\mathbf{R} + \frac{1}{2}\mathbf{n}v_0t\right)\right|F(\mathbf{R},\mathbf{n},t;\omega) = \delta(t) \\ &\left[i\omega - i\frac{d}{dt} - \frac{i}{2}v_0\mathbf{n}\cdot\nabla_R - p_0\mathbf{n}\cdot v_s\left(\mathbf{R} + \frac{1}{2}\mathbf{n}v_0t\right)\right]F(\mathbf{R},\mathbf{n},t;\omega) \\ &\quad + \left|\Delta\left(\mathbf{R} + \frac{1}{2}\mathbf{n}v_0t\right)\right|G(\mathbf{R},\mathbf{n},t;\omega) = 0 \end{aligned} \tag{12.22}$$

which can be written as

$$\begin{aligned} &\left[i\omega\hat{1} + i\hat{\sigma}_z\left(\frac{d}{dt} + \frac{1}{2}v_0t\mathbf{n}\cdot\nabla_R\right) - p_0\mathbf{n}\cdot\mathbf{v}_s\left(\mathbf{R} + \frac{1}{2}\mathbf{n}v_0t\right)\right. \\ &\quad \left. - \hat{\sigma}_x\Delta\left(\mathbf{R} + \frac{1}{2}\mathbf{n}v_0t\right)\right]G(\mathbf{R};\mathbf{n},t;\omega) = \delta(t) \end{aligned} \tag{12.23}$$

where

$$G = \frac{1}{2}\mathrm{Tr}\,(1+\hat{\sigma}_z)\hat{G}, \qquad F = -\frac{1}{2}\mathrm{Tr}\,(\hat{\sigma}_x + i\hat{\sigma}_y)\hat{G}\ . \tag{12.24}$$

If we perform the transform

$$\mathbf{R}' = \mathbf{R} - \frac{1}{2}\mathbf{n}v_0 t\,, \quad t' = t$$

and

$$\frac{\partial}{\partial R} = \frac{\partial}{\partial R'}\,; \qquad \frac{\partial}{\partial t} = \frac{\partial}{\partial t'} - \frac{1}{2}v_0\mathbf{n}\cdot\frac{\partial}{\partial \mathbf{R}'}\,, \tag{12.25}$$

Eq. (12.23) becomes

$$\left[i\omega\hat{1} - i\hat{\sigma}_z\frac{d}{dt} - p_0\mathbf{n}v_s(\mathbf{R}+\mathbf{n}v_0 t)\right]\hat{G}(\mathbf{R},\mathbf{n},t;\omega) = 0\ . \tag{12.26}$$

From (12.26), we can write the equation which describes the influence of the non-magnetic impurities as

$$\begin{aligned}\Big\{i\omega\hat{1} + i\hat{\sigma}_z\Big(\frac{d}{dt} + \frac{1}{2}\mathbf{n}v_0\nabla_R\Big) - \frac{1}{2}\Big[\Delta\Big(\mathbf{R}+\frac{1}{2}v_0\mathbf{n}t\Big)\hat{\sigma}_+ \\ +\Delta^*\Big(\mathbf{R}+\frac{1}{2}\mathbf{n}v_0 t\Big)\hat{\sigma}_-\Big] - \frac{1}{\tau}\int\frac{d\mathbf{n}}{4\pi}\hat{G}(\mathbf{R},\mathbf{n},t;\omega)\Big\}\,\hat{G}(\mathbf{R},\mathbf{n},t;\omega) = \delta(t)\end{aligned} \tag{12.27}$$

where τ is the scattering time of the electrons by impurities.
In the homogeneous state, the Green function G depends only on time and for a pure superconductor, Eq. (12.23) becomes

$$\left[i\omega\hat{1} + i\hat{\sigma}_z\frac{d}{dt} - \hat{\sigma}_x\Delta\right]\hat{G}(\omega,t) = \delta(t) \tag{12.28}$$

which can be solved by standard methods. Taking

$$\hat{G}(\omega,t) = \Big(i\omega\hat{1} - i\hat{\sigma}_z\frac{d}{dt} + \hat{\sigma}_x\Delta\Big)K(\omega,t) \tag{12.29}$$

one obtains

$$\Big(\frac{d^2}{dt^2} - \omega^2 - \Delta^2\Big)K(\omega,t) = \delta(t) \tag{12.30}$$

which gives

$$K(\omega,t) = -\frac{1}{2\pi}\int\frac{dt\,\exp(i\varepsilon t)}{\varepsilon^2+\omega^2+\Delta^2}\,,$$

and from (12.29), $G(\omega, t)$ and $F(\omega, t)$ are

$$
\begin{aligned}
G(\omega,t) &= -\frac{i}{2}\left[\frac{\omega}{\sqrt{\omega^2+\Delta^2}} + \operatorname{sign} t\right]\exp[-\sqrt{\omega^2+\Delta^2}\,|t|] \\
F(\omega,t) &= \frac{\Delta}{2\sqrt{\omega^2+\Delta^2}}\exp[-\sqrt{\omega^2+\Delta^2}\,|t|]\,.
\end{aligned}
\tag{12.31}
$$

The general equation for current in this formalism is

$$
J(\mathbf{R}) = \frac{1}{2}ev_0N(0)\pi T\sum_\omega \int d\mathbf{n}\,\mathrm{Tr}\,(1+\hat{\sigma}_z)\hat{G}(\mathbf{R},\mathbf{n},t;\omega) \tag{12.32}
$$

and if the current has the direction Oz, Eq. (12.32) can be written as

$$
J(\mathbf{R}) = 2\pi ev_0\pi T\sum_\omega \int_0^1 dx[G(\omega,x,0) - G(\omega,-x,0)] \tag{12.33}
$$

where

$$
G(\omega,x) = \frac{-i\omega + p_0v_sx}{2\sqrt{(\omega+ip_0v_sx)^2+\Delta^2}}\,. \tag{12.34}
$$

In linear approximation of v_s, Eq. (12.33) becomes

$$
\begin{aligned}
\mathbf{J} &= \frac{2}{3}\pi ev_0p_0N(0)v_s\pi T\sum_\omega \Delta^2[\omega^2+\Delta^2]^{-\frac{3}{2}} \\
&= \pi ev_s\Delta^2\pi T\sum_\omega[\omega^2+\Delta^2]^{-\frac{3}{2}}
\end{aligned}
\tag{12.35}
$$

where

$$
n = \frac{p_0^3}{3\pi^2} = \frac{2}{3}v_0p_0N(0)\,.
$$

Using now $J = en_sv_s$, (12.35) leads to

$$
\frac{n_s}{n} = \Delta^2\pi T\sum_\omega[\omega^2+\Delta^2]^{-\frac{3}{2}} \tag{12.36}
$$

which gives the density of superconducting electrons as function of temperature.

Let us consider (12.27) for the homogeneous superconducting state. These equations are

$$
\begin{aligned}
&\left[i\omega + i\frac{d}{dt} - \frac{1}{\tau_0}G(\omega;0)\right]G(\omega;t) + (\Delta + \tau_0^{-1}F(\omega,0))F(\omega,t) = \delta(t) \\
&\left[i\omega + i\frac{d}{dt} - \frac{1}{\tau_0}G(\omega;0)\right]F(\omega,t) + (\Delta^* + \tau_0^{-1}F(\omega,0))G(\omega,t) = 0
\end{aligned}
\tag{12.37}
$$

or after the substitutions

$$\omega' = \omega + \frac{i}{\tau_0} G(\omega, 0) , \quad \Delta(\omega, 0) = \Delta + \frac{1}{\tau_0} F^*(\omega, 0) \tag{12.38}$$

one gets the equations

$$\begin{aligned} &\left[i\omega' + i\frac{d}{dt}\right] G(\omega, t) + \Delta(\omega) F(\omega, t) = \delta(t) \\ &\left[i\omega' - i\frac{d}{dt}\right] F(\omega, t) + \Delta(\omega) G(\omega, t) = 0 \end{aligned} \tag{12.39}$$

with the solutions

$$\begin{aligned} G(\omega', t) &= -\frac{i}{2}\left[\frac{\omega'}{\sqrt{\omega'^2 + \Delta^2}} + \operatorname{sign} t\right] \exp[-\sqrt{\omega'^2 + \Delta^2}\,|t|] \\ F(\omega', t) &= \frac{\Delta(\omega)}{2\sqrt{\omega'^2 + \Delta^2}} \exp[-\sqrt{\omega'^2 + \Delta^2}\,|t|] \,. \end{aligned} \tag{12.40}$$

Before using these results let us go back to the Gor'kov equations (9.6) which can be written as

$$\widehat{G}(\mathbf{r}_1, \mathbf{r}_2; \omega)\left[i\omega\widehat{1} - \widehat{\sigma}_z\widehat{\varepsilon} - \frac{1}{2}(\Delta(\mathbf{r}_2)\widehat{\sigma}_+ + \Delta^*(\mathbf{r}_2)\widehat{\sigma}_-)\right] = \delta(\mathbf{r}_1 - \mathbf{r}_2) \tag{12.41}$$

where in $\hat{\varepsilon}$, the electromagnetic field is zero. Using the same method as for (12.23), one obtains

$$\begin{aligned} \widehat{G}(\mathbf{R}, \mathbf{n}, t; \omega)\Big[i\omega\widehat{1} &- i\widehat{\sigma}_z\left(\frac{d}{dt} - \frac{1}{2}v_0 t\mathbf{n}\nabla_R\right) - p_0\mathbf{n}\cdot\mathbf{v}_s(\mathbf{R} - \frac{1}{2}tv_0\mathbf{n}) \\ &- \widehat{\sigma}_x\Delta\left(\mathbf{R} - \frac{1}{2}v_0 t\mathbf{n}\right)\Big] = \delta(t) \,. \end{aligned} \tag{12.42}$$

From (12.23) and (12.42), we obtain the equation

$$iv_0\mathbf{n}\nabla_R\widehat{G}(\mathbf{R}, \mathbf{n}, 0; \omega) + \widehat{\omega}\widehat{G}(\mathbf{R}, \mathbf{n}, 0; \omega) - \widehat{G}(\mathbf{R}, \mathbf{n}, 0; \omega)\widehat{\omega}\widehat{\sigma}_z = 0 \tag{12.43}$$

where we considered the Green function $\widehat{G}(\mathbf{R}, \mathbf{n}, t; \omega)$ for $t = 0$ and

$$\widehat{\omega} = (i\omega - p_0\mathbf{n}\cdot\mathbf{v}_s(\mathbf{R}))\widehat{\sigma}_z - i\Delta(\mathbf{R})\widehat{\sigma}_y \,. \tag{12.44}$$

If we introduce the function

$$\widehat{\mathcal{G}}(\mathbf{R}, \mathbf{n}; \omega) = \widehat{G}(\mathbf{R}, \mathbf{n}, 0; \omega)\widehat{\sigma}_z \,, \tag{12.45}$$

Eq. (12.43) becomes

$$iv_0\mathbf{n}\cdot\nabla_R\widehat{\mathcal{G}}(\mathbf{R},\mathbf{n};\omega)+\widehat{\omega}\widehat{\mathcal{G}}(\mathbf{R},\mathbf{n};\omega)-\widehat{\mathcal{G}}(\mathbf{R},\mathbf{n};\omega)\widehat{\omega}=0 \tag{12.46}$$

which is known as the Eilenberger equation.[7]
In (12.48), the non-magnetic impurities can be considered by performing the transform

$$\widehat{\omega}=(i\omega-p_0\mathbf{n}\cdot\mathbf{v}_s(\mathbf{R})\widehat{\sigma}_z-\widehat{\Delta}(\mathbf{R}))-\frac{1}{\tau}\int\frac{d\mathbf{n}}{4\pi}\widehat{\mathcal{G}}(\mathbf{R},\mathbf{n};\omega)\,, \tag{12.47}$$

$$\widehat{\Delta}(\mathbf{R})=\begin{bmatrix}0 & \Delta(\mathbf{R})\\ -\Delta^*(\mathbf{R}) & 0\end{bmatrix}\,. \tag{12.48}$$

If one writes (12.46) as

$$iv_0\mathbf{n}\cdot\nabla_R\widehat{\mathcal{G}}(\mathbf{R},\mathbf{n};\omega)+[\widehat{\omega},\widehat{\mathcal{G}}]=0\,, \tag{12.49}$$

one easily obtains

$$\mathbf{n}\cdot\nabla_R\mathrm{Tr}\widehat{\mathcal{G}}(\mathbf{R},\mathbf{n};\omega)=0\,. \tag{12.50}$$

In a superconductor with a strong Meissner effect, the Green function $\widehat{\mathcal{G}}$ can be approximated as

$$\widehat{\mathcal{G}}^0(\mathbf{R},\mathbf{n};\omega)=\frac{-i\omega\widehat{\sigma}_z+i\Delta\sigma_y}{2\sqrt{\omega^2+\Delta^2}} \tag{12.51}$$

which satisfies $\mathrm{Tr}\,\widehat{\mathcal{G}}^0(\mathbf{R},\mathbf{n};\omega)=0$ and we will take as a good approximation, the same property for $\widehat{\mathcal{G}}$, then

$$\mathrm{Tr}\,\widehat{\mathcal{G}}(\mathbf{R},\mathbf{n};\omega)=0\,. \tag{12.52}$$

If we calculate $\widehat{\mathcal{G}}^2(\mathbf{R},\mathbf{n},\omega)$, we obtain

$$\mathrm{Tr}\,\widehat{\mathcal{G}}^2(\omega)=2(G^2-F\widehat{F}) \tag{12.53}$$

and

$$\widehat{\mathcal{G}}^{20}(\omega)=-\frac{1}{4}\widehat{1}+\frac{\omega\Delta[\sigma_y,\sigma_y]}{4(\omega^2+\Delta^2)} \tag{12.54}$$

which gives

$$\mathrm{Tr}\,\widehat{\mathcal{G}}^2(\omega)=-\frac{1}{2} \tag{12.55}$$

and finally for $\widehat{\mathcal{G}}^2(\omega)$, one gets

$$\widehat{\mathcal{G}}^2(\omega) = -\frac{1}{4} . \tag{12.56}$$

Equation (12.49) has been generalized by Ushadel[8] to study the influence of the impurities at high concentrations on the superconducting state. This problem is very important for anisotropic superconductors. If we introduce

$$\widehat{\mathcal{G}}^0(\mathbf{R},\omega) = \int \frac{d\mathbf{n}}{4\pi}\widehat{\mathcal{G}}(\mathbf{R},\mathbf{n};\omega) \tag{12.57}$$

for the impurities with $l \ll \xi_0$ one may consider the solution of (12.49) as

$$\widehat{\mathcal{G}}(\mathbf{R},\mathbf{n};\omega) = \widehat{\mathcal{G}}^0(\mathbf{R},\omega) + n\widehat{\Gamma}(\mathbf{R}) \tag{12.58}$$

which becomes

$$\frac{iv_0}{3}\nabla_R\widehat{\Gamma}(\mathbf{R},\omega) + [\omega, \widehat{\mathcal{G}}^0(\mathbf{R},\omega)]_- = 0 \tag{12.59}$$

and from (12.57), we get

$$iv_0\nabla_R\widehat{\mathcal{G}}^0(\mathbf{R},\omega) + [\widehat{\omega}, \widehat{\Gamma}(\mathbf{R})]_- = 0 \tag{12.60}$$

which gives

$$iv_0\nabla_R\widehat{\mathcal{G}}^0(\mathbf{R},\omega) + \left[i\omega\widehat{\sigma}_z - \Delta - \frac{1}{\tau_0}\widehat{\mathcal{G}}^0(\mathbf{R},\omega), \widehat{\Gamma}(\mathbf{R},\omega)\right]_- = 0 . \tag{12.61}$$

Using (12.56), $\widehat{\Gamma}(\mathbf{R},\omega)$ can be calculated as

$$\widehat{\Gamma}(\mathbf{R},\omega) = -2il\widehat{\mathcal{G}}^0(\mathbf{R},\omega)\nabla\widehat{\mathcal{G}}^0(\mathbf{R},\omega) \tag{12.62}$$

and (12.61) becomes

$$D\nabla_R[\widehat{\mathcal{G}}^0(\mathbf{R},\omega)\nabla_R\widehat{\mathcal{G}}^0(\mathbf{R},\omega)] + \frac{1}{2}[\widehat{\omega}, \widehat{\mathcal{G}}^0(\mathbf{R},\omega)]_- = 0 \tag{12.63}$$

where $D = lv_0/3$. If $\widehat{\mathcal{G}}^0$ is a matrix defined as

$$\widehat{\mathcal{G}}^0 = \begin{bmatrix} G & F \\ -F & -G \end{bmatrix}$$

with property (12.56) and $G(\omega,t)$, $F(\omega,t)$ are well approximated with $G(\omega,\infty)=\omega/\sqrt{\omega^2+\Delta^2}$, $F(\omega,\infty)=\Delta/\sqrt{\omega^2+\Delta^2}$, Eq. (12.63) gives the Ushadel equations

$$\omega F(\omega)+\frac{1}{2}D[F(\omega)\nabla^2 G(\omega)-G(\omega)\nabla^2 F(\omega)]=\Delta^* G(\omega)\,, \tag{12.64}$$

$$|G(\omega)|^2+|F(\omega)|^2=1 \tag{12.65}$$

which have been used to calculate the current as

$$\mathbf{j}(\mathbf{R})=\pi e n_s v_s(\mathbf{R})\frac{e\Delta}{v_0}\tanh\frac{\Delta}{2T}\,. \tag{12.66}$$

As an application of Eq. (12.23), we consider the superconductor in a constant weak magnetic field. Taking

$$\delta\widehat{G}(\mathbf{R},\mathbf{n},t;\omega)=\widehat{G}^0(\mathbf{R},\mathbf{n},t;\omega)+\delta\widehat{G}(\mathbf{R},\mathbf{n},t;\omega) \tag{12.67}$$

from (12.23), one obtains

$$\left[i\omega+i\frac{d}{dt}\widehat{\sigma}_z+\Delta\widehat{\sigma}_x\right]\delta G(\mathbf{R},\mathbf{n},t;\omega)=p_0\mathbf{n}\cdot\mathbf{v}_s(\mathbf{R}+\mathbf{n}\,v_0t)\widehat{G}(\mathbf{R},\mathbf{n}+\omega) \tag{12.68}$$

which gives the result

$$\delta\widehat{G}(\mathbf{R},\mathbf{n},t;\omega)=p_0\int dt'\widehat{G}^0(\omega,t'-t)\mathbf{n}\cdot\mathbf{v}_s(\mathbf{R}+\mathbf{n}t'v_0)\widehat{G}^0(\omega,t')$$

but we are confident with $\delta\widehat{G}(\mathbf{R},\omega)$ given by

$$\delta\widehat{G}(\mathbf{R},\mathbf{n},0;\omega)=p_0\int dt\widehat{G}^0(\omega,-t)\mathbf{n}\cdot\mathbf{v}_s(\mathbf{R}+\mathbf{n}\,tv_0)\widehat{G}^0(\omega,t) \tag{12.69}$$

and using (12.32), the current $j(\mathbf{R})$ is given by the equation

$$\mathbf{j}(\mathbf{R})=\frac{3}{4}en\pi T\sum_\omega\frac{\Delta^2}{(\omega^2+\Delta^2)}\int d\mathbf{n}\cdot\mathbf{n}\int dt \\ \times\exp[-2\sqrt{\omega^2+\Delta^2}\,|t|]\mathbf{n}\,v_s(\mathbf{R}+\mathbf{n}\,v_0t) \tag{12.70}$$

and if we introduce

$$K(t)=\frac{1}{2\pi}\int d\varepsilon K(\varepsilon)\exp(-i\varepsilon t)\,,\quad v_s(R)=\int\frac{d^3\mathbf{q}}{(2\pi)^3}v_s(\mathbf{q})e^{i\mathbf{q}\cdot\mathbf{R}}\,, \tag{12.71}$$

Eq. (12.70) becomes

$$\mathbf{j}(\mathbf{q}) = \frac{3}{4} en\pi T \sum_{\omega} \frac{\Delta}{\omega^2+\Delta^2} \int d\mathbf{n} \cdot \mathbf{n}[\mathbf{n}\, v_s(q) \widetilde{K}(v_0 \mathbf{n}\cdot\mathbf{q})] \tag{12.72}$$

where

$$\widetilde{K}(\varepsilon) = \frac{\sqrt{\omega^2+\Delta^2}}{\omega^2+\Delta^2+\varepsilon^2/4} \,. \tag{12.73}$$

In order to perform the integration over the angles in (12.72) we take q parallel to Oz and $\mathbf{v}_s$ as $\mathbf{v}_s \equiv (v_{s\|}, v_{s\perp})$. From (12.73), we see that $\widetilde{K}(\varepsilon)$ is an even function and (12.72) can be transformed as

$$\mathbf{j}(\mathbf{q}) = \frac{3\pi en}{2} \pi T \sum_{\omega} \frac{\Delta^2}{\omega^2+\Delta^2} \left[v_{s\perp}(q) \int_0^1 dx (1-x^2) \widetilde{K}(v_0 q x) \right.$$
$$\left. + 2 v_{s\|}(q) \int_0^1 dx\, x^2 K(v_0 q x) \right] \,. \tag{12.74}$$

From (12.9) and (12.24), we can calculate $\delta\Delta(q)$ using the Fourier transform of Eq. (12.65). The result has the general form

$$\delta\Delta(\mathbf{q}) = 2\pi N(0) p_0 v_0 \pi T \sum_{\omega} \frac{\Delta}{\omega^2+\Delta^2} \int_0^1 \frac{x^2 dx}{4(\omega^2+\Delta^2)+q^2x^2} \tag{12.75}$$

and this correction disappears if $\mathbf{v}_{\|}(\mathbf{q}) = 0$, a condition which can always be satisfied.
Using now

$$\mathbf{v}_s(q) = i\chi(q)\mathbf{q} - \frac{e}{m}\mathbf{A}(\mathbf{q}) \,, \tag{12.76}$$

we get from the condition

$$\chi(q) = \frac{e}{im} \frac{\mathbf{q}\cdot\mathbf{A}(\mathbf{q})}{q^2} \tag{12.77}$$

and if the Coulomb gauge is taken as $\mathbf{q}\cdot\mathbf{A}(\mathbf{q}) = 0$ one obtains $\chi = 0$. From (12.74), the current becomes

$$\mathbf{j} = eQ(\mathbf{q}) v_{s\perp}(\mathbf{q}) \tag{12.78}$$

where

$$Q(q) = \frac{3\pi}{2} n\pi T \sum_{\omega} \frac{\Delta^2}{\omega^2+\Delta^2} \int_0^1 \frac{(1-x^2)dx}{\omega^2+\Delta^2+\frac{1}{4}v_0^2 q^2 x^2} \,. \tag{12.79}$$

Using now the equation

$$\frac{dH(z)}{dz} = 4\pi j + 2H_0\delta(z)\,, \quad H(z) = -\frac{dA}{dz}\,, \quad H_0 = H(z=0) \qquad (12.80)$$

and (12.78), one obtains

$$\left[q^2 + \frac{4\pi e^2}{m}Q(q)\right] A(q) = 2H_0 \qquad (12.81)$$

which gives

$$A(z) = \frac{2H_0}{\pi}\int_0^\infty \frac{\cos qz\, dg}{q^2 + \frac{4\pi e^2}{m}Q(q)} \qquad (12.82)$$

and in this way the penetration depth is given by

$$\delta = \frac{1}{H_0}\int_0^\infty dz H(z) = \frac{A_0}{H_0}$$

and from (12.82)

$$\delta = \frac{1}{\pi}\int_0^\infty \frac{dq}{q^2 + \frac{4\pi e^2}{m}Q(q)} \qquad (12.83)$$

which gives in the local limit

$$\delta = \sqrt{\frac{m}{4\pi e^2 n_s}} = \lambda(T) \qquad (12.84)$$

which is in fact the London penetration depth.
From (12.66), we can calculate (using $\nabla \times \mathbf{v}_s = -\frac{e}{m}\mathbf{H}$) the penetration depth in the dirty limit as

$$\lambda_D(T) = \left[\frac{1}{4\pi^2 \Delta\sigma \tanh \Delta/2T}\right]^{\frac{1}{2}} \qquad (12.85)$$

where σ is the conductivity of the normal metal.
From (12.74) and (12.75), we get

$$\frac{\lambda_D(0)}{\lambda(0)} \cong \frac{\xi_0}{l}\,. \qquad (12.86)$$

In the non-local limit in (12.83), we have to take a concrete form for $Q(q)$. For $q \to \infty$, we can take the limit $x \to 0$ and neglect x^2 in the denominator of (12.79). Then we obtain

$$Q(q) = \frac{3\pi n\Delta^2}{v_0 q}\pi T \sum_\omega \frac{1}{\omega^2+\Delta^2}\tan^{-1}\frac{v_0 q}{2\sqrt{\omega^2+\Delta^2}} \qquad (12.87)$$

which in the limit $q \to \infty$ gives

$$Q(q) \cong \frac{3\pi^2 n\Delta}{4v_0 q}\left(\tanh\frac{\Delta}{2T}\right)^{-1} . \tag{12.88}$$

With this result, we calculate δ from (12.83) and

$$\delta(T) = \frac{4}{3\sqrt{3}}\left[\frac{p_0}{3ne^2\Delta\tanh\Delta/2T}\right]^{\frac{1}{3}} \tag{12.89}$$

and

$$\delta(0) \cong (\xi_0\lambda^2)^{\frac{1}{3}} \tag{12.90}$$

a result which is different from (12.84) obtained in the local limit at $T = 0$.

13. Strong-Coupling Theory of Superconductivity

The BCS theory has gone far in explaining the properties of superconducting materials. However, it must be considered as a phenomenological theory with respect to the use of an "effective potential" which describes the Coulomb and phonon-induced interactions between the electrons in a metal. The BCS interaction can be considered as an instantaneous effective potential with a strongly oscillating behaviour in the coordinate space. The BCS interaction is still adequate to describe the properties of superconductors because "g" is an adjustable parameter. On the other hand, it is difficult to see how this parameter "g" can be related to the retarded time-dependent electron-phonon interaction.

Several physical properties showed that the spatial dependence of the interaction plays an important role whereas the time retardation is of great importance in determining various cut-off phenomena. It is important to mention that the long-range part from the phonon induced interaction results primarily from the emission and absorption of very long wavelength phonons.

These phonons may contribute either through "direct" processes in which the dielectric screening of the metal is nearly complete or through "umpklapp" processes. Morel and Anderson[9] showed that these processes are not effective anymore and the important result is that the phonon-induced interaction is mediated primarily through short wavelength phonons. This part of the phonon spectrum is peaked about a few definite processes so that the Einstein model is appropriate in calculations.

The second direction leading to the same conclusion starts from the Gor'kov definition of the gap function in the coordinate space as

$$\Delta(\mathbf{r}-\mathbf{r}';t-t') = g(\mathbf{r}-\mathbf{r}';t-t')F(\mathbf{r}-\mathbf{r}';t-t')$$

where g is the retarded potential caused by phonons. The Green function $F(\mathbf{r})$ has a long range and oscillates with a wavelength of the order of $2p_0^{-1}$, while the spatial dependence of $g(\mathbf{r}-\mathbf{r}')$ is only related to the particular features of the phonon spectrum and may display an oscillating behaviour with the wavelength of the order of the inverse of the Debye momentum. The destructive interference makes the product $g(\mathbf{r})F(\mathbf{r})$ small for all but quite small spatial distances $|\mathbf{r}-\mathbf{r}'|$. Thus the main conclusion of this discussion is that only the short-range part of interaction potential is important.

Eliashberg equations

The interaction between electrons and phonons will be written as

$$\mathcal{H}_i^c = \frac{1}{2}\sum_{\mathbf{p}_1,\mathbf{p}_2}\sum_{\mathbf{p}_1',\mathbf{p}_2'}\langle\mathbf{p}_1,\mathbf{p}_2|V^c|\mathbf{p}_2',\mathbf{p}_1'\rangle(\psi^\dagger_{\mathbf{p}_1}\hat{\tau}_3\psi_{\mathbf{p}_2})(\psi^\dagger_{\mathbf{p}_2}\hat{\tau}_3\psi_{\mathbf{p}_1}) \tag{13.1}$$

and

$$\mathcal{H}_i^{e-ph} = \sum_{\mathbf{p}-\mathbf{p}'=\mathbf{q}}\sum_j g_j(q)(\psi^\dagger_{\mathbf{p}}\hat{\tau}_3\psi_{\mathbf{p}'})(b_{\mathbf{q}j}+b^\dagger_{-\mathbf{q}j}) \tag{13.2}$$

where V^c is the Coulombic electron-electron interaction and $g_j(q)$ is the matrix element of the electron-phonon interaction. Using the operators

$$\psi^\dagger_{\mathbf{p}} = (\psi^\dagger_{\mathbf{p}\uparrow},\psi_{-\mathbf{p}\downarrow}), \qquad \psi_{\mathbf{p}} = \begin{pmatrix}\psi_{\mathbf{p}\uparrow}\\ \psi^\dagger_{-\mathbf{p}\downarrow}\end{pmatrix},$$

the Hamiltonian of the free electrons is

$$\mathcal{H}_0 = \sum_{\mathbf{p}}\varepsilon(\mathbf{p})\psi^\dagger_{\mathbf{p}}\hat{\tau}_3\psi_{\mathbf{p}} \tag{13.3}$$

where $\hat{\tau}_1, \hat{\tau}_2$ and $\hat{\tau}_3$ are the Pauli matrices in the isospin space.

From (13.3), we get the Green function for the free electrons as

$$\hat{G}^{0^{-1}}(\omega,\mathbf{p}) = i\omega\hat{1} - \varepsilon(\mathbf{p})\hat{\tau}_3\,. \tag{13.4}$$

The Dyson equation for the Green function of the electrons is

$$\hat{G}^{-1}(\omega,\mathbf{p}) = \hat{G}^{0^{-1}}(\omega,\mathbf{p}) - \hat{\Sigma}(\mathbf{p};\omega) \tag{13.5}$$

where $\hat{\Sigma}$ is the self-energy written as

$$\hat{\Sigma}(\mathbf{p},\omega) = [1 - Z(\mathbf{p},\omega)]i\omega\hat{1} - \chi(\mathbf{p})\hat{\tau}_3 + \varphi(\mathbf{p})\hat{\tau}_2 \; . \tag{13.6}$$

From (13.5) and (13.6), we have

$$\frac{i\omega Z(\mathbf{p},\omega)\hat{1} + \bar{\varepsilon}(\mathbf{p})\hat{\tau}_3 + \varphi(\mathbf{p},\omega)\hat{\tau}_1}{[i\omega Z(\mathbf{p},\omega)]^2 - \varphi^2(\mathbf{p},\omega) - \bar{\varepsilon}^2(\mathbf{p},\omega)} \tag{13.7}$$

where

$$\bar{\varepsilon}(\mathbf{p},\omega) = \varepsilon(\mathbf{p}) + \chi(\mathbf{p},\omega) \; .$$

The self energy $\hat{\Sigma}$ will be calculated in the Hartree-Fock approximation. We will neglect the Hartree diagrams of the type

which gives only some contributions to the renormalization of the kinetic energy. The first important diagrams are

$\hat{\Sigma}_1(\mathbf{p},\omega) =$

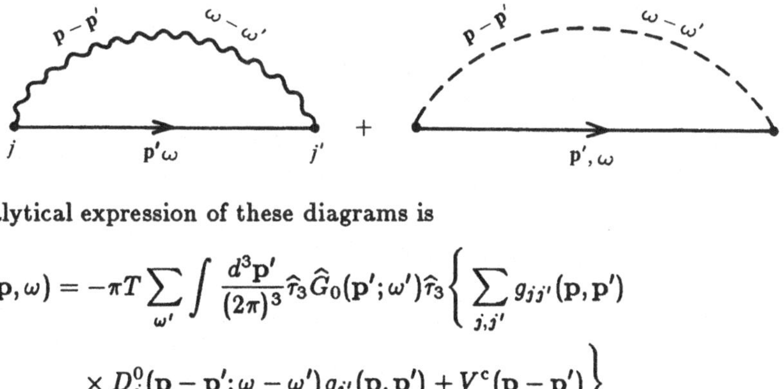

and the analytical expression of these diagrams is

$$\hat{\Sigma}_1(\mathbf{p},\omega) = -\pi T \sum_{\omega'} \int \frac{d^3\mathbf{p}'}{(2\pi)^3} \hat{\tau}_3 \hat{G}_0(\mathbf{p}';\omega')\hat{\tau}_3 \left\{ \sum_{j,j'} g_{jj'}(\mathbf{p},\mathbf{p}') \times D_j^0(\mathbf{p}-\mathbf{p}';\omega-\omega') g_{j'}(\mathbf{p},\mathbf{p}') + V^c(\mathbf{p}-\mathbf{p}') \right\} \tag{13.8}$$

where the first contribution is given by the electron-phonon interaction $\hat{\Sigma}_{e-ph}$ and the second $\hat{\Sigma}_c$ by the electron-electron interaction. We consider now these two contributions

$$\hat{\Sigma}_{e-ph} = -\pi T \sum_{\omega'} \int \frac{d^3\mathbf{p}'}{(2\pi)^3} \hat{\tau}_3 \hat{G}(\mathbf{p}';\omega')\hat{\tau}_3 \sum_j |g_j(\mathbf{p},\mathbf{p}')|^2 \times D_j^0(\mathbf{p}-\mathbf{p}';\omega-\omega') \tag{13.9}$$

and

$$\widehat{\Sigma}_c = -\pi T \sum_{\omega'} \int \frac{d^3\mathbf{p}'}{(2\pi)^3} \widehat{\tau}_3 \widehat{G}(\mathbf{p}',\omega)\widehat{\tau}_3 V^c(\mathbf{p}-\mathbf{p}') \tag{13.10}$$

where we used $g_{jj'} = g_j \delta_{jj'}$ and $g_{jj'} = g^*_{j'j}$.
Using now the relation

$$\widehat{\tau}_3 \widehat{\tau}_1 \tau_3 = -\widehat{\tau}_1$$

we obtain

$$\widehat{\tau}_3 \widehat{G} \widehat{\tau}_3 = \frac{i\omega\tau_3 Z(\mathbf{p},\omega) + \bar{\varepsilon}(p)\widehat{\tau}_3 - \varphi(\mathbf{p},\omega)\widehat{\tau}_1}{(i\omega Z(\mathbf{p},\omega))^2 - \varphi^2(\mathbf{p},\omega) - \bar{\varepsilon}^2(\mathbf{p},\omega)} \ . \tag{13.11}$$

In order to perform the summation in (13.9) and (13.10), we write the Green functions $\widehat{G}(\omega;\mathbf{p})$ and $D_j(\omega,\mathbf{p})$ as

$$G(\omega,\mathbf{p}) = \frac{1}{2\pi}\int_{-\infty}^{\infty} dz' \frac{a(\mathbf{p},z')}{i\omega - z'} \tag{13.12}$$

$$D_j(\omega,\mathbf{p}) = \frac{1}{2\pi}\int_{-\infty}^{\infty} dz \frac{b(\mathbf{p},z)}{i\omega - z} \ . \tag{13.13}$$

Using now

$$\pi T \sum_{\omega'} \frac{1}{(i\omega' - z)[i(\omega-\omega') - z]} = -\frac{1}{2}\frac{\tanh\frac{z'}{2T} + \coth\frac{z}{2T}}{i\omega - z - z'}$$

$$\pi T \sum_{\omega'} \frac{1}{(i\omega - z)} = -\frac{1}{2}\tanh\frac{z'}{2T} \tag{13.14}$$

and the relation

$$a(\mathbf{p},z) = -2\,\mathrm{Im}\,G(\mathbf{p},z) \ , \tag{13.15}$$

Eqs. (13.9) and (13.10) become

$$\widehat{\Sigma}_{\mathrm{e-ph}} = -\int \frac{d^3\mathbf{p}'}{(2\pi)^3} \sum_j |g_j(\mathbf{p},\mathbf{p}')|^2 \int \frac{dz}{2\pi}\int\frac{dz'}{2\pi} b_j(z-z')$$

$$\times\, \widehat{\tau}_3\,\mathrm{Im}\,\widehat{G}(z',\mathbf{p}')\widehat{\tau}_3 \frac{\tanh z'/2T + \coth z/2T}{\omega - z - z' + i\delta} \tag{13.16}$$

$$\widehat{\Sigma}_c(\mathbf{p},\omega) = -\int \frac{d^3\mathbf{p}'}{(2\pi)^3} V^c_{(\mathbf{p}-\mathbf{p}')} \int_{-\infty}^{\infty} \frac{dz'}{2\pi} \widehat{\tau}_3\,\mathrm{Im}\,\widehat{G}(\mathbf{p}',z')\widehat{\tau}_3 \tanh\frac{z'}{2T} \ . \tag{13.17}$$

Equation (13.11) will be re-written as

$$\widehat{\tau}_3 \widehat{G} \widehat{\tau}_3 = \frac{Z(\mathbf{p},\omega)\omega + \bar{\varepsilon}(\mathbf{p})\widehat{\tau}_3 - \varphi(\mathbf{p},\omega)\widehat{\tau}_1}{[Z(\mathbf{p},\omega)\omega]^2 - \varphi^2(\mathbf{p},\omega) - \bar{\varepsilon}^2(\mathbf{p},\omega)} \tag{13.18}$$

and the self-energy Σ will be considered as

$$\begin{aligned}\widehat{\Sigma} &= \widehat{\Sigma}_{e-ph}(\mathbf{p},\omega) + \widehat{\Sigma}_c(\mathbf{p};\omega) \\ &= (1 - Z(\mathbf{p},\omega))\omega\widehat{1} + \chi(p)\widehat{\tau}_3 + \varphi(\mathbf{p},\omega)\widehat{\tau}_1 \ . \end{aligned} \tag{13.19}$$

In the next calculations we will consider $\chi(\mathbf{p},\omega) = 0$ because this term can give rise to a shift in the chemical potential, and we put $\bar{\varepsilon}(\mathbf{p},\omega) = \varepsilon'(\mathbf{p},\omega)$.

If we consider that the electrons which participate in the superconducting condensation are near the Fermi surface, we may write

$$\int \frac{d^3\mathbf{p}'}{(2\pi)^3} \cdots \longrightarrow \int d\varepsilon \int_{S(\varepsilon_0)} \frac{d^2p'}{v_{\mathbf{p}'}} \tag{13.20}$$

where $v_{p'}$ is the velocity at the surface of the energy $S(\varepsilon_0)$. The integral over the energy can be performed using

$$\int_{-\infty}^{\infty} \frac{d\varepsilon}{\varepsilon^2 - A^2} = \frac{i\pi}{\sqrt{A}} \operatorname{sign} A \ . \tag{13.21}$$

In order to perform further calculations in (13.16) and (13.17) we have to calculate the integral over the directions of $\mathbf{p}$ on the Fermi surface. If we introduce

$$\widehat{\Sigma}_{e-ph} = \int_{S(\varepsilon_0)} \frac{d^2\mathbf{p}}{v_\mathbf{p}} \widehat{\Sigma}_{e-ph} \Big/ \int_{S(\varepsilon_0)} \frac{d^2\mathbf{p}}{v_\mathbf{p}} \ , \tag{13.22}$$

Eq. (13.16) becomes

$$\begin{aligned}\widehat{\Sigma}_{e-ph} = &\int_{-\infty}^{\infty} \frac{dz'}{2\pi} \int_{-\infty}^{\infty} \frac{dz}{2\pi} \left(\int_{S(\varepsilon_0)} \frac{d^2\mathbf{p}}{v_\mathbf{p}} \right)^{-1} \int \frac{d^2\mathbf{p}}{v_\mathbf{p}} \int \frac{d^2\mathbf{p}'}{v_{\mathbf{p}'}} \\ &\times \sum_j |g_j|^2 b_j(\mathbf{p} - \mathbf{p}') \operatorname{sign} z' \left\{ \operatorname{Re} \frac{Z(\mathbf{p}', z') z' \widehat{\tau}_0 - \varphi(\mathbf{p}', z')\tau_1}{\sqrt{z'^2 Z^2(\mathbf{p}, z') - \varphi^2(\mathbf{p}', z')}} \right. \\ &\times \left. \left[\frac{\tanh z'/2T + \coth z/2T}{\omega - z - z' + i\delta} \right] \right\} . \end{aligned} \tag{13.23}$$

If the Fermi surface is anisotropic and we cannot neglect the anisotropy functions, Z and φ, of $\mathbf{p}$ and z. For a perfect isotropy, we can take

$Z(\mathbf{p}', z') = Z(z'), \varphi(\mathbf{p}, z') = \varphi(z')$ and (13.23) becomes

$$\hat{\Sigma}_{\mathrm{e-ph}} = \int_{-\infty}^{\infty} dz' \int_{-\infty}^{\infty} dz \frac{1}{4\pi} \int_{S(\epsilon_0)} \frac{d^2\mathbf{p}'}{v_{\mathbf{p}'}} \int_{S(\epsilon_0)} \frac{d^2\mathbf{p}}{v_{\mathbf{p}'}} \\ \times \sum_j |g_j(\mathbf{p}, \mathbf{p}')|^2 b_j(\mathbf{p}, \mathbf{p}'; z) \left(\frac{d^2\mathbf{p}'}{v_{\mathbf{p}'}}\right)^{-1} \\ \times \frac{\tanh z'/2T + \coth z/2T}{\omega - z - z' + i\delta} \mathrm{Re} \left\{ \frac{z' Z(z')\hat{1} - \varphi(z')\hat{\tau}_1}{\sqrt{z'^2 Z^2(z') - \varphi^2(z')}} \right\} . \tag{13.24}$$

If we write down (13.19) only for the phonon contribution as

$$\hat{\Sigma}_{\mathrm{e-ph}} = [1 - Z_{\mathrm{ph}}(\omega)]\omega \hat{1} + \varphi_{\mathrm{ph}}(\omega)\hat{\tau}_1 , \tag{13.25}$$

we get from (13.24) two equations

$$[1 - Z_{\mathrm{ph}}(\omega)]\omega = - \int_{-\infty}^{\infty} dz' K_{\mathrm{ph}}(z', \omega) \frac{z' \operatorname{sign} z'}{\sqrt{z'^2 - \Delta^2(z')}} \tag{13.26}$$

$$Z_{\mathrm{ph}}(\omega)\Delta_{\mathrm{ph}}(\omega) = \int_{-\infty}^{\infty} dz' K_{\mathrm{ph}}(z', \omega) \,\mathrm{Re}\, \frac{\Delta(z')}{\sqrt{z'^2 - \Delta^2(z')}} \tag{13.27}$$

where we substituted

$$\Delta_{\mathrm{ph}}(\omega) = \frac{\varphi_{\mathrm{ph}}(\omega)}{Z_{\mathrm{ph}}(\omega)}, \quad \Delta(\omega) = \frac{\varphi(\omega)}{\Delta(\omega)} . \tag{13.28}$$

The kernel of Eqs. (13.26)–(13.27) is

$$K_{\mathrm{ph}}(z', \omega) = \int_0^{\infty} dz\, \alpha^2(z) F(z) \frac{1}{2} \left\{ \frac{\tanh z'/2T + \coth z/2T}{z' + z - \omega - i\delta} \right. \\ \left. - \frac{\tanh z'/2T - \coth z/2T}{z' - z - \omega - i\delta} \right\} \tag{31.29}$$

where

$$\alpha^2(z)F(z) \\ = \int_{S(\epsilon_0)} \frac{d^2\mathbf{p}}{v_{\mathbf{p}}} \int_{S(\epsilon_0)} \frac{d^2\mathbf{p}'}{v_{\mathbf{p}'}} \sum_j |g_j(\mathbf{p}, \mathbf{p}')|^2 \delta[z - \omega_j(q)] \Big/ \int_{S(\epsilon_0)} \frac{d^2\mathbf{p}}{v_{\mathbf{p}}} . \tag{13.30}$$

Equations (13.26), (13.27) and (13.30) have been obtained using for the spectral density $b_j(z, \mathbf{q})$, the simple form

$$b_j(z, \mathbf{q}) = \operatorname{Im} D_j(\mathbf{q}, z) = 2\pi[\delta(z - \omega_j(q)) + \delta(z - \omega_j(q))] \ . \qquad (13.31)$$

Let us calculate the Coulombic contribution in the self energy. In (13.17), the integration over z which has to be performed on $-\infty$ to ∞ can be transformed to an integration over $z' > 0$. Using (13.18) it is easy to see that the contribution proportional to $Z(z)$ vanishes and

$$\varphi_c(\mathbf{p}, \omega) = \frac{1}{\pi} \int_0^\infty dz' \tanh \frac{z'}{2T} V^c(\mathbf{p} - \mathbf{p}') \times \operatorname{Im} \frac{\varphi(\mathbf{p}', z)}{z'^2 Z^2(z') - \varphi^2(\mathbf{p}, z') - \varepsilon^2(\mathbf{p}')} \ . \qquad (13.32)$$

This equation contains integrals which depend on $\mathbf{p}$ and z. If we neglect the influence of the direction of $\mathbf{p}$ in $\varphi(\mathbf{p}, z)$, Eq. (13.32) becomes

$$\varphi_c(\mathbf{p}, z) = \frac{1}{\pi} \int_0^\infty dz' \tanh \frac{z'}{2T} \int d\varepsilon N(0) V^c_{\mathbf{p}-\mathbf{p}'} \times \operatorname{Im} \frac{\varphi(\mathbf{p}', z)}{z'^2 Z^2(z') - \varphi^2(\mathbf{p}', z') - \varepsilon^2(\mathbf{p}')} \qquad (13.33)$$

where

$$N(0) V^c_{(\mathbf{p}-\mathbf{p}')} = \int_{S(\varepsilon_0)} \frac{d^2\mathbf{p}}{v_\mathbf{p}} \int_{S(\varepsilon_0)} \frac{d^2\mathbf{p}'}{v_{\mathbf{p}'}} V^c_{(\mathbf{p}-\mathbf{p}')} \Big/ \int_{S(\varepsilon_0)} \frac{d^2\mathbf{p}}{v_\mathbf{p}} \qquad (13.34)$$

and $N(0) = \int_{S(\varepsilon_0)} d^2\mathbf{p}/v_\mathbf{p}$ is the density of states of the electrons on the Fermi surface. The energy of the electrons which interact by V_c is of the order

$$\omega_D \leq \omega_F \leq \varepsilon_0$$

and then the integral (13.32) will be performed from 0 to ω_c and from ω_c to ∞.

In the interval $z' > \omega_c$, the electron-phonon interaction is not important and $\varphi(\mathbf{p}', z') \ll \omega_c$. We also have $T_c \ll \omega_c$ and $\tanh z'/2T \to 1$. The integration over z' becomes

$$\frac{1}{\pi} \int_{\omega_c}^\infty dz' \operatorname{Im} \frac{\varphi(\mathbf{p}')}{z'^2 - \varphi^2(\mathbf{p}') - \varepsilon^2(\mathbf{p}')} = \frac{\varphi(\mathbf{p}')}{\pi} \int_{\omega_c}^\infty dz' \operatorname{Im} \frac{1}{(z' + i\delta)^2 - E^2(p)} = \frac{\varphi(\mathbf{p}')}{2E(\mathbf{p}')} \theta(E(\mathbf{p}') - \omega_c) \qquad (13.35)$$

where

$$E^2(\mathbf{p}) = \varepsilon^2(\mathbf{p}) + \varphi_c^2(\mathbf{p}) \,. \tag{13.36}$$

Equation (13.32) is now written as

$$\varphi_c(\mathbf{p}) + \int \frac{d\varepsilon'}{2E'}\theta(E(\mathbf{p}') - \omega_c)N(0)V^c_{(\mathbf{p}-\mathbf{p}')}\varphi_c(\mathbf{p},\mathbf{p}')$$
$$= \int d\varepsilon V^c_{(\mathbf{p}-\mathbf{p}')}F(\mathbf{p}') \tag{13.37}$$

where

$$F(\mathbf{p}') = N(0)\int_0^{\omega_c} dz' \frac{1}{\pi}\frac{\varphi(\mathbf{p}',z)}{z'^2Z^2(z') - \varphi^2(p)} \tanh\frac{z'}{2T} \,. \tag{13.38}$$

The solution of the integral equation (13.37) is of the form

$$\varphi_c(\mathbf{p}) = \int d\varepsilon\, U_c(\mathbf{p},\mathbf{p}')F(\mathbf{p}') \tag{13.39}$$

where $U_c(\mathbf{p},\mathbf{p}')$ is the solution of the integral equation

$$U_c(\mathbf{p},\mathbf{p}') + \int \frac{d\varepsilon''}{E''}\theta(E'' - \omega_c)N(0)V^c_{(\mathbf{p},\mathbf{p}'')} = V^c_{(\mathbf{p},\mathbf{p}')} \tag{13.40}$$

and is in fact a pseudopotential describing the electron-electron interaction near the Fermi surface. In this approximation, we may consider $p \simeq p_0$ and

$$\varphi_c(\mathbf{p}) \cong U_c \int_{-\infty}^{\infty} d\varepsilon'\, F(\mathbf{p}') \tag{13.41}$$

where $U_c = U_c(p_0, p_0)$.
Performing the integral over ε', we get

$$\varphi_c(\mathbf{p}) = -U_c N(0)\int_0^{\omega_c} dz' \tanh\frac{z'}{2T}\,\mathrm{Re}\,\frac{\Delta(z')}{\sqrt{z'^2 - \Delta^2(z')}} \,. \tag{13.42}$$

If now we take

$$V^c_{(\mathbf{p},\mathbf{p}')} = \begin{cases} V^c; & \varepsilon < \varepsilon_0 \\ 0; & \varepsilon > \varepsilon_0 \end{cases} ,$$

Eq. (13.40) becomes

$$U_c + N(0)V_c \int_{-\infty}^{\infty} \frac{d\varepsilon}{\sqrt{\varepsilon^2 + \varphi_c^2}}\theta[\sqrt{\varepsilon^2 + \varphi_c^2} - \omega_c]U_c = V_c \tag{13.43}$$

and because $\varphi_c \sim T_c$, we get from (13.43)

$$U_c = \frac{V_c}{1 + N(0)\ln \varepsilon_0/\omega_c} . \tag{13.44}$$

If we now write

$$Z(\omega) = Z_{ph}(\omega) , \qquad Z(\omega)\Delta(\omega) = Z_{ph}(\omega)\Delta_{ph}(\omega) + \varphi_c , \tag{13.45}$$

we get from (13.26), (13.27) and (13.42) the equations

$$[1 - Z(\omega)]\omega = -\int_{-\infty}^{\infty} dz' K_{ph}(z', \omega) \operatorname{Re} \frac{z' \operatorname{sign} z'}{\sqrt{z'^2 - \Delta^2(z')}} \tag{13.46}$$

$$Z(\omega)\Delta(\omega) = \int_{-\infty}^{\infty} dz' K_{ph}(z', \omega) \operatorname{Re} \frac{\Delta(z')}{\sqrt{z'^2 - \Delta^2(z')}} \operatorname{sign} z' - U_c N(0) \int_0^{\omega_c} dz' \tanh \frac{z'}{2T} \operatorname{Re} \frac{\Delta(z')}{\sqrt{z'^2 - \Delta^2(z')}} . \tag{13.47}$$

Using (13.29), we can transform these equations to

$$\begin{aligned} [1 - Z(\omega)]\omega = \frac{1}{2} \int_0^{\infty} dz\, \alpha^2 F(z) \int_0^{\infty} dz' \operatorname{Re} \frac{z'}{\sqrt{z'^2 + \Delta^2}} \\ \times \left\{ \left[\tanh \frac{z'}{2T} + \coth \frac{z}{2T}\right] \right. \\ \times \left[\frac{1}{z' + z + \omega + i\delta} - \frac{1}{z' + z - \omega - i\delta}\right] \\ - \left[\tanh \frac{z'}{2T} - \coth \frac{z}{2T}\right] \left[\frac{1}{z' - z - \omega + i\delta} \right. \\ \left. \left. - \frac{1}{z' - z - \omega - i\delta}\right] \right\} , \end{aligned} \tag{13.48}$$

$$\begin{aligned} Z(\omega)\Delta(\omega) = \frac{1}{2} \int_0^{\infty} dz\, \alpha^2(z) F(z) \int_0^{\infty} dz' \operatorname{Re} \frac{\Delta(z')}{\sqrt{z'^2 - \Delta^2(z')}} \\ \times \left\{ \left[\tanh \frac{z'}{2T} + \coth \frac{z}{2T}\right] \left[\frac{1}{z' + z + \omega + i\delta} \right. \right. \\ \left. - \frac{1}{z' + z - \omega - i\delta}\right] - \left[\tanh \frac{z'}{2T} - \coth \frac{z}{T}\right] \\ \left. \times \left[\frac{1}{z' - z - \omega + i\delta} - \frac{1}{z' - z - \omega - i\delta}\right] \right\} . \end{aligned} \tag{13.49}$$

From these equations known as the Eliashberg equations,[10] we can calculate the critical temperature if we know the phonon spectrum for the superconducting material. Other properties of superconductors in this model have been studied by Scalapine *et al.*[11] in connection with the tunnelling experiments.

The critical temperature

In order to calculate the critical temperature, we linearize the Eliashberg equations. Near the critical temperature, $\Delta(\omega)$ is small and from (13.48) we obtain

$$[1 - Z(\omega)]\omega = \int_{-\infty}^{\infty} dz' K_{\rm ph}(z';\omega)\,, \tag{13.50a}$$

$$Z(\omega)\Delta(\omega) = \int_{-\infty}^{\infty} dz' K_{\rm ph}(z',\omega) \operatorname{Re} \frac{\Delta(z')}{z'} - U_c N(0) \int_0^{\omega_c} dz' \tanh \frac{z'}{2T_c} \operatorname{Re} \frac{\Delta(z')}{z'}\,. \tag{13.50b}$$

From these equations we calculate the critical temperature in the following approximations.

a) *Weak coupling*

In the limit of weak electron-phonon coupling we have to obtain from the Eliashberg equations (13.48)–(13.49) the basic equations of the BCS theory. Indeed, the kernel $K_{\rm ph}(z',\omega)$ of (13.29) decreases sharply at the values of arguments $|z'|, |\omega| \gtrsim \omega_{\rm D}$. Then we can adopt the following approximation: the kernel $K_{\rm ph}(z',\omega)$ is taken as constant in the interval of variation of the argument from 0 to the frequency $\omega_{\rm D}$ and zero elsewhere.

The quantity

$$K_{\rm ph}(z' \to 0) = \frac{\lambda}{2} \tanh \frac{z'}{2T} \tag{13.51}$$

where

$$\lambda = 2 \int_0^{\infty} dz \frac{\alpha^2(z) F(z)}{z}$$

is now the electron-phonon coupling constant.

Using (13.50), $K_{\rm ph}(z',\omega)$ can be written (in this approximation) as

$$K_{\rm ph}(z',\omega) = \frac{\lambda}{2} \tanh \frac{z'}{2T} \theta(\omega_{\rm D} - |z'|)\theta(\omega_{\rm D} - |\omega|) \tag{13.52}$$

which gives for $Z(\omega)$, in (13.50a), the result

$$Z(\omega) = 1 \, . \tag{13.53}$$

In this case

$$\begin{aligned}\Delta(\omega) &= \lambda\theta(\omega_D - |\omega|) \int_0^{\omega_D} \frac{dz'}{z'} \tanh \frac{z'}{2T} \Delta(z') \\ &\quad - U_c N(0) \int_0^{\omega_c} \frac{dz'}{z'} \tanh \frac{z'}{2T} \Delta(z')\end{aligned} \tag{13.54}$$

and this equation can be solved if we suppose

$$\Delta = \begin{cases} \Delta_{ph} \, ; & |\omega| < \omega_D \\ \Delta_c \, ; & |\omega| < \omega_c \end{cases} \tag{13.55}$$

and we get

$$\begin{aligned}&[1 - (\lambda - N(0)U_c)J]\Delta_{ph} + U_c N(0) \ln \frac{\omega_c}{\omega_D} \Delta_c = 0 \\ &U_c N(0) J \Delta_{ph} + \left[1 + U_c N(0) \ln \frac{\omega_c}{\omega_D} \Delta_c\right] = 0\end{aligned} \tag{13.56}$$

where

$$J = \frac{1}{2} \int_0^{\omega_D/2T_c} \frac{dx}{x} \tanh x \, .$$

From the conditions

$$T_c = \frac{2\gamma}{\pi} \omega_D \exp\left[- \frac{1}{\lambda - \mu^*} \right] \tag{13.57}$$

where

$$\mu^* = \frac{N(0)V^c}{1 + N(0)V^c \ln \varepsilon_0/\omega_D} \, .$$

b) *Intermediate coupling*

We will consider a more rigorous solution of the linearized Eliashberg equations which gives corrections to the BCS formula for the critical temperature T_c, and thus corresponds to the case of intermediate electron-phonon coupling. This will give us the possibility to obtain a criterion for the validity of the weak coupling approximation, just as the strong coupling yields a criterion for the validity of the intermediate coupling.

We consider $z \gg T_c$, which is equivalent to the approximation of $N(z) \to 0$. Then in Eqs. (13.48) and (13.49), $\coth z/2T \to 1$ and $\tanh z'/2T \to 1$. In these approximations, Eqs. (13.48) and (13.49) become

$$Z(\omega)\Delta(\omega) = \int_0^\infty \frac{dz'}{z'} \tanh \frac{z'}{2T} \left[\int_0^\infty dz\, \alpha^2(z)F(z)K_+(z',z;\omega) - \mu(z',\omega) \right] \tag{13.58}$$

and

$$[1 - Z(\omega)]\omega = \int_{-\infty}^\infty dz' \int_0^\infty dz\, \alpha^2(z)F(z)f(-z')K_-(z',z;\omega) \tag{13.59}$$

where

$$K_\pm = \frac{1}{z'+z+\omega+i\delta} \pm \frac{1}{z'+z-\omega-i\delta} \tag{13.60}$$

and using

$$\tanh \frac{z'}{2T} = 1 - 2f(z') = -1 + 2f(z') \,,$$

we get the critical temperature T_c as

$$T_c = \frac{2\gamma}{\pi}\omega_D \exp\left[-\frac{1+\lambda}{\lambda - \mu^*(1+\lambda A)/(1+\lambda)} \right] \tag{13.61}$$

which contains the influence of the phonons on the critical temperature. Here A is a constant. If we consider

$$\begin{aligned} \alpha^2(z)F(z) &= \frac{\lambda\Omega_0}{2}\delta(z-\Omega_0) \\ T_c &= \frac{\Omega_0}{1.143} \exp\left[-\frac{1+\lambda}{\lambda - \mu^*(1+0.5\lambda)/(1+\lambda)} \right] . \end{aligned} \tag{13.62}$$

From (13.62) we see that the weak coupling limit can be reobtained for $\lambda \ll 1$. This condition determines the validity of the weak coupling theory. In contrast to strong coupling which corresponds to $\lambda \simeq 1$, it is rather difficult to formulate a criterion for the validity of the intermediate coupling.

c) *Strong coupling*

The intermediate coupling approximation has been obtained as a result of neglecting the thermal phonons, and replacing $\tanh z'/2T_c$ by unity in the Eliashberg equations. This corresponds to neglecting of temperature effects when they are of the order of $(T_c/\omega_D)^2$ in Im $Z(\omega)$ and keeping the terms of the order of $(T_c/\omega_D)^2 \ln \omega_D/T_c$ in $\mathrm{Re}\, Z(\omega)$. The presence of such a temperature dependence in $\mathrm{Re}\, Z(\omega)$ may lead to a deviation of the ratio $2\Delta(0)/T_c$ from the value (3.52) which results from the BCS model.

The contribution of the thermal phonons ($\mathrm{Im}\,\Delta(\omega) \neq 0$ and $\mathrm{Re}\,\Delta(\omega) \neq 0$) in the Eliashberg equations ensure an accuracy that permits the maintaining of the terms $(T_c/\omega_D)^2 \ln \omega_D/T_c$ and $(T_c/\omega_D)^2$. In the strong coupling case, which becomes realistic at $T_c \gg \omega_D$, it is very difficult to get an analytic formula for T_c. However, there are some approximations which one can obtain an analytical equation for T_c. We shall present in the following the McMillan[12] approach. If we take $\Delta' \to 0$ in (13.48) and (13.49), these equations can be solved using the approximation

$$\Delta_t = \begin{cases} \Delta_0\,; & 0 < \omega < \omega_0 \\ \Delta_\infty\,; & \omega > \omega_0\,. \end{cases} \tag{13.63}$$

The equation for $\Delta_0^{(1)}$ will be obtained from (13.49) by neglecting the thermal phonons

$$\Delta^{(1)}(0) \cong \frac{\Delta(0)}{Z(0)} \int_0^{\omega_D} \frac{dz'}{z'} \tanh \frac{z'}{2T_c} 2 \int_0^{\omega_D} dz\, \frac{\alpha^2(z)F(z)}{z} \tag{13.64}$$

and if we denote

$$\lambda = 2 \int_0^{\omega_D} \frac{dz}{z} \alpha^2(z)F(z) \tag{13.65}$$

where λ is the correspondent of $N(0)g$ from BCS.
Further we have

$$\Delta^{(2)}(0) = \frac{1}{Z(0)} \int_{\omega_0}^{\infty} \frac{dz'}{z'} \Delta_\infty \int_0^{\infty} dz\, \alpha^2 F(z) \frac{2}{z'+z} \cong \frac{\Delta_\infty}{Z(0)} \frac{\langle\omega\rangle}{\omega_0} \tag{13.66}$$

where

$$\langle\omega\rangle = \int_0^{\infty} dz\, \alpha^2(z)F(z) \Big/ \int_0^{\infty} \frac{dz}{z} \alpha^2(z)F(z) \cong 0.5\omega_0 \tag{13.67}$$

and we have neglected z' as compared with z.

The Coulomb interaction contribution $\Delta^{(3)}$ will be

$$\Delta^{(3)}(0) = -\frac{N(0)V^c}{Z(0)} \left[\Delta_0 \ln \frac{\omega_D}{T_c} + \Delta_\infty \ln \frac{\varepsilon_B}{\omega_D}\right] \tag{13.68}$$

where ε_B is the electronic bandwidth. At high energies, the only contribution is from Coulomb interaction and $Z(0) = 1 + \lambda, Z(\infty) = 1$. The consistency requirements at low and high energies give

$$\Delta(0) \equiv \Delta_0 = \left[\frac{\Delta_0\lambda}{Z(0)}\ln\frac{\omega_0}{T_c} + \frac{\Delta_\infty}{Z(0)}\right]\frac{\langle\omega\rangle}{\omega_0}\lambda$$

$$- \frac{N(0)Z_c}{Z(0)}\left[\Delta_0 \ln\frac{\omega_0}{T_c} + \Delta_\infty \ln\frac{\varepsilon_B}{\omega_0}\right], \tag{13.69}$$

$$\Delta(\infty) \equiv \Delta_\infty = \frac{-N(0)V^c\Delta_0 \ln\frac{\omega_0}{T_c}}{1 + N(0)V^c \ln\frac{\varepsilon_B}{\omega_c}}. \tag{13.70}$$

From (13.69) and (13.66), we get

$$\Delta_0 = \frac{\Delta_0[1 - \mu^* - \mu^*\lambda(\langle\omega\rangle/\omega_0)]\ln\omega_c/T_c}{1+\lambda}$$

and the critical temperature T_c is

$$T_c = \omega_0 \exp\left[-\frac{(1+\lambda)}{\lambda - \mu^* - \langle\omega\rangle/\omega_0\lambda\mu^*}\right]. \tag{13.71}$$

The relation which is usually used for the comparison with experimental results has been obtained by McMillan[12]

$$T_c = \frac{\omega_D}{1.45}\exp\left[-\frac{1.04(1+\lambda)}{\lambda - \mu^*(1+0.62\lambda)}\right]. \tag{13.72}$$

This formula gives a good fit to experimental data even when μ^* and λ are taken from independent experiments. However, these results have been obtained using many properties of strong coupling superconductor; the frequency $\langle\omega\rangle$ in (13.67) was averaged with respect to $F(\omega)$ of the Niobium spectrum which was obtained from inelastic neutron scattering.

The McMillan formula (13.72) gives a good agreement with the numerical solutions of the Eliashberg equations for $\lambda \leq 1.5$. Allen and Dynes[13] performed similar calculations for large values of λ and found a discrepancy between the exact results and the data obtained using (13.72) even for $\lambda = 2$. Instead of a maximum in $T_c(\lambda)$ in the vicinity of $\lambda = 2$ as predicted from (13.72), they obtained a monotonically increasing T_c and found that in the ultrastrong coupling limit ($\lambda \geq 10$ and $\mu^* = 0.1$), the critical temperature becomes

$$T_c \cong 0.15\sqrt{\langle\omega^2\rangle\lambda}\,. \tag{13.73}$$

Allen and Dynes have also investigated the independence of T_c on the form of phonon spectrum. They solved the Eliashberg equations for Pb and Hg phonon spectra. On the other hand, if in the McMillan formula (13.72), we use for $\langle\omega\rangle$ the expression

$$\omega_{\ln} = \exp\left[\frac{2}{\lambda}\int_0^\infty \frac{d\omega}{\omega}\alpha^2(\omega)F(\omega)\ln\omega\right] , \tag{13.74}$$

the λ dependence of T_c remains the same for the same three metals: Pb, Hg and Nb. Then the dependence of T_c on the form of the phonon spectrum manifests itself only when $\lambda \gg 2$.

The quantities $\alpha^2(\omega)F(\omega)$ and $\mu^*(\omega_c)$ are contained in the current I and voltage V characteristics of a tunnel junction. The $I-V$ data can be inverted to get these microscopic parameters using the McMillan-Rowel inversion procedure.[14] The validity of the procedure proposed by Allen and Dynes has been proved by Mitrovic *et al.*[15] by the calculation of T_c in the strong coupling regime. This expression has been calculated first by Geilikman and Kresin[16] as

$$\frac{2\Delta(0)}{T_c} = 3.53\left[1 + 5.3\left(\frac{T_c}{\tilde{\omega}}\right)^2 \ln\frac{\tilde{\omega}}{T_c}\right] \tag{13.75}$$

where $\tilde{\omega}$ is some phonon energy which is not sharply specified. In fact this equation shows the deviation of the strong coupling theory from the BCS theory. In order to recalculate (13.75), the Eliashberg equations have been considered at $T = 0$, which can be written as

$$\Delta(\omega)Z_s(\omega) = \int_0^{\omega_c} dz' \,\mathrm{Re}\left[\frac{\Delta(z')}{\sqrt{z'^2 - \Delta^2(z')}}\right][K_+(\omega, z') - \mu^*(\omega_c)] \tag{13.76}$$

$$[1 - Z_s(\omega)]\omega = \int_0^\infty dz' \,\mathrm{Re}\left[\frac{z'}{\sqrt{z'^2 - \Delta^2(z')}}\right] K_-(\omega, z') \tag{13.77}$$

where

$$K_\pm(\omega, z') = \int_0^\infty d\Omega\left[\frac{1}{z' + \Omega + \omega + i0} \pm \frac{1}{z' + \Omega - \omega - i0}\right] .$$

If we consider the real part of (13.76) at the gap edge $\omega = \Delta_0$, take the cutoff ω_c to be equal to the maximum phonon frequency and approximate $\Delta(\omega)$ by Δ_0, we get from (13.76), the simple equation

$$\Delta_0 Z_s(\Delta_0) = \left[-\bar{\lambda} + (\lambda - \mu^*)\ln\frac{\omega_c}{\Delta_0}\right]\Delta_0 \tag{13.78}$$

where λ and $\bar{\lambda}$ are given by

$$\lambda = 2\int_0^{\omega_c} \frac{d\Omega}{\Omega}\alpha^2(\Omega)F(\Omega), \quad \bar{\lambda} = 2\int_0^{\omega_c} \frac{d\Omega}{\Omega}\alpha^2(\Omega)F(\Omega)\ln\left(1+\frac{\omega_c}{\Omega}\right). \tag{13.79}$$

In obtaining (13.76)–(13.77) we followed the procedure given by Leavens and Carbotte[17] and we also considered that the important phonon frequencies from the density $\alpha^2(\Omega)F(\Omega)$ satisfy $\Omega \gg \Delta_0$. In this approximation Eq. (13.77), in the limit $\omega \to 0$, reduces to

$$Z_s(\omega) = 1 + \lambda + \Delta_0^2\int_0^{\infty} d\Omega\frac{2\alpha^2(\Omega)F(\Omega)}{\Omega^3}\ln\frac{\Delta_c}{2\Omega}. \tag{13.80}$$

The last integral in (13.80) can be approximated by

$$\frac{\alpha\lambda}{\bar{\Omega}^2}\ln\left(\frac{\Delta_0\beta}{\bar{\Omega}}\right) \tag{13.81}$$

where α and β are constants to be determined by phenomenological fit (or experiment) and $\bar{\Omega}$ is a suitably defined average phonon frequency, we then get from (13.78) the expression

$$\Delta_0 = 2\omega_c\exp\left[-\frac{1+\lambda+\bar{\lambda}}{\lambda-\mu^*} - \alpha\left(\frac{\Delta_0}{\bar{\Omega}}\right)^2\frac{\lambda}{\lambda-\mu^*}\ln\left(\frac{\Delta_0\beta}{\bar{\Omega}}\right)\right].$$

In the same approximation, Leavens and Carbotte[17] calculated the critical temperature as

$$T_c = 1.143\omega_D\exp\left[-\frac{1+\lambda(T_c)+\bar{\lambda}}{\lambda-\mu^*}\right] \tag{13.82}$$

where

$$\lambda(T) = 2\int_0^{\infty} d\omega\left(-\frac{\partial f}{\partial\omega}\right)\int_0^{\infty} d\Omega\frac{\alpha^2(\Omega)F(\Omega)}{\Omega+\omega} \tag{13.83}$$

and taking $\lambda(T_c) = \lambda$, Eq. (13.82) becomes

$$\frac{2\Delta(0)}{T_c} = 3.53\left[1-\alpha\left(\frac{\Delta_c}{\bar{\Omega}}\right)^2\frac{\lambda}{\lambda-\mu^*}\ln\frac{\Delta_0\beta}{\bar{\Omega}}\right], \tag{13.84}$$

and assuming $\Delta_0 \ll \bar{\Omega}$, the quantity $\lambda-\mu^*$ can be approximated by λ, and we get

$$\frac{2\Delta(0)}{T_c} = 3.53\left[1+\alpha\left(\frac{\Delta_0}{\bar{\Omega}}\right)\ln\frac{\bar{\Omega}}{\beta\Delta_0}\right]. \tag{13.85}$$

Taking now for $\bar{\Omega}$ the expression (13.74), we get

$$\frac{2\Delta(0)}{T_c} = 3.53\left[1 + 12.5\left(\frac{T_c}{\omega_{\ln}}\right)\ln\frac{\omega_{\ln}}{T_c}\right] \tag{13.86}$$

which gives a good agreement with the experimental results.

For materials with a wide valence s-p electron band and a narrow d band (which contains the Fermi level) Hopfield[18] calculated the expression for λ as

$$\lambda = \frac{1}{M\langle\omega^2\rangle}\left(\frac{dU}{dz}\right)^2 N_p(0) \tag{13.87}$$

which mainly contains the atomic constants: the potential gradient (dU/dz) (U is the electron-ion interaction) and the partial density of p-type states on the Fermi surface. Since T_c is determined mainly by the electron-phonon coupling constant λ, the result (13.87) shows that the critical temperature in the transition metals is determined by parameters of atomic nature.

Another important problem for the theory of superconductivity which has been analysed using the Eliashberg equations is the effect of the thermal phonons on the critical temperature. Anderson[19] suggested that it might be the most important barrier to the achievement of high-temperature superconductivity. In fact Allen and Dynes[13] showed that there is nothing in the Eliashberg theory to prevent T_c from growing indefinitely as the strength of the electron-phonon coupling increases without limit.

The first attempt to solve this problem has been made by Bergmann and Rainer,[20] who showed that $\delta T_c/\delta\alpha^2(\omega)F(\omega) \geq 0$ but in their calculation there is no distinction between thermal and virtual phonons. An interesting result was obtained by Leavens and Talbot,[21] who showed by accurate numerical methods applied to the Eliashberg equations that for Nb_3Sn, the thermal phonons depress the critical temperature by only 2%. In the very strong electron-phonon coupling, this suppression saturates at about 50%.

We may conclude that the electron-phonon coupling has to be very important in T_c high temperature superconductor.

14. Spin Fluctuations in Superconductors

In metals which do not present a magnetic ordering in the itinerant-electron system (ferro or antiferromagnetism) we can have strong spin fluctuations (paramagnons) which may, in some cases, help or suppress the different phase transitions. In such metals, the occurrence of superconductivity is less favourable. In fact, the transition metals at the end of the period, such as Pd and Pt, are not superconductors, although the atomic magnetic moments are absent. The strong electronic correlations due to the Coulomb interaction is supposed to be responsible for the occurrence of the spin fluctuations.

In order to study the influence of the paramagnons on the strong-coupling superconductors, Berk and Schrieffer[22] analysed the system of interacting electrons on the superconducting transition line. The Coulomb interaction described by the Hamiltonian

$$\mathcal{H}_c = U \sum_j n_{j\uparrow} n_{j\downarrow} \tag{14.1}$$

where U is the effective Coulombic coupling constant, will be analysed by two-particle Green function

$$K(\mathbf{r}_j - \mathbf{r}_{j'}; \tau) = \langle T_\tau c^\dagger_{j\uparrow}(\tau) c_{j\downarrow}(\tau) c^\dagger_{j'\downarrow}(0) c_{j'\uparrow} \rangle \ , \tag{14.2}$$

with the Fourier transform

$$K(\mathbf{q}, \tau) = \sum_{\mathbf{p},\mathbf{p}'} \langle T c^\dagger_{\mathbf{p}+\mathbf{q}\uparrow} c^\dagger_{\mathbf{p}\downarrow} c_{\mathbf{p}'-\mathbf{q}\downarrow} a_{\mathbf{p}'\uparrow} S(\infty) \rangle \ . \tag{14.3}$$

In the zero-order theory of perturbations, $K_0(\mathbf{q}, \omega)$ is given by

$K_0(\mathbf{q}, \omega) \equiv$ 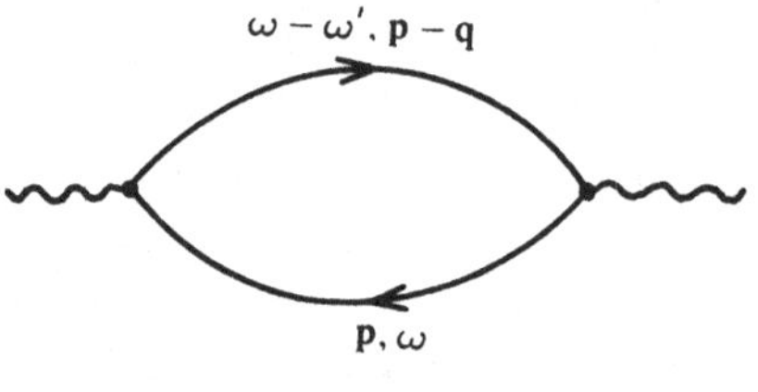

$$= -\pi T \sum_{\omega'} \sum_{\mathbf{p}} G^0(\omega; \mathbf{p}) G^0(\omega + \omega'; \mathbf{p} + \mathbf{p}') \ . \tag{14.4}$$

On the other hand, K can be calculated in the Random Phase Approximation (RPA) as

$$K_{\mathrm{RPA}}(\mathbf{q}, \omega) = \frac{K_0(\mathbf{q}, \omega)}{1 - K_0(\mathbf{q}; \omega)} \tag{14.5}$$

and in the same approximation, the dynamical susceptibility $\chi(\mathbf{q}, \omega)$ is

$$\chi(\mathbf{q}, \omega) = \frac{\chi_0 U_{\mathrm{L}}(\mathbf{q}, \omega)}{1 - N(0) U_{\mathrm{L}}(\mathbf{q}, \omega)} \tag{14.6}$$

where χ_0 is the Pauli susceptibility and $U_L(\mathbf{q},\omega)$ the Lindhard function, which in the limit $q/2p_0 \ll 1$ and $\omega/qv_0 \ll 1$, can be expanded as

$$U_L(\mathbf{q},\omega) = 1 - \frac{1}{3}\left(\frac{q}{2p_0}\right)^2 + \frac{i\pi\omega}{qv_0} . \tag{14.7}$$

In the limit $\omega = 0$, we have

$$\chi(\mathbf{q},0) = \frac{\chi_0}{1 - N(0)U\frac{1}{3}(q/2p_0)^2} . \tag{14.8}$$

With these results, the self energy of the electrons can be calculated from the diagram:

$\hat{\Sigma}_m(\mathbf{p}) =$

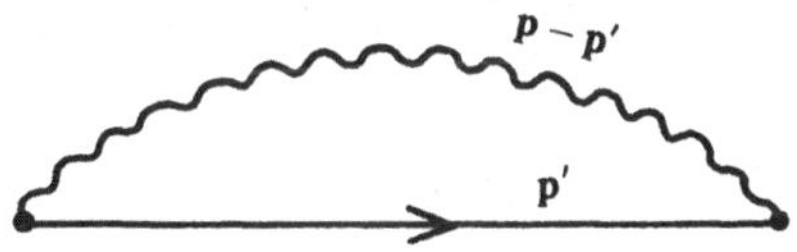

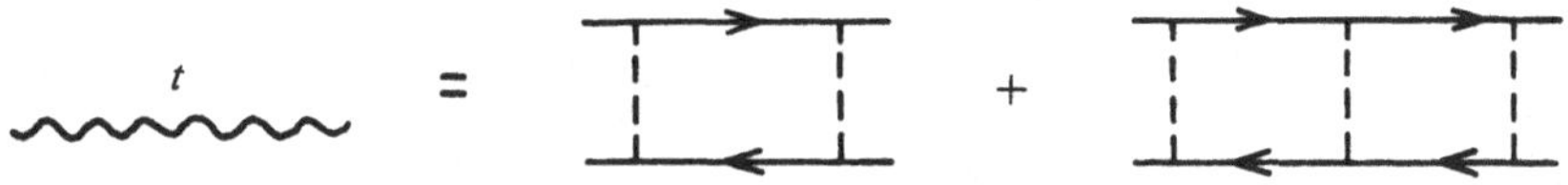

where $t(\mathbf{q},\omega)$ can be calculated as

$$t(\mathbf{q},\omega) = \frac{K_0(\mathbf{q},\omega)U}{1 - UK_0(\mathbf{q},\omega)} = -U^2 K_{\mathrm{RPA}} . \tag{14.9}$$

The analytical expression for $\hat{\Sigma}_m(\mathbf{p},\omega)$ is

$$\hat{\Sigma}_m(\mathbf{p},\omega) = \pi T \sum_{\omega'} \int \frac{d^3\mathbf{p}'}{(2\pi)^3} G^0(\omega';\mathbf{p}')t(\mathbf{p}-\mathbf{p}',\omega-\omega') .$$

Let us consider the case of superconductor with strong-spin fluctuations. The Coulombic interaction (14.1) can be written as

$$\mathcal{H}_i = U\sum_{\mathbf{p},\mathbf{q}} (\psi^+_{\mathbf{p}+\mathbf{q}}(\tau)\hat{\tau}^+\psi_{\mathbf{p}})(\psi^+_{\mathbf{p}'-\mathbf{q}}\hat{\tau}^-\psi_{\mathbf{p}'}) \tag{14.10}$$

where

$$\hat{\tau}^+ = \begin{bmatrix} 1 & 0 \\ 0 & 0 \end{bmatrix} , \quad \hat{\tau}^- = \begin{bmatrix} 0 & 0 \\ 0 & -1 \end{bmatrix} .$$

The electron-paramagnon interaction can be described by the self-energy $\widehat{\Sigma}_{\rm m}$ expressed as

$$\widehat{\Sigma}_{\rm m}(\mathbf{p},\omega) = -\pi T \sum_{\omega'} \int \frac{d^3\mathbf{p}'}{(2\pi)^3} t(\mathbf{p}-\mathbf{p}',\omega-\omega')\widehat{\tau}_1 \widehat{G}(\omega';\mathbf{p}')\widehat{\tau}_1 \tag{14.11}$$

and the Eliashberg equations (13.48)–(13.49) at $T = 0$ becomes

$$[1 - Z(\omega)]\omega = \int dz' [K_{\rm ph}(z',\omega) + K_{\rm m}(z',\omega)] \mathrm{Re}\, \frac{z'}{\sqrt{z'^2 - \Delta^2(z')}} \tag{14.12a}$$

and

$$Z(\omega)\Delta(\omega) = \int dz' [K_{\rm ph}(z',\omega) - K_{\rm m}(z',\omega)] \mathrm{Re}\, \frac{\Delta(z')}{\sqrt{z'^2 + \Delta^2(z')}} \tag{14.12b}$$

where the electron-paramagnon kernel $K_{\rm m}$ is given by

$$\begin{aligned} K_{\rm m}(z',\omega) = \int_0^\infty dz\, g_{\rm m}(z) \frac{1}{2} \Bigg\{ & \frac{\tanh z'/2T + \coth z/2T}{z' + z - \omega - i\delta} \\ & + \frac{\tanh z'/2T - \coth z/2T}{z' + z - \omega - i\delta} \Bigg\} \end{aligned} \tag{14.13}$$

with $g_{\rm m}(z)$ represents the average of the spectral density of paramagnons on the Fermi surface

$$g_{\rm m}(z) = \int_{S(\epsilon_0)} \frac{d^2\mathbf{p}}{v_{\mathbf{p}}} \int \frac{d^2\mathbf{p}'}{v_{\mathbf{p}'}} \left[-\frac{1}{\pi} \mathrm{Im}\, t(\mathbf{p}-\mathbf{p}'; z - z') \right] \Bigg/ \int_{S(\epsilon_0)} \frac{d^2\mathbf{p}}{v_{\mathbf{p}}} \, . \tag{14.14}$$

If we linearize Eqs. (14.12), we get

$$\begin{aligned} & \Delta(\omega) \left[1 + \frac{1}{\omega} A_{\rm ph}(\omega) + \frac{1}{\omega} A_{\rm m}(\omega) \right] \\ & = -\int_{-\infty}^{\infty} dz' [K_{\rm ph}(z',\omega) - K_{\rm m}(z',\omega)] \mathrm{Re}\, \frac{\Delta(z')}{z'} \end{aligned} \tag{14.15}$$

where

$$\begin{aligned} A_{\rm ph}(\omega) &= \int_{-\infty}^{\infty} dz' K_{\rm ph}(z';\omega) \\ A_{\rm m}(\omega) &= \int_{-\infty}^{\infty} dz\, K_{\rm m}(z;\omega) \, . \end{aligned} \tag{14.16}$$

From (14.16), $A_{\rm ph}(\omega) \simeq \lambda\omega$ and $A_{\rm m}(\omega) \cong \lambda_{\rm m}\omega$ where

$$\lambda_{\rm m} = 2\int_0^\infty dz\,\frac{g_{\rm m}(z)}{z}\,. \tag{14.17}$$

Near the critical temperature T_c

$$K_{\rm m}(z',\omega) = \frac{\lambda_{\rm m}}{2}\theta(\omega_{\rm m} - |z|)\theta(\omega_{\rm m} - |\omega|)\tanh\frac{z'}{2T} \tag{14.18}$$

and using (13.51) for $K_{\rm ph}(z',\omega)$, we get

$$T_c = \frac{2\gamma}{\pi}\bar{\omega}\exp\left[-\frac{1+\lambda+\lambda_{\rm m}}{1-\lambda_{\rm m}}\right] \tag{14.19}$$

where

$$\bar{\omega} = \omega_{\rm m}\left(\frac{\omega_{\rm D}}{\omega_{\rm m}}\right)^{\nu},$$

$$\nu = \frac{\lambda^2}{(\lambda-\lambda_{\rm m})[\lambda-\lambda_{\rm m}+\lambda\lambda_{\rm m}/(1+\lambda+\lambda_{\rm m})]\ln(\omega_{\rm D}/\omega_{\rm m})}\,.$$

For the quadratic dispersion law of electrons, (14.14) becomes

$$g_{\rm m}(z) = N(0)\int_0^{2p_0} dq\,\frac{q}{2p_0}\left[-\frac{1}{\pi}\,{\rm Im}\,t(q,z)\right] \tag{14.20}$$

and

$$-\frac{1}{\pi}\,{\rm Im}\,t(\mathbf{q};z) = \frac{U}{\pi}\,\frac{(\pi/2)N(0)U(z/qv_0)}{[1-N(0)U-N(0)Uq^2/(12p_0)^2+((\pi/2)N(0)Uz/qv_0)^2]}\,. \tag{14.21}$$

Substituting (14.20) and (14.21) in (14.16), we get

$$\lambda_{\rm m} = aUN(0)\ln\frac{1}{1-N(0)U} \tag{14.22}$$

where $a \simeq 2$ if $N(0) = U \cong 1$ and $a \cong 1$ if $N(0)U \ll 1$.
We can see that near the ferromagnetic instability $N(0)U \cong 1$, the electron-paramagnon coupling constant is increasing and the superconductivity will be suppressed if $\lambda_{\rm m} > \lambda$. This model has also been applied to explain

the low critical temperature in some non-magnetic alloys of the transition metals.

15. Triplet Pairing

The idea of the p-states pairing appeared a short time after the s-pairing had been proposed and gained particular interest when it was realized that He^3 is a Fermi liquid. The theory of the p-pairing for the superfluid He^3 has been developed by Leggett[23] but the possibility of the occurrence of the superconducting state with p-states pairing has been developed by Anderson and Morel,[24] Gor'kov and Galitskii.[25]

The electron-electron interaction has been considered as

$$g(\mathbf{p}-\mathbf{p}') = \sum_l g_l P_l(\theta) \tag{15.1}$$

where l is the quantum orbital number and $P_l(\theta)$ the Legendre polynomials.

The anomalous averages $\langle N+2, l, m|\psi^\dagger\psi^\dagger|N,0\rangle$ have $2l+1$ components and depend on the quantum magnetic number m. The equations for the Green functions are

$$\begin{aligned} &[\omega-\varepsilon(\mathbf{p})]\hat{G}(\omega;\mathbf{p}) - i\sum_m \hat{\Delta}_m(\mathbf{p})F_m^\dagger(\omega;\mathbf{p}) = 1 \\ &[\omega+\varepsilon(\mathbf{p})]F_m(\mathbf{p};\omega) + i\sum_m \hat{\Delta}_m^\dagger(\mathbf{p})\hat{G}(\omega;\mathbf{p}) = 0 \end{aligned} \tag{15.2}$$

where

$$\hat{\Delta}_m(\mathbf{p}) = \frac{1}{(2\pi)^4}\int d^3\mathbf{p}' g(\mathbf{p}-\mathbf{p}')\hat{F}_m(\mathbf{p}';\omega) \tag{15.3}$$

$$\hat{\Delta}_m^\dagger(\mathbf{p}) = \frac{1}{(2\pi)^3}\int d^3\mathbf{p}' g(\mathbf{p}-\mathbf{p}')\hat{F}_m^\dagger(\mathbf{p}';\omega)\,. \tag{15.4}$$

The Green functions $\hat{G}_{\alpha\beta}$ and $\hat{F}_{\alpha\beta}$ have the structure

$$\hat{G}_{\alpha\beta} = \delta_{\alpha\beta}G(\omega,\mathbf{p})\,,\quad \hat{F}_{m;\alpha,\beta} = -FI_{l,m}Y_{l,m}(\theta,\varphi) \tag{15.5}$$

with

$$I_{\alpha,\beta} = -I_{\beta,\alpha} \quad \text{and} \quad I_{\alpha\beta}^2 = 1\,.$$

Using the definition

$$Y_{l,m} = \left[\frac{2l+1}{4\pi}\frac{(l-|m|)!}{(l+|m|)!}\right]^{\frac{1}{2}} P_l^m(\theta)e^{im\varphi}$$

and the relation

$$P_l(\theta,\theta';\varphi,\varphi') = \sum_m \frac{(l-|m|)!}{(l+|m|)!} P_l^m(\theta) P_{l'}(\theta) e^{im(\varphi-\varphi')} ,$$

one obtains the equation for Δ_m^* as

$$\Delta_m^* = \left[\frac{g_l N(0)}{2l+1}\right] \int \frac{d\omega}{2\pi} \int_{-\infty}^{\infty} d\varepsilon F_m^*(\omega,\mathbf{p}) \tag{15.6}$$

where the Green function is

$$F_m^*(\omega,\mathbf{p}) = \frac{-i\Delta_m^*}{\omega^2 - E^2(\mathbf{p})}$$

with

$$E^2(\mathbf{p}) = \varepsilon^2(\mathbf{p}) + |\Delta(\mathbf{p})|^2 \tag{15.7}$$

and

$$|\Delta(\mathbf{p})|^2 = \sum_m |\Delta_m(\mathbf{p})|^2 \frac{(l-|m|)!}{(l+|m|)!} |P_l(\theta)|^2 . \tag{15.8}$$

If one takes G as isotropic, using $P_l(\theta,\theta')$, Eq. (15.8) becomes

$$|\Delta|^2 = \frac{1}{2} |\Delta_m|^2 (2l+1)(2s+1) P_l(\theta) \tag{15.9}$$

and from (15.6), the order parameter is

$$|\Delta| = 2\tilde{\omega}_{\rm D} \exp\left[-\frac{(2l+1)}{N(0)V_l}\right] . \tag{15.10}$$

The spin states are important in this case and we have to consider the spin structure of the order parameter. If one considers

$$c(\mathbf{p}) = \begin{bmatrix} c_{\mathbf{p}\uparrow} \\ c_{\mathbf{p}\downarrow} \end{bmatrix} , \qquad c^\dagger(\mathbf{p}) = \begin{bmatrix} c^\dagger_{\mathbf{p}\uparrow} \\ c^\dagger_{\mathbf{p}\downarrow} \end{bmatrix} ,$$

$$c(\mathbf{p}) = [c_{\mathbf{p}\uparrow}, c_{\mathbf{p}\downarrow}] , \qquad c^\dagger(\mathbf{p}) = [c^\dagger_{\mathbf{p}\uparrow}, c^\dagger_{\mathbf{p}\downarrow}]$$

the gap can be expressed by the correlation

$$\hat{x}(\mathbf{p}) = \langle c(\mathbf{p}) \tilde{c}(-\mathbf{p}) \rangle \tag{15.11}$$

and the analysis of the symmetry properties showed that (15.11) can be expressed, using a vector $\widehat{x}$, as

$$\widehat{x}(\mathbf{p}) = (\mathbf{x}(\mathbf{p})\widehat{\sigma})\widehat{\sigma}_y \tag{15.12}$$

which gives

$$\mathbf{x}(\mathbf{p}) = \frac{1}{2}\mathrm{Tr}\,\{\mathbf{x}(\mathbf{p})\widehat{\sigma}_y\widehat{\sigma}\}\ . \tag{15.13}$$

This vector can be expressed as

$$\sum_{\mathbf{p}} \mathbf{x}(\mathbf{p}) = \psi \mathbf{d}(\mathbf{n}) \tag{15.14}$$

where ψ is a scalar function and

$$\int \frac{d\Omega}{4\pi} |\mathbf{d}(\mathbf{n})|^2 = 1\ . \tag{15.15}$$

The order parameter for the triplet state depends on the spin states as the correlation (15.11) and one may take a similar structure as (15.12), namely

$$\widehat{\Delta}(\mathbf{p}) = \Delta[\mathbf{d}(\mathbf{n})\widehat{\sigma}]\widehat{\sigma}_y\ . \tag{15.16}$$

For the p-pairing ($l = 1$), there are four states which have been discussed in connection with the p-pairing in He^3.[23]

1. Balian-Werthamer State

In this case

$$\mathbf{d}(\mathbf{n}) = \mathbf{n} \tag{15.17}$$

which satisfies

$$|\mathbf{d}(\mathbf{n})|^2 = 1 \quad \text{and} \quad \langle |\mathbf{d}(\mathbf{n})|^2 \rangle = \int \frac{d\Omega}{4\pi} |d(\mathbf{n})|^2\ .$$

The anisotropy of the order parameter can be characterized by

$$\chi = \int \frac{d\Omega}{4\pi} |\mathbf{d}(\mathbf{n})|^2 .$$

2. Anderson-Brinkmann-Morel States

In this case, the vector $\mathbf{d}$ is given by[27]

$$\mathbf{d}(\mathbf{n}) = \begin{cases} 0 \\ 0 \\ \sqrt{3}/2(n_x + i n_y) \end{cases} \tag{15.18}$$

where $\mathbf{d}$ and χ satisfy

$$|d(\mathbf{n})|^2 = \frac{3}{2}(n_x^2 + n_y^2)\,, \quad \langle|\mathbf{d}(\mathbf{n})|\rangle = 1\,, \quad \chi = \langle|\mathbf{d}(\mathbf{n})|^4\rangle = \frac{6}{5}\,.$$

3. Planar States

In this case

$$d_x = \sqrt{\frac{3}{2}}\,n_x\,, \quad d_y = \sqrt{\frac{3}{2}}\,n_y\,, \quad d_z = 0\,, \quad \chi = \frac{6}{5}\,. \tag{15.19}$$

4. Polar States

For the polar states

$$d_z = \sqrt{3}\,n_z\,, \quad d_x = d_y = 0\,, \quad \chi = \frac{9}{5}\,. \tag{15.20}$$

The free energy has been calculated as function of the parameter Δ and the most stable state is the Balian-Werthamer one followed by the other states in the order presented above. However, at temperature much lower than the critical temperature T_c, the Anderson-Brinkmann-Morel phase is more stable than the Balian-Werthamer state. In the presence of an external magnetic field, the most stable phase (called A_1) is defined by $S_z = +1$ and the parameters

$$d_x = -i/\sqrt{2}\,, \quad d_y = i/\sqrt{2}\,, \quad d_z = 0\,.$$

The magnetic susceptibility χ_{ik} defined as

$$\chi_{ik} = \mu_0 \sum_{\alpha,\beta,\gamma,\delta} (\sigma_i)_{\alpha\beta}(\sigma_k)_{\gamma\delta} \int d^4x' \langle T\psi_\gamma^\dagger(x')\psi_\delta(x')\psi_\alpha^\dagger(x)\psi_\beta(x)\rangle \tag{15.21}$$

can be calculated by the standard methods using the results obtained by Gor'kov and Galitskii.[25]

The average from (15.21) can be expressed as

$$\begin{aligned}&\langle T\psi_\alpha(x)\psi_\beta^\dagger(x')\psi_\gamma(y)\psi_\delta^\dagger(y')\rangle \\ &= G_{\alpha\beta}(x-x')G_{\gamma\delta}(y-y') - G_{\alpha\delta}(x-y')G_{\gamma\delta}(y-x') \\ &\quad - \sum_m F_{m;\alpha\gamma}(x-y)F^\dagger_{m;\beta\delta}(x'-y')\end{aligned} \tag{15.22}$$

where the Fourier transform of the Green functions G and F are given by (15.21). Using these results, (15.21) becomes

$$\chi_{ik} = \frac{\mu_0^2}{(2\pi)^3}\pi T \sum_{\omega} \sum_{\alpha,\beta,\gamma,\delta} (\sigma_i)_{\alpha\beta}(\sigma_k)_{\gamma\delta} \int d^3\mathbf{p} \times \left[-G_{\delta\alpha}(\mathbf{p},\omega)G_{\beta\gamma}(\mathbf{p};\omega) + \sum_m F_{m;\delta\beta}(\mathbf{p};\omega)F^{\dagger}_{m;\alpha\gamma}(\mathbf{p};\omega)\right]. \tag{15.23}$$

If we introduce the notations χ^{s} and χ^{t} for the susceptibilities in the singlet and triplet states, then

$$\chi^{s} = \chi_{\parallel} = -\frac{2\mu_0^2}{(2\pi)^3}\pi T \sum_{\omega} \int d^3\mathbf{p}\left[G^2(\mathbf{p};\omega) + \frac{2l+1}{4\pi}F^2(\mathbf{p};\omega)\right] \tag{15.24}$$

$$\chi^{t} = \chi_{\perp} = -\frac{2\mu_0^2}{(2\pi)^3}\pi T \sum_{\omega} \int d^3\mathbf{p}\left[G^2(\mathbf{p};\omega) - \frac{2l+1}{4\pi}F^2(\mathbf{p};\omega)\right] \tag{15.25}$$

we get

$$\chi^{s} \equiv \chi_{\parallel} = \frac{\mu_0^2}{(2\pi T)^3}\frac{T}{2}\int \frac{d^3\mathbf{p}}{\cosh^2 E(\mathbf{p})/2T}, \tag{15.26}$$

$$\chi^{t} \equiv \chi_{\perp} = \frac{\mu_0^2}{(2\pi)^3}\int d^3\mathbf{p}\left[\frac{\varepsilon^2(\mathbf{p})}{2TE(\mathbf{p})}\frac{1}{\cosh^2 E(\mathbf{p})/2T} + \frac{\Delta^2}{E^3(\mathbf{p})}\tanh\frac{E(\mathbf{p})}{2T}\right]. \tag{15.27}$$

For a constant density of states, (15.26) and (15.27) become

$$\chi^{s}(T) = \chi_0 Y(T) \tag{15.28}$$

$$\chi^{t}(T) = \chi_0 \tag{15.29}$$

where

$$Y(T) = \frac{1}{2T}\int_0^{\infty} \frac{d\varepsilon}{\cosh^2(1/2T)\sqrt{\varepsilon^2+\Delta^2}} \tag{15.30}$$

and χ_0 is the Pauli paramagnetic susceptibility.
For a polycrystalline sample

$$\chi_{\text{pol}}^{\perp} = \frac{1}{3}\chi_{\parallel} + \frac{2}{3}\chi_{\perp} = \chi_0\left[\frac{2}{3} + \frac{1}{3}Y(T)\right], \tag{15.31}$$

which is identical to that obtained by Balian and Werthamer[26] for the first state.

The anisotropic Knight shift suggested a strong dependence of the susceptibility on the spin state. Let us consider **d** and **H** as two parallel vectors and $\mathbf{d}_1 = \mathbf{d}_2 = 0$. In this case, the electron pairs have $S_z = 0$ and

$$\hat{\Delta} = i\Delta_3\hat{\sigma}_x \, . \tag{15.32}$$

The magnetic susceptibility in this case is

$$\chi_{\|} = \chi_0 Y(\mathbf{n}, T) \tag{15.33}$$

which contains $Y(\mathbf{n}, T)$ dependent on $\Delta(\mathbf{n}, T)$.
If the vector $\mathbf{d}(\mathbf{n})$ is perpendicular to $\mathbf{H}$ and $\mathbf{d}_3 = 0$, the pairs are superpositions of states with $S_z = \pm 1$ and

$$\chi_{\perp} = \chi_0 \, . \tag{15.34}$$

A more accurate analysis can be performed if one calculates the magnetization M defined as

$$\mathbf{M} = \chi_{\|}\mathbf{H}_{\|} + \chi_{\perp}\mathbf{H}_{\perp} \tag{15.35}$$

where $\mathbf{H}_{\|}$ and $\mathbf{H}_{\perp}$ are the components of the magnetic field H parallel and perpendicular to the vector **d** respectively.
Then we have

$$\mathbf{H}_{\|} = (\mathbf{u}\cdot\mathbf{H})\mathbf{u}\,, \quad \mathbf{H}_{\perp} = \mathbf{H} - (\mathbf{u}\cdot\mathbf{H})\mathbf{u} \tag{15.36}$$

where **u** is the unit vector of **d**.
The magnetization (15.35) will be given by

$$\mathbf{M}_i = \chi_{\perp}\mathbf{H}_i + (\chi_{\|} - \chi_{\perp})\sum_j u_i u_j H_j = \sum_j \chi_{ij} H_j \tag{15.37}$$

and the susceptibility tensor has the components

$$\chi_{ij} = \chi_{\perp}\delta_{ij} + (\chi_{\|} - \chi_{\perp})u_i u_j$$

which can be transformed using (15.33) and (15.34) as

$$\chi_{ij} = \chi_0 \left\{ \delta_{ij} - [1 - Y(\mathbf{n}, T)]\frac{d_i(\mathbf{n})d_j^*(\mathbf{n})}{|\mathbf{d}(\mathbf{n})|^2} \right\} \, . \tag{15.38}$$

For the Balian-Werthnamer state if one takes

$$\left\langle \frac{d_i(\mathbf{n})d_j^*(\mathbf{n})}{|\mathbf{d}(\mathbf{n})|^2} \right\rangle = \int \frac{d\Omega}{4\pi} d_i^*(\mathbf{n})d_j(\mathbf{n}) \, ,$$

Eq. (15.38) becomes

$$\chi_{ij} = \chi_0 \delta_{ij} \left[\frac{2}{3} + \frac{1}{3} Y(T) \right] \tag{15.39}$$

which is in fact identical to (15.31).
If one considers the Anderson-Brinkmann-Morel state with

$$\mathbf{d} = \sqrt{\frac{3}{2}}\, \mathbf{d}(\alpha_1 \mathbf{n} + \alpha_2 \mathbf{n})$$

we get

$$\chi^{ij}_{\mathrm{ABM}} = \chi_0 [\delta_{ij} - (1 - Y(T)) d_i d_j] \tag{15.40}$$

where

$$Y(T) = \frac{1}{4\pi} \int d\Omega \, Y(\mathbf{n}, T) \, .$$

The triplet pairs may be responsible for the superconductivity in the heavy fermion superconductors, which explains the behaviour of some materials at low temperatures.

Appendix

In this appendix we discuss shortly the occurrence of the attractive electron-electron interaction due to the electron-phonon interaction.

We start with the simple Debye model for the electron-lattice interaction. In the second quantization, the Hamiltonian describing this model is

$$\mathcal{H}_{\mathrm{e-ph}} = \gamma \int d^3\mathbf{r} \, \psi_\alpha^\dagger(\mathbf{r}) \psi_\alpha(\mathbf{r}) \varphi(\mathbf{r}) \tag{A.1}$$

where $\psi_\alpha^\dagger(\mathbf{r})$, $\psi_\alpha(\mathbf{r})$ are the operators describing the creation and annihilation of an electron with the spin α, and the phonon field is

$$\varphi(\mathbf{r}) = \frac{1}{V} \sum_{\mathbf{q}} \frac{\omega(\mathbf{q})}{2} [b_{\mathbf{q}} e^{i\mathbf{q}\cdot\mathbf{r}} + b_{\mathbf{q}}^\dagger e^{-i\mathbf{q}\cdot\mathbf{r}}] \tag{A.2}$$

where ω_D is the Debye frequency, $\omega(\mathbf{q})$ is the phonon energy and V is the volume of the system.

In the theory of superconductivity, the electron-phonon interaction mediates the effective electron-electron interaction. In order to obtain the concrete form of this interaction, we can apply the theory of perturbation which, in fact, is limited by the existence of the small vertex correction.

If the perturbation theory cannot be applied, we have to use the method of canonical transform which is exact and nonperturbative.

a) Perturbative treatment

The self consistent study of the electron-phonon interaction implies the use of the Dyson equations for the Green functions of electrons and phonons. For the electron-phonon interaction in the normal metals, such an approach can be simplified by the result obtained by Migdal[28] which is called Migdal's theorem. Migdal's theorem states that the vertex corrections are at least of the order $\gamma\sqrt{m_e/M_i}$ (m_e is the mass of the electron, M_i is the mass of ion) and to a good approximation, we can replace the vertex corrections Γ by the point value γ. A complete demonstration is contained in Ref. 29 and here we confine ourselves to show the physics beyond this approximation. In fact, the argument is that the electrons and phonons wave vectors have the same scale, while the energy scales are very different. This makes $\partial\Sigma/\partial k$ (where Σ is the electronic self-energy) a very small quantity anywhere on the Fermi surface, while $\partial\Sigma/\partial\omega$ is very large. In the simplest diagram

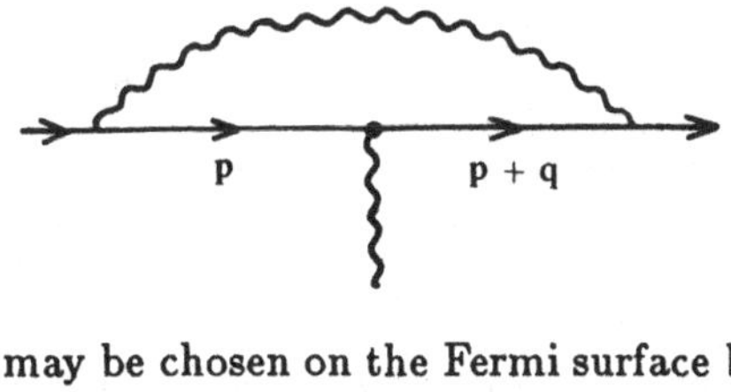

the wave vector $\mathbf{p}$ may be chosen on the Fermi surface but for any $\mathbf{q}$ so that $qv_0 \gg \omega_D$, the wave vector is far from the Fermi surface; this introduces an energy denominator of the order of the electronic energy and hence a factor proportional to m_e/M_i.
Englesberg and Schrieffer[30] noted that for $qv_0 \ll \omega$, the Migdal approximation is not valid, but this region has little influence on most physical properties.

For a two-band electronic system with nesting, the Migdal theorem is no more valid. The nesting condition is expressed by $\varepsilon(\mathbf{p}) = \varepsilon(\mathbf{p}+\mathbf{Q})$, and if we calculate the self-energy of the electrons expressed by the diagram

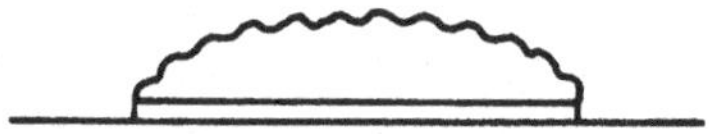

we obtain for the real part of the self-energy a logarithmic behaviour taking Einstein phonons. If one considers the Ward-identity for $\mathbf{q} = \mathbf{Q}$, we get

$$\Gamma(\mathbf{p},\omega;\mathbf{Q},\Omega) = 1 - \frac{\gamma^2 N(0)}{\omega(q)} \frac{\Sigma(\mathbf{p}+\mathbf{Q},\Omega) - \Sigma(\mathbf{p},\omega)}{\Omega} \tag{A.3}$$

and because $\gamma^2 N(0)/\Omega \sim 1$, Eq. (A.3) shows the breakdown of the Migdal approximation.

The high temperature anomalies are connected with the failure of the Migdal approximation which is in fact a restatement of the adiabatic approximation. More physically, if the temperature is not low enough, the electrons and phonons cannot be considered as independent entities whose residual interaction may be calculated using the perturbation theory. However, for usual superconductors we consider that the Migdal theorem is valid and an electron-electron attraction can appear if we treat the electron-phonon interaction by the perturbation theory.

We will consider that the Migdal theorem can be applied for an interacting system of electrons and phonons and that the electron-electron interaction can occur due to the phonon exchange. In this case, we can perform the identification

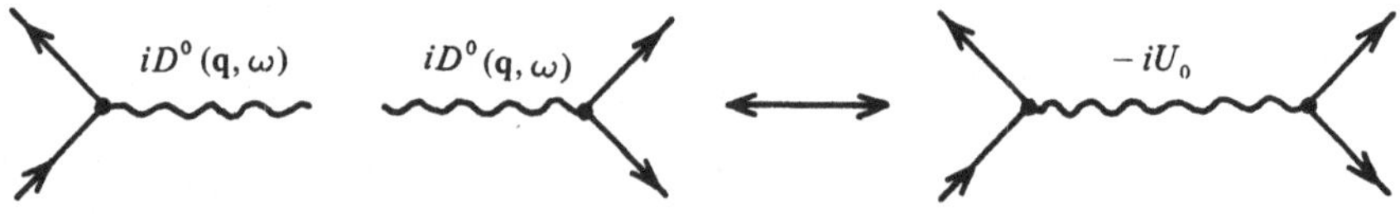

where $D^0(\mathbf{q},\omega)$ is the Green function for the phonons

$$D^0(\mathbf{q},\omega) = \frac{\omega^2(q)}{\omega^2 - \omega^2(q)} \,. \tag{A.4}$$

We can write the effective interaction $U_0(\mathbf{q},\omega)$ as

$$U_0(\mathbf{q},\omega) = g^2 D^0(\mathbf{q},\omega) = g^2\theta(\omega_D - \omega(\mathbf{q}))\frac{\omega^2(q)}{\omega^2 - \omega^2(q)} \,. \tag{A.5}$$

This potential is frequency dependent even in the lowest order. In the static limit (A.5) reduces to

$$U_0(\mathbf{q},0) = -g^2\theta(\omega_D - \omega(q)) \equiv \int d^3x\, e^{-i\mathbf{q}\cdot\mathbf{x}} V_a(x) \ . \tag{A.6}$$

If we assume $U_0(\mathbf{q},0) = -g^2$ for all $\mathbf{q}$, then the electron-electron potential becomes an attractive one, of the form

$$V_a(x) = -g^2\delta(x) \ . \tag{A.7}$$

In Eq. (A.5), the potential $U_0(\mathbf{q},\omega)$ will cease to be attractive when the energy transfer ω satisfies $|\omega| \gg \omega(q)$. Since $\omega(q) < \omega_D$, we get

$$U_0(\mathbf{q},\omega) > 0 \quad \text{if} \quad |\omega| > \omega_D \tag{A.8}$$

an equation which shows that in the noninteracting ground state where the electrons form a filled Fermi sea $U_0(\mathbf{q},\omega)$ can be attractive only for the electrons lying within an energy shell of thickness ω_D below the Fermi surface, because only these electrons can be excited to unoccupied levels with energy transfer less than ω_D. This argument is used when we perform the integrals over electronic energy on the interval $0,\omega_D(-\omega_D,\omega_D)$. However, if the electrons are strongly localized, and the electron-phonon interaction is strong, the perturbation theory does now work anymore. A well-known problem is that of the excitons in insulators or semiconductors. In this case, the electronic energy is drastically affected by the phonons due to a strong electron-phonon interaction. In this case, the perturbations theory cannot be applied and for the treatment of this problem, it was developed an exact method called the canonical transformation. This method has been used to describe a polaron-type representation of the exciton by a canonical transformation of the Hamiltonian.

b) Canonical transformation method

If the system consisting of electrons and phonons is described by the Hamiltonian

$$\mathcal{H} = \mathcal{H}_0 + \mathcal{H}_i \tag{A.9}$$

where $\mathcal{H}_0$ is the Hamiltonian for the free electrons and phonons and $\mathcal{H}_i$ is the Hamiltonian describing the electron-phonon interaction, we can transform (A.9) in a Hamiltonian containing only the electron-electron interaction. Let us introduce the unitary transformation S which transforms the Hamiltonian (A.9) as

$$\mathcal{H}' = \exp(-iS)\mathcal{H}\exp(iS) \tag{A.10}$$

where S will be developed after the powers of the coupling constant g as

$$S = S_1 + S_2 + \dots \tag{A.11}$$

and

$$\exp(iS) = 1 + iS_1 + iS_2 - \frac{1}{2}S_1^2 + \dots . \tag{A.12}$$

The Hamiltonian (A.10) can be written as

$$\mathcal{H}' = \mathcal{H}_0 + \mathcal{H}_i - i[S_1, \mathcal{H}_0] - i[S_1, \mathcal{H}_i] - i[S_2, \mathcal{H}_0] + \dots . \tag{A.13}$$

The contribution S_1 can be obtained from the condition that the nondiagonal contributions in the Hamiltonian (which are proportional to g) has to be avoided, and S_2 imposing the same constrain on the term containing g^2. In this way, for S_1 we have the condition

$$\mathcal{H}_i - i[S_1, \mathcal{H}_0] = 0 . \tag{A.14}$$

This equation can be solved and S_1 is

$$\langle n|S_0|m\rangle = i\frac{\langle n|\mathcal{H}_0|m\rangle}{E_n^0 - E_m^0} \tag{A.15}$$

where $|n\rangle$ and $|m\rangle$ are eigenstates of H_0 with the eigenvalues E_n^0 and E_m^0. If the function $(E_n^0 - E_m^0)^{-1}$ can be considered as a principal part, we obtain

$$\begin{aligned}\langle n|S_1|m\rangle &= \frac{1}{2}\langle n|\mathcal{H}_i|m\rangle\left\{\int_0^\infty dt\exp[i(E_n^0 - E_m^0 + i\eta)t]\right.\\ &\qquad\left. - \int_{-\infty}^0 dt\exp[i(E_n^0 - E_m^0 - i\eta)t]\right\}\\ &= \frac{1}{2}\left\{\int_0^\infty dt\, e^{-\eta t}\langle n|\exp(i\mathcal{H}_0 t)\mathcal{H}_i\exp(-i\mathcal{H}_0 t)|m\rangle\right.\\ &\qquad\left. - \int_{-\infty}^0 dt\, e^{\eta t}\langle n|\exp(i\mathcal{H}_0 t)\mathcal{H}_i\exp(-i\mathcal{H}_0 t)|m\rangle\right\}\end{aligned} \tag{A.16}$$

where we used the identity

$$\begin{aligned}P\frac{1}{x} &= \frac{1}{2}\left(\frac{1}{x+i\eta} - \frac{1}{x-i\eta}\right)\\ &= \frac{i}{2}\left[\int_{-\infty}^0 dt\, e^{i(x-i\eta)t} + \int_0^\infty dt\, e^{i(x+i\eta)t}\right] .\end{aligned}$$

In the second integral of (A.16), we take $t \to -t$ and in the limit $\eta \to 0$, S_1 is

$$S_1 = \frac{1}{2}\int_0^\infty dt\, e^{-\eta t}[\mathcal{H}_i(t) - \mathcal{H}_i(-t)]e^{-\eta t} \tag{A.17}$$

where

$$\mathcal{H}_i(t) = \exp(i\mathcal{H}_0 t)\mathcal{H}_i \exp(i\mathcal{H}_0 t)\ .$$

Using now for the operators $c_{\mathbf{k}}(t)$ and $b_{\mathbf{k}}(t)$, the expressions

$$\begin{aligned} c_{\mathbf{k}}(t) &= \exp(i\mathcal{H}_0 t)c_{\mathbf{k}} \exp(-i\mathcal{H}_0 t) = \exp(-i\varepsilon(\mathbf{k})t) \\ b_{\mathbf{k}}(t) &= \exp(i\mathcal{H}_0 t)b_{\mathbf{k}} \exp(-i\mathcal{H}_0 t) = \exp(-i\omega(\mathbf{q})t) \end{aligned}$$

the integral over t in (A.17) can be performed and we get

$$S_1 = i\sum_{\mathbf{k},\mathbf{q},\alpha} \frac{g(\mathbf{q})}{\varepsilon(\mathbf{k}) - \varepsilon(\mathbf{k}+\mathbf{q}) + \omega(q)}(c^\dagger_{\mathbf{k},\alpha}c_{k+q,\alpha}b_{\mathbf{q}} + h.c.) \tag{A.18}$$

where $g(q) = g[\omega(q)/2V]^{\frac{1}{2}}$.

Let us consider terms of the second order from (A.13). Using (A.14), we have

$$i[S_1, \mathcal{H}_i] + \frac{1}{2}[S_1, [S_1, \mathcal{H}_0]] = \frac{i}{2}[S_1, \mathcal{H}_0] \tag{A.19}$$

and the Hamiltonian $\mathcal{H}'$ becomes

$$\mathcal{H}' = \mathcal{H}_0 - \frac{i}{2}[S_1, \mathcal{H}_0]_{\mathrm{d}}$$

where "d" indicates that we have to keep only the diagonal terms in the phononic operators. Using this approximation (A.18) for the electron-phonon interaction, we get a new Hamiltonian

$$\mathcal{H}' = \mathcal{H}_0 - \frac{g^2}{2V}\sum_{\substack{\mathbf{k},\mathbf{k}',\mathbf{q}\\ \alpha,\alpha'}} \frac{\omega^2(\mathbf{k}-\mathbf{k}')}{\omega^2(\mathbf{k}-\mathbf{k}') - (\varepsilon(\mathbf{k}) - \varepsilon(\mathbf{k}'))^2} c^\dagger_{\mathbf{k},\alpha}c^\dagger_{\mathbf{k}-\mathbf{q},\alpha'}c_{\mathbf{k}-\mathbf{q},\alpha'}c_{\mathbf{k},\alpha'} \tag{A.20}$$

with the second term giving the effective electron-electron interaction. For $|\varepsilon(\mathbf{k}) - \varepsilon(\mathbf{k}')| < \omega(\mathbf{k}-\mathbf{k}')$, the interaction contribution at the Hamiltonian (A.20) becomes attractive.

References

1. H. Fröhlich, *Proc. Roy. Soc.* **215** (1952) 291.
2. L. N. Cooper, *Phys. Rev.* **104** (1956) 1189.
3. J. Bardeen, L. N. Cooper and J. R. Schrieffer, *Phys. Rev.* **108** (1957) 1175.
4. N. N. Bogoliubov, *Sov. Phys. JETP* **7** (1951) 41.
5. L. P. Gor'kov, *Zh. Eksp. Fiz.* **34** (1958) 735; *Sov. Phys. JETP* **7** (1958) 505.
6. I. M. Khalatnikov and A. A. Abrihosov, *Advn. Phys.* **8** (1959) 45.
7. G. Eilenberger, *Z. Phys.* **B214** (1968) 195.
8. K. Usadell, *Phys. Rev. Lett.* **25** (1970) 507.
9. P. Morel, P. W. Anderson, *Phys. Rev.* **125** (1962) 1263.
10. G. M. Eliashberg, *Sov. Phys. JETP* **12** (1960) 696, 1000.
11. D. J. Scalapino, J. R. Schrieffer, J. W. Wilkins, *Phys. Rev.* **148** (1966) 263.
12. W. L. MacMillan, *Phys. Rev.* **167** (1968) 331.
13. P. B. Allen and R. C. Dynes, *Phys. Rev.* **B12** (1975) 905.
14. W. L. McMillan and J. M. Rowell, *Superconductivity*, R. D. Parks (Ed. Vol. 1, M. Dekker, New-York 1969, Chap. 11, p. 561).
15. B. Mitrovic, H. G. Zarate and J. P. Carbotte, *Phys. Rev.* **B29** (1984) 184.
16. B. T. Geilikman and V. Z. Kresin, *Fiz. Tverd. Tela* (Leningrad) **1** (1965) 3294. [*Sov. Phys. - Solid State* **7** (1966) 2659].
17. C. R. Leavens and J. P. Carbotte, *Can. J. Phys.* **49** (1971) 724.
18. J. J. Hopfield, *Phys. Rev.* **186** (1969) 443.
19. P. W. Anderson, *Physics* **2** (1966) 151.
20. G. Bergman and D. Reiner, *Z. Phys.* **263** (1973) 59.
21. C. R. Leavens and G. Talbot, *Phys. Rev.* **B28** (1983) 1304.
22. N. F. Berk and J. R. Schrieffer, *Phys. Rev. Lett.* **17** (1966) 433.
23. A. J. Leggett, *Rev. Mod. Phys.* **47** (1975) 331.
24. P. Morel and P. W. Anderson, *Phys. Rev.* **125** (1962) 1263.
25. L. P. Gor'kov and V. M. Galitskii, *Sov. Phys. JETP* **13** (1961) 792.
26. R. Balian and N. R. Westhamer, *Phys. Rev.* **131** (1963) 1553.
27. P. W. Anderson and V. W. F. Brinkmann, *The Helium Liquids*, (Academic, New York, 1975, pp. 315–416).
28. A. B. Migdal, *Zh. Eksperim. Teor. Fiz.* **34** (1958) 1438; *Sov. Phys. JETP* **7** (1958) 996.
29. A. L. Fetter and J. D. Walecka, *Quantum Theory of Many-Particle System*, (McGraw-Hill, Book Comp. New York, 1971, p. 406).
30. S. Engelsberg and J. R. Schrieffer, *Phys. Rev.* **131** (1963) 993.

III
THEORY OF SUPERCONDUCTING ALLOYS

The problem of magnetic impurities in metals has received a great amount of attention in solid state theory since Fridel and Anderson proposed a model to explain the occurrence of the suppression of magnetic moment in the normal dilute alloys. When atoms of rare-earths or transition metals are dissolved in non-magnetic hosts, the localized moment may appear in the alloy or may be suppressed due to the interaction between conduction electrons and magnetic moments.

The experimental measurements showed that the magnetic impurities have a drastic effect on superconductivity. If the concentration of these impurities is low, the system is called dilute superconducting alloy or simply superconducting alloys. The magnetic impurities dissolved in a superconductor give rise to a decreasing in the critical temperature. The pair breaking effect can be given by the exchange interaction between conduction electrons and magnetic impurity atoms. Another possible effect is the admixture between localized and conduction electrons which gives rise to a repulsion in the electron-electron interaction. These simple mechanisms have been developed later taking into consideration the effect of the resonant levels, the spin correlations, and in the last period the Anderson localization.

The Kondo effect extensively developed for the normal metals may appear in the superconducting alloys. In the normal metals, the Kondo effect is given by the electron-spin interaction and is in fact an example of strong interaction in the solid state physics. The problem has been solved by Wilson using the Renormalization Group method. In the superconducting metal the Kondo problem was not completely solved, and more than that, the occurrence of the superconductivity in a Kondo lattice becomes more exciting and more difficult at the same time.

All these problems will be treated using the Green function method with an additional but essential approach: the average on the impurities. One of the most important results of this theory was the prediction of the decrease of the critical temperature with the concentration of the impurities, the occurrence of the gapless superconductivity, the quenching of the superconductivity and the re-entrance effects.

We also reconsidered the decreasing of the critical temperature in connection with the Anderson localization. The main result obtained in this framework is that even the non-magnetic impurities may give rise to a decrease in the critical temperature because the Anderson localization is a quantum effect and in the Abrikosov-Gor'kov theory, the scattering of electrons on the non-magnetic impurities has been treated classically.

Another point which has been attentively discussed is the physical justification of different approximations. The average on the impurities presented in Appendix is an approximation generally accepted in theory of dilute alloys and, in fact, it is an important approach of the disordered systems.

16. The Influence of Impurities on the Superconducting State

The interaction between electrons and impurities in a superconductor is described by the Hamiltonian

$$\mathcal{H} = \mathcal{H}_s + \sum_{\alpha,\beta} \int d\mathbf{r} \psi_\alpha^\dagger \widehat{U}(\mathbf{r}) \psi_\beta(\mathbf{r}) \tag{16.1}$$

where $\mathcal{H}_s$ describes the pure superconductor and is given by (9.2) and $\widehat{U}(\mathbf{r})$ is the potential due to the impurity.

Let us consider the simple case of non-magnetic impurity. In this case

$$\widehat{U}(\mathbf{r}) = \sum_i V(\mathbf{r} - \mathbf{R}_i) = \int d\mathbf{r}' \rho(\mathbf{r}') V(\mathbf{r} - \mathbf{r}') \tag{16.2}$$

where $\mathbf{R}_i$ is the impurity position, and $\rho(\mathbf{r})$ is the density of the impurities.

The Green function (9.6) calculated from (16.1) is

$$[i\omega\hat{1} - \tilde{\varepsilon} - \hat{U}(\mathbf{r}_1)\sigma_z]\hat{G}(\omega;\mathbf{r}_1,\mathbf{r}_2) = \delta(\mathbf{r}_1 - \mathbf{r}_2) \tag{16.3}$$

where

$$\tilde{\varepsilon} = \hat{\sigma}_z\varepsilon + \frac{1}{2}(\Delta(\mathbf{r}_1)\sigma^+ + \Delta^*(\mathbf{r}_1)\sigma^-) \ . \tag{16.4}$$

The interaction term is

$$V(\mathbf{r}_1)\hat{\sigma}_z\hat{G}(\omega;\mathbf{r}_1,\mathbf{r}_2) = \int d\mathbf{r}V(\mathbf{r}-\mathbf{r}_1)\rho(\mathbf{r}_1)\hat{G}(\omega;\mathbf{r}_1,\mathbf{r}_2) \tag{16.5}$$

and can be re-written as

$$\int d\mathbf{r}V(\mathbf{r})\hat{G}(\omega)$$

where $\hat{G}(\omega)$ is the Green function (16.5) and $\hat{V}(\mathbf{r})$ is the diagonal part of

$$(\hat{V}(\mathbf{r}))_{\mathbf{r}_1,\mathbf{r}_2} = \sigma_z V(\mathbf{r}-\mathbf{r}_1)\delta(\mathbf{r}_1 - \mathbf{r}_2) \ . \tag{16.6}$$

Using these notations we write Eq. (16.3) as

$$[i\omega\,\hat{1} - \tilde{\varepsilon} - \int d\mathbf{r}\rho(\mathbf{r})\hat{V}(\mathbf{r})] = \hat{G}(\omega) = 1 \ . \tag{16.7}$$

The impurities are randomly distributed and the *average on the impurities* consists of performing all integrals on the impurities co-ordinates and to divide these integrals by V^N where V is the volume of the system and N the number of impurities. The average on the density is

$$\rho(\mathbf{r}) = \overline{\sum_i \delta(\mathbf{r} - R_i)} = \frac{N}{V} = c \tag{16.8}$$

where c is the concentration of the impurities. The product

$$\overline{\rho(\mathbf{r})\rho(\mathbf{r}')} = c\delta(\mathbf{r}-\mathbf{r}') + c^2 \tag{16.9}$$

and if (15.8) is finite (in the thermodynamic limit this is correct) and the alloy is dilute, the second term of (16.9) can be neglected.

The Green function for the system of electrons interacting with the impurities will be determined from the Dyson equation

$$[i\omega\,\hat{1} - \tilde{\varepsilon} - \hat{\Sigma}(\omega)]\hat{G}(\omega) = 1 \ . \tag{16.10}$$

The self-energy $\widehat{\Sigma}$ can be calculated in the Born approximation as

$$\widehat{\Sigma}(\omega) = c\int d\mathbf{r}\widehat{V}(\mathbf{r})\overline{\widehat{G}}\widehat{V}(\mathbf{r}) \tag{16.11}$$

or using the same method as for (16.7), we get

$$[i\omega\,\widehat{1} - \widetilde{\varepsilon} - c\int d\mathbf{r}\widehat{V}(\mathbf{r})\widehat{G}(\mathbf{r})\widehat{V}(\mathbf{r})]\widehat{G}(\mathbf{r}) = 1\,. \tag{16.12}$$

If $\widehat{V}(\mathbf{r})$ is expressed by the Eq. (16.6), Eq. (16.12) becomes

$$[i\omega - \widetilde{\varepsilon} - c|V|^2\widehat{G}(\omega;\mathbf{r}_1,\mathbf{r}_2)\sigma_z]\widehat{G}(\omega;\mathbf{r}_1,\mathbf{r}_2) = \delta(\mathbf{r}_1 - \mathbf{r}_2)\,. \tag{16.13}$$

Using now the transforms

$$\omega' = \omega + ic|V|^2 G(\omega;\mathbf{r},\mathbf{r}) \tag{16.14}$$

$$\Delta(\omega) = \Delta + c|V|^2 G(\omega;\mathbf{r},\mathbf{r}) \tag{16.15}$$

Eq. (16.13) can be re-written as

$$[i\omega' - \widetilde{\varepsilon}]G(\omega;\mathbf{r}_1,\mathbf{r}_2) + \Delta(\omega)F(\omega;\mathbf{r}_1,\mathbf{r}_2) = \delta(\mathbf{r}_1 - \mathbf{r}_2) \tag{16.16}$$

$$[i\omega' + \widetilde{\varepsilon}]F(\omega;\mathbf{r}_1,\mathbf{r}_2) + \Delta^*(\omega)G(\omega;\mathbf{r}_1,\mathbf{r}_2) = 0\,. \tag{16.17}$$

The average on the impurities restored the homogeneity of the system and we can perform the Fourier transforms on (16.14)–(16.17). The transforms (16.14) and (16.15) can be re-written as

$$\widetilde{\omega}' = \omega - ic|V|^2\int\frac{d\mathbf{p}}{(2\pi)^3}\frac{i\omega' + \varepsilon}{\omega'^2 + \varepsilon^2(\mathbf{p}) + \Delta^2(\omega)} \tag{16.18}$$

$$\widetilde{\Delta} = \Delta + c|V|^2\int\frac{d\mathbf{p}}{(2\pi)^3}\frac{\Delta(\omega')}{\omega'^2 + \varepsilon^2(\mathbf{p}) + \Delta^2(\omega)}\,. \tag{16.19}$$

Using the standard notation for the scattering time

$$\frac{1}{\tau} = 2\pi c N(0)|V|^2 \tag{16.20}$$

Eqs. (16.18) and (16.19) become

$$\widetilde{\omega} = \eta\omega \qquad \widetilde{\Delta} = \eta\Delta \tag{16.21}$$

where

$$\eta = 1 + \frac{1}{\tau\sqrt{\omega^2 + \Delta^2}} . \tag{16.22}$$

If we now write the self-consistent equation for Δ using (16.16)–(16.17), we get

$$\Delta = g\pi T \sum_{\omega} \int \frac{d^3\mathbf{p}}{(2\pi)^3} \frac{\widetilde{\Delta}(\omega)}{\widetilde{\omega}^2 + \varepsilon^2(\mathbf{p}) + \widetilde{\Delta}^2} = N(0)\pi T \sum_{\omega} \frac{\widetilde{\Delta}}{\sqrt{\widetilde{\omega}^2 + \widetilde{\Delta}^2}} . \tag{16.23}$$

From this equation and from (16.21) we see that Δ is not affected by the non-magnetic impurities. This result is known as the Anderson[1] theorem and is in fact only an approximation. However, it is relevant if we compare the influence of magnetic impurities with that of the non-magnetic impurities.

Let us consider, instead of the simple interaction (16.2), the interaction

$$\hat{U}(\mathbf{r}) = \sum_i V(\mathbf{r} - \mathbf{r}_i) + \sum_i J(\mathbf{r} - \mathbf{R}_i)\mathbf{S}_i\sigma \tag{16.24}$$

which describes the interaction between the superconducting electrons and the paramagnetic impurities. The first term from (16.24) is the independent spin contribution (which was discussed above) and the second describes the exchange interaction which is spin dependent. If we perform the average on the impurities — approximation proposed by Abrikosov and Gor'kov[2] the new Green functions are

$$G(\omega;\varepsilon) = \frac{i\widetilde{\omega} + \widetilde{\varepsilon}}{\widetilde{\omega}^2 + \widetilde{\varepsilon}^2 + \widetilde{\Delta}^2} \qquad F(\omega,\varepsilon) = \frac{\widetilde{\Delta}}{\widetilde{\omega}^2 + \widetilde{\varepsilon}^2 + \widetilde{\Delta}^2} \tag{16.25}$$

where

$$\widetilde{\omega} = \omega + \frac{\widetilde{\omega}}{2\tau_1\sqrt{\widetilde{\omega}^2 + \widetilde{\Delta}^2}} \tag{16.26a}$$

$$\widetilde{\Delta} = \Delta + \frac{\widetilde{\Delta}}{2\tau_2\sqrt{\widetilde{\omega}^2 + \widetilde{\Delta}^2}} \tag{16.26b}$$

with the scattering times

$$\tau_1^{-1} = 2\pi c N(0)[|V|^2 + \frac{1}{4}S(S+1)|J|^2] \tag{16.27a}$$

$$\tau_2^{-1} = 2\pi c N(0)[|V|^2 - \frac{1}{4}S(S+1)|J|^2] . \tag{16.27b}$$

The equation for the order parameter $\tilde{\Delta}$ is

$$\tilde{\Delta} = |g|\pi T \sum_{\omega} \int \frac{d^3\mathbf{p}}{(2\pi)^3} \frac{\tilde{\Delta}}{\tilde{\omega}^2 + \tilde{\varepsilon}^2(\mathbf{p}) + \tilde{\Delta}^2} \tag{16.28}$$

and can be transformed using the substitution $i\tilde{\omega} = z$. Relations (16.26a)–(16.26b) become

$$\tilde{z} = z + \frac{1}{2\tau_1} \frac{\tilde{z}}{\sqrt{\tilde{z}^2 - \tilde{\Delta}^2}} \tag{16.29a}$$

$$\tilde{\Delta} = \Delta + \frac{1}{2\tau_2} \frac{\tilde{\Delta}}{\sqrt{\tilde{z}^2 - \tilde{\Delta}^2}} . \tag{16.29b}$$

Defining now the new variable

$$u(z) = \frac{\tilde{z}}{\tilde{\Delta}} \tag{16.30}$$

we get

$$\frac{z}{\Delta} = u(z)\left[1 - \frac{i}{2\tau_s} \frac{u(z)}{\sqrt{u^2(z) - 1}}\right] \tag{16.31}$$

where

$$\frac{1}{2\tau_s} = \frac{1}{\tau_1} - \frac{1}{\tau_2} . \tag{16.32}$$

Equation (16.28) gives

$$\ln \frac{T}{T_c} = 2x_0 \sum_{n=0}^{\infty} \left[\frac{1}{(1+u^2)^{\frac{1}{2}}} - \frac{1}{x}\right] \tag{16.33}$$

where $x_n = (zn+1)T/\Delta$. If we consider now the approximation $x_0 \gg 1$ $(T \gg \Delta)$ Eq. (16.33) becomes

$$\ln \frac{T_c}{T_{c0}} = \sum_{\omega} \left[\frac{1}{\alpha} - \frac{1}{\alpha}\frac{x_n}{u} - \frac{1}{x_n + \alpha}\right] - \chi(\rho) \tag{16.34}$$

where

$$\chi(\rho) = \Psi\left(\frac{1}{2} + \rho\right) - \Psi\left(\frac{1}{2}\right) \tag{16.35}$$

and $\Psi(\rho) = d \ln \Gamma(\rho)/d\rho$. These results have been obtained by different authors[3,4] and generalize the Abrikosov-Gor'kov result.

Equation (16.34) can be written in the form

$$\ln \frac{T_c}{T_{c0}} = 2\pi T_c \sum_{\omega} \left[\frac{1}{\Delta}\frac{\tilde{\Delta}}{\tilde{\omega}} - \frac{1}{\tilde{\omega}}\right] \tag{16.36}$$

where for $T \to T_c$, $\Delta \to 0$. In this case, (16.34) is the classical Abrikosov-Gor'kov[1] result

$$\ln \frac{T_c}{T_{c0}} = \Psi\left(\frac{1}{2} + \rho_{AG}\right) - \Psi\left(\frac{1}{2}\right) \tag{16.37}$$

where the pair-breaking parameter ρ_{AG} is given by

$$\rho_{AG} = \frac{1}{\pi T_c \tau_s} \; . \tag{16.38}$$

An important problem in the theory of superconducting dilute alloys is the spectrum of the elementary excitations. At $T = 0$, Eq. (16.31) is

$$\frac{\omega}{\Delta} = u(\omega)\left[1 - \frac{\gamma}{(1 - u(\omega))^{\frac{1}{2}}}\right] \tag{16.39}$$

where

$$\gamma = (\tau_s \Delta)^{-1} \; .$$

If the solutions of this equation are real, the Cooper pairs do not decay. In the case of non-magnetic impurities $\tau_1 = \tau_2 = \tau$ and we get

$$\tilde{\omega} = \eta\omega \, , \quad \tilde{\Delta} = \eta\Delta \, , \quad \eta = 1 + \frac{i}{2\tau}\frac{1}{\sqrt{\omega^2 - \Delta^2}} \tag{16.40}$$

which shows that the number of quasiparticles near $\varepsilon = 0$ and $\Delta = 0$ do not change and the density of states will have the same form.

For the magnetic scattering ($V = 0, \tau_s^{-1} = 0$), we get from (16.29)

$$\tilde{\Delta} = \Delta(1 - u^2)^{\frac{1}{2}} = i\varepsilon/\Delta \tag{16.41}$$

and the energy $\tilde{\omega}$ becomes

$$\tilde{\omega} = \omega - \frac{i(\varepsilon^2 + \Delta^2)^{\frac{1}{2}}}{\tau_s \varepsilon} \tag{16.42}$$

which can be approximated as

$$\omega = \tilde{\omega} + i\Gamma \, , \quad \Gamma \cong \frac{1}{\varepsilon} \tag{16.43}$$

and for $\varepsilon \to 0$, Γ becomes important, which means that electrons which are near the gap suffer a strong scattering. If the energy of these particles satisfies

$$\omega_0 \leq \omega \leq \Delta \tag{16.44}$$

where

$$\omega_0 = \Delta[1 - \gamma^{\frac{2}{3}}]^{\frac{3}{2}} \tag{16.45}$$

the scattering on impurities introduces new states inside the gap which suffer a decay.

We give in Fig. 15 the function

$$f(u) = u\left[1 - \frac{\gamma}{\sqrt{1-u^2}}\right] \tag{16.46}$$

for (a) $\gamma \ll 1$ and (b) $\gamma \gg 1$. We see that the solution is real for $\gamma \ll 1$ and $\omega < \omega_0$. For $\omega > \omega_0$ there are no real solutions but there are only decay states. In particular for $\gamma = 1$ the gapless superconducting state may appear.

In the approximations (a) and (b) we can calculate the order parameter from (16.28). At $T = 0$ we get

$$\ln \frac{\Delta_0(0)}{\Delta(0)} = \frac{\pi\gamma}{4}, \qquad \gamma > 1 \tag{16.47}$$

$$\ln \frac{\Delta_0(0)}{\Delta(0)} = \ln \frac{1+\sqrt{1-\gamma^2}}{\gamma} - \frac{1}{2}\sqrt{1-\gamma^2} - \frac{\sin^{-1}\gamma}{2\gamma}, \qquad \gamma \ll 1\,. \tag{16.48}$$

From (16.48) we see that at $\tau_s^c = \pi T_{c0}/2\gamma = \Delta(0)/2$, the gap $\Delta(0)$ vanishes. When $\gamma = 1$, we have $\Delta(0) = \Delta_0 \exp(-\pi/4)$ which shows that there is some value of concentration which is lower than the critical one for which the gap in energy (16.45) vanishes. This concentration is

$$c' = 2\exp(-\pi/4)c^c \simeq 0.91 c_r$$

where c_r is given by the condition $\tau_s = 2/\Delta(0)$ and for the concentration $c' < c < c_r$, the alloy presents gapless superconductivity.

The localized excited states in the energy gap of a superconductor containing magnetic impurities appear even if the magnetic impurity is treated as a *classical spin*.

This problem has been treated by Shiba[5] beyond the lowest Born approximation, for one impurity and for a finite impurity concentration. If in (16.24) we take $V = 0$, the "s-d" Hamiltonian becomes

$$\mathcal{H}_{\text{s-d}} = -\frac{J}{N}\sum_{\mathbf{k},\mathbf{k}'} c^{\dagger}_{\mathbf{k}}\sigma c_{\mathbf{k}'}\mathbf{S} \tag{16.49}$$

and the "classical limit" means that $J \to 0, S \to \infty$ but JS = finite.

In this case, the quantum mechanical properties of spins are neglected and the localized spin is equivalent to a local magnetic field.

The superconductor containing magnetic impurities described by the Hamiltonian

$$\mathcal{H} = \mathcal{H}_{\text{s}} + \mathcal{H}_{\text{s-d}} \tag{16.50}$$

can be analysed using the method developed by Abrikosov-Gor'kov. Shiba treated the scattering of electrons on a classical spin beyond the lowest Born approximation. In the case of one impurity problem (one classical spin located in the superconductor) the Green function

$$\widehat{G}(\mathbf{p},\mathbf{p}';\omega) = \widehat{G}^0(\mathbf{p};\omega)\delta_{\mathbf{p},\mathbf{p}'} + \widehat{G}^0(\mathbf{p};\omega)t(\omega)\widehat{G}^0(\mathbf{p}';\omega) \tag{16.51}$$

will give all the information on the single particle excitation spectrum. Using now for $\widehat{G}^0(\mathbf{p};\omega)$ the expression

$$\widehat{G}^0(\mathbf{p};\omega) = [\omega\,\widehat{1} - \varepsilon(\mathbf{p})\widehat{\sigma}_z - \Delta(0)\widehat{\sigma}_y]^{-1} \tag{16.52}$$

we get

$$t(\omega) = \frac{1}{N}\frac{[-(JS)^2]F(\omega)}{1-[-(JS)F(\omega)]^2} \tag{16.53}$$

where

$$F(\omega) = \frac{1}{N}\sum_{\mathbf{p}} \widehat{G}^0(\mathbf{p};\omega)\ . \tag{16.54}$$

The localized excited states in the energy gap ($\omega < \Delta_0$) will be obtained as the poles of

$$F(\omega) = -\pi N(0)\frac{\omega\,\widehat{1} + \Delta(0)\widehat{\sigma}_y}{\sqrt{\Delta(0)^2 - \omega^2}} \tag{16.55}$$

which gives the poles of the t-matrix (16.53)

$$\omega_0 = \pm\Delta_0\frac{1-(JS\pi N(0))^2}{1+(JS\pi N(0))^2} \tag{16.56}$$

an expression which gives the low-lying excited state of one spin in a superconductor. In fact, the energy ω_0 is temperature dependent due to Δ_0. From (16.56), we see that the occurrence of the bound state is independent on the magnitude of $|\pi N(0)JS|^2$ and this is because the density of states of the BCS superconductor is sharp at the gap edge. On the other hand, a direct calculation showed that the ordinary potential $V(J=0)$, does not give bound states.

The finite concentration problem can be treated using these results. The Green function averaged over the random spatial distribution of the impurities and on the random orientation of spins $\bar{G}(\mathbf{p},\mathbf{p}';\omega)$ can be obtained from

$$\bar{G}(\mathbf{p},\mathbf{p}';\omega) = G^0(\mathbf{p};\omega) + G^0(\mathbf{p};\omega)\sum(\mathbf{p}',\omega)\bar{G}(\mathbf{p},\mathbf{p}';\omega) \tag{16.57}$$

where Σ is the self-energy. The self-energy can be calculated as

$$\Sigma(\omega) = c\frac{[-(JS)]^2\bar{F}(\omega)}{1-[-(JS)\bar{F}(\omega)]^2} \tag{16.58}$$

where

$$\bar{F}(\omega) = \frac{1}{N}\sum_{\mathbf{p}}\bar{G}(\mathbf{p},\mathbf{p}';\omega)\ . \tag{16.59}$$

The function (16.59) can be calculated using

$$\bar{G}(\mathbf{p},\mathbf{p}';\omega) = [\tilde{\omega}\hat{1} - \tilde{\varepsilon}(\mathbf{p})\hat{\sigma}_z - \tilde{\Delta}\hat{\sigma}_y]^{-1} \tag{16.60}$$

and we get

$$\bar{F}(\omega) = -\pi N(0)\frac{\tilde{\omega}+\tilde{\Delta}\hat{\sigma}_y}{\sqrt{\tilde{\Delta}^2-\tilde{\omega}^2}}\ . \tag{16.61}$$

In the lowest Born approximation, the self-energy is

$$\Sigma_{\mathrm{AG}} = -\frac{1}{\tau_s}\frac{\omega\hat{1}+\tilde{\Delta}\hat{\sigma}_y}{\sqrt{\tilde{\Delta}^2-\tilde{\omega}^2}} \tag{16.62}$$

and if we introduce $u=\tilde{\omega}/\tilde{\Delta}$ as in (16.30), the variable ω/Δ can be written as

$$\frac{\omega}{\Delta} = u\left[1 - \frac{1}{\tau_s\Delta\sqrt{1-u^2}}\frac{u^2-1}{u^2-\gamma^2}\right] \tag{16.63}$$

where

$$\gamma = \left|\frac{1-\varsigma^2}{1+\varsigma^2}\right|, \quad \varsigma = \pi N(0)JS \tag{16.64}$$

and

$$\tau_s^{-1} = \tau_{0s}^{-1}(1+\varsigma^2)^{-2} \,. \tag{16.65}$$

The scattering time τ_{0s}^{-1} is different from the Abrikosov-Gor'kov scattering time (16.32), and is defined as

$$\tau_{0s}^{-1} = 2\pi N(0)\pi c(JS)^2 \,. \tag{16.66}$$

Now, we can see the existence of the "impurity band" if we analyse the function

$$\varphi(u) = u\left[1 - \frac{1}{\tau_s\Delta\sqrt{1-u^2}}\frac{u^2-1}{u^2-\gamma^2}\right] \tag{16.67}$$

which gives information about the density of states. If the concentration c is small and $(\tau_s\Delta)^{-1} \ll 1$, the function (16.67) has two extrema in the region, $\gamma < u < 1$ and Eq. (16.63) has complex solutions for $\varphi_1 < \omega/\Delta < \varphi_2$ or $\varphi_3 < \omega/\Delta$. In this case, the density of states of a single particle excitations consists of two parts: the "impurity band" ($\varphi_1 < \omega/\Delta < \varphi_2$) and the continuum $\omega/\Delta > \varphi_1$.

If $(\tau_s\Delta)^{-1}$ is so large that $\varphi(u)$ has no extremum in the region $\gamma < u < 1$, the impurity band cannot be separated from the continuum and in fact when $\varphi_1 = 0$, the gapless superconductivity begins. The concentration dependence of the width W of the "impurity band" can be obtained by solving (16.63), and if $(\tau_s\Delta)^{-1} \ll 1$ we get for W

$$W = \Delta[8\alpha(1-\gamma^2)^{\frac{1}{2}}]^{\frac{1}{2}} \tag{16.68}$$

where

$$\alpha = (\tau_s\Delta)^{-1} \tag{16.69}$$

is the pair-breaking parameter.

The critical temperature can be obtained from the equation

$$1 = |g|N(0)\int_0^{\omega_D} d\omega\frac{\omega}{\omega^2+\tau_s^{-2}}\tan\frac{\omega}{2T_c} \tag{16.70}$$

and at $T=0$ we get the critical concentration

$$\left(\tau_{s0}^{-1}\right)_c = \frac{1}{2}\Delta_0(0) \tag{16.71}$$

identical with the result from the Born approximation (16.48).

The concentration at which gapless superconductivity begins, c' is defined as $\omega(c) = 0$, and is obtained from $\varphi'(0) = 0$ as

$$\frac{c'}{c_r} = 2\exp\left[-\frac{\pi}{4}\right]\exp\left[\frac{\pi}{2}\frac{(1+2\gamma)(1-\gamma)}{1+\gamma}\right] < 2\exp\left[-\frac{\pi}{4}\right] . \tag{16.72}$$

These results show that using a simple model for dilute alloys but improving the approximation, one obtains an "impurity band". Then the results obtained in the Born approximation should be revised. In fact, there are more effects which have to be considered in the dilute alloys.

Gor'kov and Rusinov[6] considered the magnetization of the impurities due to the polarization of the conduction electrons as well as the spin-orbit scattering due to the non-magnetic impurities. A general equation taking into consideration all effects on the conduction electrons was given by Keller and Benda.[7] The Green function for the electrons is given by

$$\hat{G}^{-1}(\mathbf{p};\omega) = \hat{G}^{0^{-1}}(\mathbf{p},\omega) - \hat{\Sigma}(\mathbf{p};\omega) \tag{16.73}$$

where

$$\hat{G}^{0^{-1}}(\mathbf{p},\omega) = i\omega - \varepsilon(\mathbf{p}) \mp I . \tag{16.74}$$

I being the field which polarizes the conduction electrons. The self energy considered in Ref. 7 contains

— the exchange scattering on the magnetic impurities
— the potential scattering
— the spin-orbit interaction

and from these contributions, we can calculate the scattering times.

$$\tau^{-1} = c_1\frac{N(0)}{2}\int d\Omega|V|^2 \tag{16.75a}$$

$$\tau_s^{-1} = c_2\frac{N(0)}{2}\int d\Omega|J|^2 \tag{16.75b}$$

$$\tau_{s0}^{-1} = c_1\frac{N(0)}{g}4\pi|V_{s0}|^2 \tag{16.75c}$$

where V_{s0} is the spin-orbit scattering coupling constant. If we calculate all these contributions in the self-energy, the new parameters $\tilde{\omega}_\pm$ and $\tilde{\Delta}_\pm$ will be

$$\begin{aligned}\tilde{\omega}_\pm = \omega \pm i(I-\delta_\pm) + (2\tau^{-1} + 2\tau_{s0}^{-1} + 2\tau_s^{-1}\langle S_z^2\rangle)\tilde{\omega}_\pm \Big/ W_\pm \\ + \tau_s\langle S_z\rangle\pi T\sum_{\omega'} D_\pm(\omega,\omega')\frac{\tilde{\omega}_\pm}{W_\mp} + \tau_{s0}^{-1}\frac{\tilde{\omega}_\mp}{W_\mp}\end{aligned} \tag{16.76}$$

$$\widetilde{\Delta}_{\pm} = \Delta_{\pm} + (2\tau_0^{-1} + 2\tau_{s0} - 2\tau_s\langle S_z^2\rangle)\widetilde{\Delta}_{\pm}\Big/W_{\pm}$$
$$- \tau_s\langle S_z\rangle\pi T\sum_{\omega'} D_{\pm}(\omega,\omega')\frac{\widetilde{\Delta}_{\mp}}{W_{\mp}} + \tau_{s0}^{-1}\frac{\widetilde{\Delta}_{\mp}}{W_{\mp}} \tag{16.77}$$

where $D(\omega,\omega')$ is the propagator for the impurities polarized by the field K.

$$D_{\pm}(\omega-\omega') = [\pm i(\omega-\omega') - K]^{-1} \tag{16.78}$$

and

$$W_{\pm} = \sqrt{\widetilde{\omega}_{\pm}^2 + \widetilde{\Delta}_{\pm}^2} \quad \text{with} \quad \mathrm{Re}\, W_{\pm} > 0\ . \tag{16.79}$$

We also introduced the notations

$$\Delta_{\pm} = gN(0)\pi T\sum_{\omega'}\frac{\widetilde{\Delta}_{\pm}}{W_{\pm}} \tag{16.80a}$$

$$\delta_{\pm} = \pm gN(0)\pi T\sum_{\omega'}\frac{i\widetilde{\omega}_{\pm}}{W_{\pm}} \tag{16.80b}$$

and $\Delta_+ \sim -\langle\psi_\uparrow\psi_\downarrow\rangle$, $\Delta_- \sim \langle\psi_\downarrow\psi_\uparrow\rangle$.
In the limit $\Delta \to 0$, from (16.80) one obtains

$$\ln\frac{T_c}{T_{c0}} = 2\pi T_c\sum_{\omega>0}\left[\mathrm{Re}\frac{1}{y} - \frac{1}{\omega}\right] \tag{16.81}$$

where

$$y = y_{\pm} = \lim_{\Delta\to 0}\widetilde{\Delta}_{\pm}\Big/\widetilde{\omega}_{\pm}\ . \tag{16.82}$$

In Eqs. (16.76) and (16.77), for $\Delta \to 0$ we can perform the transform

$$A_{\pm}y_{\pm} + \tau_{s0}\mathrm{sign}\,\omega(y_{\pm} - y_{\mp}) = 1 - \tau_s\langle S_z\rangle\sum_{\omega'} D_{\pm}(\omega,\omega')\,\mathrm{sign}\,\omega' y_{\mp} \tag{16.83}$$

with

$$A_{\pm} = \omega \pm I + \tau_s^{-1}\langle S_z\rangle\mathrm{sign}\,\omega + \tau_s^{-1}\langle S_z\rangle\sum_{\omega'} D_{\pm}(\omega,\omega')\,\mathrm{sign}\,\omega\ . \tag{16.84}$$

These calculations show that the potential scattering effect drops out from the final equation which is in fact in agreement with the Anderson theorem.

Let us consider the following cases

a) $I \to 0$, $K \to 0$, $\langle S_z \rangle = \frac{1}{3} S(S+1)$ which correspond to the paramagnetic impurities and unpolarized electrons.

In this case, we get from (16.81) the Abrikosov-Gor'kov equation (16.37).

b) $I \neq 0$, $K \neq 0$ correspond to the case of polarized electrons and polarized impurities. From (16.81), we get

$$\ln \frac{T_c}{T_{c0}} = \Psi\left(\frac{1}{2}\right) - \frac{1}{2}\left\{\left[1 - \frac{F_2}{\sqrt{F_2^2 - I^2}}\right] \Psi\left(\frac{1}{2} + \rho_+\right) + \left[1 - \frac{F_2}{\sqrt{F_2^2 - I^2}}\right] \Psi\left(\frac{1}{2} + \rho_-\right)\right\} \tag{16.85}$$

where

$$F_1 = \tau_s^{-1}\langle S_z^2 \rangle + (2\tau_s)^{-1}(\langle S_x^2 \rangle + \langle S_y^2 \rangle) + \tau_{s0}^{-1}$$
$$F_2 = \tau_{s0}^{-1} - (2\tau_s)^{-1}(\langle S_x^2 \rangle + \langle S_y^2 \rangle)$$

and

$$\rho_\pm = (2\pi T_c)^{-1}[F_1 \pm \sqrt{F_2^2 - I_1^2}] \,. \tag{16.86}$$

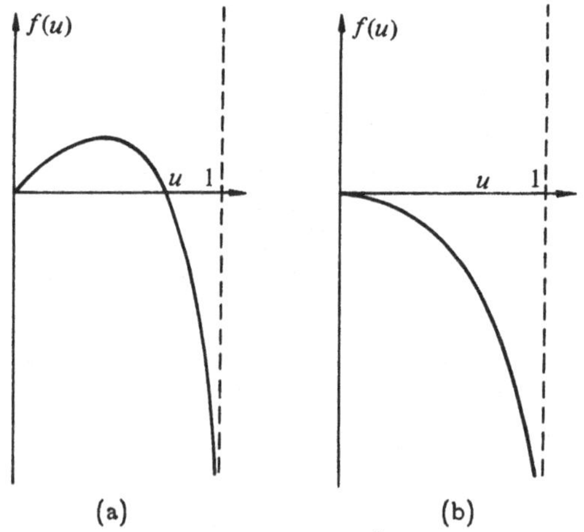

Fig. 15. The function $f(u) = u[1 - \gamma(1-u^2)^{-\frac{1}{2}}]$ for (a) $\gamma \ll 1$ and (b) $\gamma \gg 1$.

This is the Gor'kov-Rusinov[6] result, and Fig. 16 shows the existence of a region with $\Delta \neq 0$ and $\langle S_z \rangle \neq 0$ which is in homogeneous superconducting state rather than in a coexistence region.

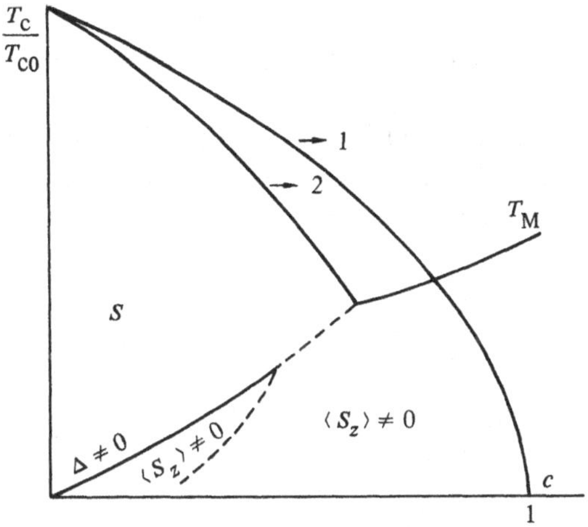

Fig. 16. The dependence of the critical temperature on the impurities concentration. (1) Abrikosov-Gor'kov result (16.37) (2) Gor'kov-Rusinov-Keller-Benda result (16.91).

c) $K = 0, I \neq 0$ correspond to the strong polarization of the conduction electrons. In this case, the critical temperature is:

$$\ln \frac{T_c}{T_{c0}} = \Psi\left(\frac{1}{2}\right) - \operatorname{Re}\Psi\left(\frac{1}{2} + i\frac{I}{2\pi T_c}\right) . \qquad (16.87)$$

This case has been treated by Smith and Vertoghen[8] and Crisan and Jones.[9]

On the other hand, de Gennes and Sarma[10] showed that if the exchange coupling between electrons and impurities is small, the "second order effects" (the variations of the critical temperatures which depend on J^2) are small and the first order effects become important in an external field of the order

$$H = cJ\langle S_z \rangle \simeq T_c \qquad (16.88)$$

if the impurity polarization is independent of the superconducting order parameter.

17. Influence of the Correlated Spins on the Critical Temperature of a Superconductor

In the Abrikosov-Gor'kov theory, the pair-breaking effect is given by the scattering of the electrons on the uncorrelated spins. If the correlations between spins is considered, the spin system can undergo a magnetic ordered phase if these correlations are strong.

Toxen, Kwok and Gambino[11] considered the influence of the spatial spin correlations of antiferromagnetic type. Rainer[12] analysed the characteristic difference in the concentration dependence of the superconducting transition temperature in the cases of the ferromagnetic and antiferromagnetic spin correlations.

The simple model considered in Refs. 2–5 is described by the Hamiltonian which contains the interaction (16.24). The self-energy which considers the effect of the correlations is

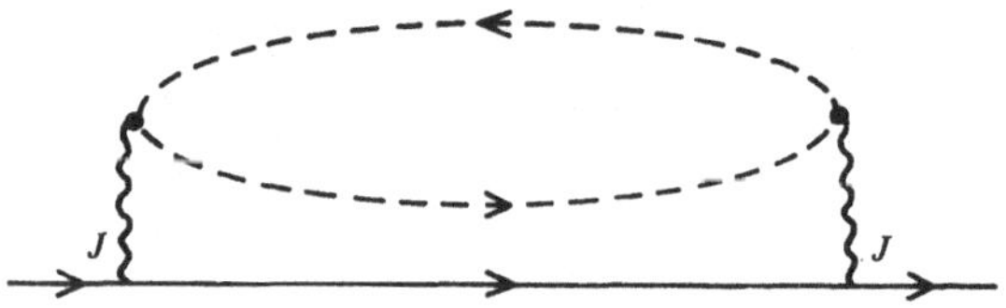

which can be written analytically as

$$\Sigma_1(\mathbf{p};\omega) = \frac{3}{2}J^2\pi T\sum_{\omega'}\frac{V}{(2\pi)^3}\int d^3\mathbf{p}' G(\mathbf{p}',\omega')\chi(\mathbf{p}-\mathbf{p}';\omega-\omega') \qquad (17.1)$$

where $\chi(\mathbf{q};\omega)$ is the spin susceptibility and G is the Green function of the electrons.

For $|\mathbf{p}| \simeq p_0$, we approximate the self-energy (17.1) by its imaginary part at the Fermi surface, averaged over all $\mathbf{p}$ directions and

$$\Sigma_1(\mathbf{p};\omega) = -i\int\frac{d\Omega}{4\pi}\mathrm{Im}\,\Sigma_1(p_0;\omega) \equiv -i\Sigma_s(\omega)\,. \qquad (17.2)$$

Equation (17.1) can be approximated for a dirty system as

$$\Sigma_s(\omega) = \frac{3}{2}J^2\sum_{\omega'}\frac{V}{(2\pi)^3}\int d^3\mathbf{p}'\frac{(2\tau_0)^{-1}\mathrm{sign}\,\omega'}{\varepsilon^2(\mathbf{p}')+(2\tau_0)^{-2}}\bar{\chi}(|\mathbf{p}_0-\mathbf{p}|;\omega) \qquad (17.3)$$

which becomes

$$\Sigma_s(\omega) = \frac{3}{2}J^2N(0)\omega\sum_{\omega'}\frac{\theta(|\omega|-\omega')}{|\omega|}\pi T\int d^3\mathbf{q}\,g(\mathbf{q};\tau_0)\bar{\chi}(\mathbf{q};\omega') \qquad (17.4)$$

where

$$\bar{\chi}(\mathbf{q};\omega) = \int\frac{d\Omega}{4\pi}\chi(\mathbf{q};\omega)\,.$$

Equation (17.4) has been obtained using the formula

$$\frac{V}{(2\pi)^3}\int d^3\mathbf{p}'\ldots \longrightarrow N(0)\int\frac{d\varepsilon}{2p_0}\int_{\varepsilon/v_0}^{\sqrt{4p_0^2+\varepsilon^2/v_0^2}} q\,dq\ldots \qquad (17.5)$$

where $q = |\mathbf{p}_0 - \mathbf{p}|$ and the function $g(\mathbf{q}; x)$ is defined as

$$g(\mathbf{q}; x) = \begin{cases} \frac{1}{4p_0^2 q} \tan^{-1} \frac{v_0 q}{|x|}\,; & q < 2p_0 \\ \frac{1}{4p_0^2 q} \left[\tan^{-1} \frac{v_0 q}{|x|} - \tan^{-1} \frac{v_0\sqrt{q^2-4p_0^2}}{x}\right]; & q > 2p_0 \end{cases} . \tag{17.6}$$

In order to calculate the critical temperature we shall use (16.87), and the self energy (17.4) to obtain $\widetilde{\omega}$ and $\widetilde{\Delta}$. Following the formalism from the strong coupling theory, we get

$$\widetilde{\omega} = \omega + (2\tau_0)^{-1}\text{sign}\,\omega + 3(2\tau)^{-1}\sum_{\omega'}\int d^3\mathbf{q} g(\mathbf{q})\chi(\mathbf{q}; \omega-\omega')\text{sign}\,\omega' \tag{17.7}$$

$$\widetilde{\Delta} = \Delta + (2\tau_0)^{-1}\text{sign}\,\omega - 3(2\tau)^{-1}\sum_{\omega'}\int d^3\mathbf{q} g(\mathbf{q}; \omega-\omega')\frac{\widetilde{\Delta}}{\widetilde{\omega}}\text{sign}\,\omega' \tag{17.8}$$

where $\tau = \tau_s S(S+1)$.
In (17.7)–(17.8), the non-magnetic scattering time τ_0^{-1} appears in both equations with the same sign, while the magnetic scattering time τ_s enters with opposite sign. This sign changes is due to the lack of time-reversal invariance of the magnetic interaction and gives rise to pair-breaking or gapless superconductivity. However, if these two equations are combined in order to calculate T_c, τ^{-1} no longer appeared and this is in agreement with the Anderson theorem.

The function $g(\mathbf{q})$ is normalized, so that

$$\int d^3\mathbf{q}\, g(\mathbf{q}) = 1 \tag{17.9}$$

and for small q it has a maximum. This behaviour shows that the ferromagnetic correlations are weighted more strongly than the antiferromagnetic correlations (which correspond to a finite q) and these imply that $\chi(\mathbf{q}; \omega)$ is the largest for $\mathbf{q} = 0$.

On the other hand, in (17.7)–(17.8) the terms with low frequencies give the largest contribution, and we obtained an important result: the spin system with low excitations energies will be very effective in the pair-breaking mechanism. The susceptibility $\chi(\mathbf{q}; \omega)$ satisfies the sum rule

$$\pi T\sum_{\omega}\int d^3\mathbf{q}\chi(\mathbf{q}; \omega) = \frac{1}{3}S(S+1)\int_{B.Z.} d^3\mathbf{q} \tag{17.10}$$

where the integral extends over the first Brillonin zone. We will consider now two simple cases of impurities correlation namely the static correlations and the dynamic correlations.

a) Static approximation of the correlations

When the characteristic frequency of the spin motion is small compared to T_c^{-1}, one may use the approximation

$$\chi(\mathbf{q};\omega) = \chi(\mathbf{q})\delta_{\omega,0} \, . \tag{17.11}$$

In this case, the theory reduces to the Abrikosov-Gor'kov results but the scattering time

$$\tau_s^{-1} = 3\tau^{-1}\int d^3\mathbf{q}\, g(\mathbf{q})\chi(\mathbf{q};T) \tag{17.12}$$

became temperature dependent.
Equation (16.87) gives

$$\ln\frac{T_c}{T_{c0}} = \Psi\left(\frac{1}{2}\right) - \Psi\left(\frac{1}{2} + \frac{1}{2\pi T_c\tau_s}\right) \tag{17.13}$$

Entel and Klose[13] calculated the influence of the short-range magnetic correlations using (17.11) for $\chi(\mathbf{q};\omega)$ the Ornstein-Zernike expression which is correct near the magnetic phase transition temperature denoted by T_M. Taking then

$$\chi(q,T,\omega) = \frac{\frac{1}{3}T\,S(S+1)}{A^2(c,T,T_M) + B^2(c,T,T_M)q^2}\delta_{\omega,0} \tag{17.14}$$

we have to mention that A defines the magnetic coherence length as

$$A^{-1} = [\tilde{T}/(\tilde{T} - T_M)]^{\frac{1}{2}} \tag{17.15}$$

and B can be determined from the sum rule (17.10). If we define a critical concentration c_M, the instability is ensured for $q \to 0$, if

$$\tilde{T} = \left[\frac{T^2 + T_{M0}^2}{T_{M0}}\right]^{\frac{1}{2}} \qquad \tilde{T}_M = \left[\frac{T_M^2 + T_{M0}^2}{T_{M0}^2}\right]^{\frac{1}{2}} \tag{17.16}$$

and the magnetic critical temperature is

$$T_M = T_{M0}[(c/c_M)^2 - 1]^{\frac{1}{2}} \tag{17.17}$$

where

$$T_{M0} = \frac{2}{3} J^2 N(0) S(S+1) .$$

Using (17.14) and taking $g(q) \cong 1$ in (17.12) the critical temperature T_c obtained in Ref. 13 is given by the equation

$$\ln \frac{T_c}{T_{c0}} = \Psi\left(\frac{1}{2}\right) - \Psi\left(\frac{1}{2} + \frac{1}{2T_c} \frac{S(S+1)}{\tau_c} \frac{1}{4p_0^2 B^2} \ln \frac{A^2 + 4p_0^2 B^2}{A^2}\right) \quad (17.18)$$

where the scattering time τ_c is

$$\frac{1}{\tau_c} = \frac{cN(0)\pi}{p_0^2} \int_0^{2p_0} q dq |J(q)|^2 \quad (17.19)$$

and $J(q)$ is the electron-spin exchange interaction. However, the nature of the phase transition is not clear and Keller[14] showed the existence of a large jump in the specific heat which suggests that the normal-superconducting phase transition may be a first-order one.

The antiferromagnetic correlations have a reverse effect[14,15] because of the weighting factor $g(\mathbf{q})$. In this case, the approximations used in Ref. 12 do not work anymore and we have to consider the result from Ref. 14 as reasonable.

b) Dynamic correlations

If the characteristic energy of the fluctuating spins is of the order or less than T, then we have to use a frequency-dependent form for $\chi(\mathbf{q}; \omega)$.

Soukoulis and Grest[16] used the effective scattering time

$$\frac{1}{\tau_{eff}} = \frac{3}{\tau} \int d^3\mathbf{q}\, g(\mathbf{q}) \frac{1}{\pi} \int_{-\infty}^{\infty} d\omega \frac{\exp(\beta\omega)}{(\exp(\beta\omega) - 1)^2} \operatorname{Im} \chi(\mathbf{q}; \omega) \quad (17.20)$$

where $\beta = 1/T$, and for $\chi(\mathbf{q}, \omega)$ it was supposed a diffusion type expression

$$\chi(\mathbf{q}; \omega) = \chi(\mathbf{q}) \frac{iDq^2}{\omega + iDq^2} . \quad (17.21)$$

However, it was pointed out[14] that for realistic values of D, the effects are too small to give rise to any changes in T_c compared with the case of the static correlations.

If the crystalline field effects are important (at low temperatures and high energies of the magnetic excitations), the magnetic excitations with energy δ (is the crystal field split energy) larger than T are not strong enough to suppress the superconductivity. Let us consider the influence of high energy excitations. In this case, the ω-dependence from $\chi(\mathbf{q};\omega)$ can be completely neglected. This approach looks strange but this is exactly as in the case of the high-frequency phonons which contribute to the order parameter Δ but cause only a negligible frequency dependence of the order parameter. Equations (17.7) and (17.8) reduce to

$$\tilde{\omega} = \omega(1 + Z_s) + (2\tau_0)^{-1}\,\mathrm{sign}\,\omega_c \tag{17.22}$$

$$\tilde{\Delta} = \Delta[1 - Z_s/gN(0)] + (2\tau_0)^{-1}\tilde{\Delta}/\tilde{\omega}\,\mathrm{sign}\,\omega \tag{17.23}$$

with

$$Z_s = 3(2\pi\tau)^{-1}\int d^3\mathbf{q}\, g(\mathbf{q})\chi(\mathbf{q};0)\,. \tag{17.24}$$

The transition temperature has been calculated as

$$T_c = 1.43\omega_D \exp\left[-\frac{1 + Z_s}{N(0)g - Z_s}\right] \tag{17.25}$$

where ω_D is the Debye frequency, and g the electron-phonon coupling constant. This formula for T_c shows that the spin-fluctuations of high energy give a pair-weakening effect instead of a pair-breaking effect.

Using in fact the same approximation, Entel and Klose neglected the frequency dependence in y from Eq. (16.87). In this approximation, the critical temperature obtained[13] by the consideration of the dynamic correlation is

$$T_c = 1.143\omega_D \exp\left[-\frac{1}{N(0)g}\frac{\tau_{\mathrm{eff}}^{-1}}{1 + \tau_{\mathrm{eff}}^{-1}}\right] \tag{17.26}$$

where

$$\tau_{\mathrm{eff}}^{-1} = \frac{1}{\pi T_c}\frac{S(S+1)}{\tau_c}\frac{1}{4p_0^2B^2}\ln\frac{A^2 + 4p_0^2B^2}{A^2}\,.$$

The relations (17.25) and (17.26) are in principle equivalent but cannot be discussed in connection with the Berk and Schrieffer formula. Even accidentally the analytical form of these two relations are similar, the spin fluctuations considered in (14.19) are given by the electron-hole scattering and the ferromagnetic instability appears in the electronic system and is of Stoner type.

An important case of dynamic correlations has been use by Crisan *et al.*[17] to describe the *superconducting spin-glass*. The scattering time

$$\frac{1}{\tau_{\mathrm{eff}}} = \frac{mp_0}{2\pi^2}\left(\frac{J}{2}\right)^2\frac{c}{T}\int d\varepsilon d\varepsilon' f(\varepsilon)(1 - f(\varepsilon'))K(\varepsilon - \varepsilon') \tag{17.27}$$

where $K(\omega)$ is the dynamic correlation function

$$K(\omega) = \int_{-\infty}^{\infty} dt\, e^{-i\omega t} \langle \mathbf{S}(0)\mathbf{S}(t) \rangle \tag{17.28}$$

has been used to describe the scattering of the electrons on the randomly distributed frozen spins.

The most important point of this approximation is that the characteristic energy of spins is of the order of T, namely

$$SH(r_c) \sim T \tag{17.29}$$

and then we can apply Eq. (17.27) which is typical for the dynamical correlations. In (17.29), $H(\mathbf{r})$ is the molecular field

$$H(\mathbf{r}) = A \frac{\cos 2p_0 r}{r^3} \exp(-r/\xi) \tag{17.30}$$

given by the short-range spin-spin interaction

$$\mathcal{H} = \sum_{i \neq j} I(\mathbf{R}_i - \mathbf{R}_j) \mathbf{S}_i \mathbf{S}_j \tag{17.31}$$

where $I(x) \sim J^2[\exp(-x/\xi) \cos 2p_0 x]/x^3$.
The scattering time (17.27) has been calculated by Abrikosov[18] as

$$\tau^{-1}(T) = \frac{mp_0}{\pi} \left(\frac{J}{2}\right)^2 c \left[S(S+1) - SB_s(H/T) \frac{\sinh \frac{H}{T} - \frac{H}{T}}{\cosh \frac{H}{T} - 1} \right] \tag{17.32}$$

where $B(x)$ is the Brillonin function and H is a random molecular field.

This scattering time was averaged using the probability

$$P(r,q) = 4\pi r^2 c \exp\left[-\frac{4\pi}{3} c r^3 \right] \frac{2}{\pi} \frac{1}{(1-q^2)^{\frac{1}{2}}} \tag{17.33}$$

where

$$q = |\cos 2p_0 r|$$

and the critical temperature has been calculated using relation (16.45) with

$$\frac{1}{\tau_s} = \left\langle \frac{1}{\tau_s} \right\rangle_{\text{conf}} = \iint dr dq \frac{1}{\tau_s} P(r,q) \ . \tag{17.34}$$

In the high concentration approximation

$$\Psi\left(\frac{1}{2}+z\right) \simeq \Psi\left(\frac{1}{2}\right) + \ln 4\gamma z + \frac{1}{24z^2} + \ldots$$

and with these approximations, T_c has been calculated in the different domains of concentrations. This model implies many approximations which are clearly justified in Refs. 17 and 18 but the main point of the superconducting spin-glass is the existence of a rigid cluster containing the frozen impurities. This cluster is destroyed by the temperature or magnetic field at a critical concentration. A reasonable mechanism considered by Abrikosov[18] and Crisan *et al.*[17] is the percolation.

Finally, we have to mention that for high temperature (17.32) gives for the scattering time the same result as the Abrikosov-Gor'kov theory. This result is expected because for $T \gg H$, the spins are randomly distributed and uncorrelated. However, a strong depairing effect appears for $T \ll H$ when the scattering time is

$$\tau_s^{-1} = \frac{mp_0}{\pi}\left(\frac{J}{2}\right)^2 cS^2$$

which corresponds to the scattering of electrons on the frozen spins.

This model appears to be close enough to that given by Bennemann[19] but he did not consider the spin as being frozen and the short-range spin-spin interaction. The validity of this model is confirmed by the experimental data and by the agreement with the similar calculation of the upper critical field H_{c2} for the superconducting spin-glass.

18. Non-magnetic Localized States in Superconducting Alloys

In the previous paragraph, we studied the influence of the paramagnetic impurities on the superconducting state and the main result obtained was that these impurities destroy the superconductivity because of the exchange interaction between conduction electrons and magnetic impurity atoms.

On the other hand, the Anderson theorem states that the non-magnetic impurities have no effect on superconductivity because of time reversal invariance. An intermediary case occurs in the alloys containing transition metal impurities. Before discussing the influence of this kind of impurities on the superconducting alloys, we will present this problem for the normal metal shortly. In the normal dilute alloys, the magnetic moment of a transition metal can be suppressed. This effect has been discussed first by Anderson[20] on the basis of a simple model. The transition metal impurity has been considered as an isolated electronic level placed into the zone of itinerant-electron states of the diluent metal. The important interactions for the behaviour of the alloy are the "s-d" hybridization of the electronic states and the Coulomb interaction between the "d" electrons of the impurities. This model has been treated[20] in the mean field approximation and the main result is the occurrence of a resonance of the finite width Γ which is proportional to the hybridization coupling constant $V_{\text{s-d}}$. The Coulomb interaction U of the electrons with opposite spins splits the resonance state with respect to the spin and if the magnitude of this splitting is such that one level passes the Fermi level, a localized magnetic moment may appear on the impurity. Concretely, when $U \gg \Gamma$, a localized magnetic moment exists and this is called localized magnetic state. In the opposite limit $U \ll \Gamma$, a localized moment is not expected and this state is called non-magnetic localized state. This last case will be considered for the superconducting alloys containing transition metals impurities because the case of magnetic states is in fact equivalent to the magnetic impurities.

We start this problem using the Anderson Hamiltonian for the superconducting alloy which is

$$\mathcal{H} = \mathcal{H}_0 + \mathcal{H}_{\text{s-d}} + \mathcal{H}_{\text{s}} + \mathcal{H}_{\text{c}} \tag{18.1}$$

where

$$\mathcal{H}_0 = \sum_{\mathbf{p},\alpha} \varepsilon(\mathbf{p}) c^\dagger_{\mathbf{p}\alpha} c_{\mathbf{p}\alpha} + \sum_{\alpha} \varepsilon_{\text{d}} d^\dagger_\alpha d_\alpha \tag{18.2}$$

where $c^\dagger_{\mathbf{p}\alpha}(c_{\mathbf{p}\alpha})$ are the creation (annihilation) operators for the s-electrons with energy $\varepsilon(\mathbf{p})$ and spin "α", and $d^\dagger_\alpha$ (d_α) are the creation (annihilation) operators for d-electrons of energy ε_{d} and spin α. The s-d hybridization can be described by

$$\mathcal{H}_{\text{s-d}} = \sum_{\mathbf{p},\alpha} (V_{\text{pd}} c^\dagger_{\mathbf{p}\alpha} d_\alpha + V_{\text{dp}} d^\dagger_\alpha c_{\mathbf{p}\alpha}) \tag{18.3}$$

where V_{pd} is the hybridization matrix element. The Cooper pairing will be described by the term

$$\mathcal{H}_{\text{s}} = -\sum_{\mathbf{p}} (\Delta c^\dagger_{\mathbf{p}\uparrow} c^\dagger_{-\mathbf{p}\downarrow} + \Delta^\dagger c_{-\mathbf{p}\downarrow} c_{\mathbf{p}\uparrow}) \tag{18.4}$$

and the "d-d" interaction will be taken as a Hubbard interaction

$$\mathcal{H}_{\text{c}} = U \sum_{i} n_{i\uparrow} n_{i\downarrow} \tag{18.5}$$

where U is the Coulomb interaction.

In the mean-field approximation, the interaction (18.5) will be approximated using

$$U n_{i\uparrow} n_{i\downarrow} \cong U\langle n_{i\uparrow}\rangle n_{i\downarrow} + U\langle n_{i\downarrow}\rangle n_{i\uparrow} + U\langle d_{i\uparrow}^{\dagger} d_{i\downarrow}^{\dagger}\rangle d_{i\downarrow} d_{i\uparrow} + U\langle d_{i\downarrow} d_{i\uparrow}\rangle d_{i\uparrow}^{\dagger} d_{i\downarrow}^{\dagger} . \tag{18.6}$$

The simplest case for the study of the influence of the non-magnetic states is to take $U = 0$. In this case, Machida and Shibata[21] suggested the occurrence of the impurity band within the BCS gap.

Using the same method as Shiba for the "classical spin" in superconductor showed that the t-matrix has two poles inside the gap and these poles correspond to the energies which are the roots of the equation:

$$z^2\left(1 + \frac{\Gamma}{\sqrt{\Delta^2 - z^2}}\right) = \varepsilon_{\mathrm{d}}^2 + \Gamma^2 \tag{18.7}$$

where

$$\Gamma = \langle |V_{\mathrm{pd}}|^2\rangle N(0)\pi .$$

This result has been obtained at $T = 0$ in the one impurity approximation, but for the finite concentration, an impurity band appears inside the BCS gap. At $T \neq 0$, the critical temperature has been calculated as

$$\ln \frac{T_{\mathrm{c}}}{T_{\mathrm{c}0}} = c\frac{\Gamma}{N(0)}\pi T_{\mathrm{c}} \sum_{\omega} \omega^{-1}[(|\omega| + \Gamma^2)^2 + \varepsilon_{\mathrm{d}}^2]^{-1} . \tag{18.8}$$

We have to point out that the growing of the impurity band is the same as that of the "classical spin" treated by Shiba. Finally, the gapless superconductivity appears.

This is a quite surprising physical result, but we can see that the properties of the system are determined by the parameters Γ and Δ even in the simple case $\varepsilon_{\mathrm{d}} = 0$, which corresponds to a total localized impurity.

In the following we will consider the influence of the Coulombic interaction between d-electrons on the properties of the system. Because of the admixture between localized and conduction electrons, a part of repulsive interaction is present in the interaction of the two electrons of a Cooper pair. In dilute alloys, the exact one-electron functions

$$\psi_n = \sum_{\mathbf{p}} \langle n|\mathbf{p}\rangle \Phi_{\mathbf{p}} + \langle n|\mathrm{d}\rangle \Phi_{\mathrm{d}}$$

is used to describe the superconducting state. The normalization of this function requires

$$\sum_{\mathbf{p}} |\langle n|\mathbf{p}\rangle|^2 + \sum_{\mathrm{d}} |\langle \mathrm{d}|n\rangle|^2 = 1$$

which is not rigorously satisfied because there is an overlapping between "d" and "p" states and these states form an "overcomplete" set. Even with this approximation, the final results are correct. Rato and Blandin,[22] Takanaka and Takano[23] reconsidered the pioneering study of this problem done by Zuckermann[24] in order to classify the importance of the Coulombic interaction in dilute superconducting alloys containing magnetic impurities with d-electrons. The Coulombic correlation has been considered using the Schrieffer-Mattis[25] procedure and performing the average on the impurities. The t-matrix has been calculated as

$$t \equiv U_{\mathrm{eff}} = \frac{U}{1 + U/\pi E \tanh^{-1} E/\Gamma} \tag{18.9}$$

where $E = \varepsilon_{\mathrm{d}} + U\langle n\rangle$ is the energy of the non-magnetic state for which $\langle n\rangle = \langle n_\uparrow\rangle = \langle n_\downarrow\rangle$. This expression has been obtained in the approximation

$$\Gamma \gg T_{\mathrm{c}} \, . \tag{18.10}$$

Using this result, the critical temperature T_{c} as function of the impurities concentration has been calculated as

$$\ln \frac{T_{\mathrm{c}}}{T_{\mathrm{c}0}} = -cN(0)[1 + AN_{\mathrm{d}}(0)U_{\mathrm{eff}}] \tag{18.11}$$

where

$$A = \ln\left[\frac{2\gamma}{T_{\mathrm{c}}}\sqrt{\varepsilon_{\mathrm{d}}^2 + \Gamma^2}\right] - \frac{\Gamma}{E}\tanh^{-1}\frac{1}{E} \tag{18.12}$$

and the density for the d-electrons at the Fermi level is

$$N_{\mathrm{d}}(0) = \frac{\Gamma/\pi}{\varepsilon_{\mathrm{d}}^2 + \Gamma^2} \, . \tag{18.13}$$

Equation (18.11) has been obtained using the usual approximation in the theory of superconductivity

$$\omega_{\mathrm{D}} \gg T_{\mathrm{c}} \, . \tag{18.14}$$

From Eqs. (18.10) and (18.14), we get

$$\Gamma \ll T_{\mathrm{c}} \tag{18.15}$$

which appears to be essential in these calculations.

However, the condition (18.15) is not generally satisfied, and the case

$$\Gamma \sim T_c \tag{18.16}$$

presents a real interest.

Using the Anderson model with $U = 0$, Okaba and Nagi[26] considered the Green function which describes the superconducting electrons interacting with d-electrons as

$$G^{-1}(\mathbf{p},\omega) = (i\widetilde{\omega}\,\widehat{1} - \varepsilon(\mathbf{p})\widehat{\sigma}_z + \widetilde{\Delta}\widehat{\sigma}_x)^{-1} \tag{18.17}$$

and the parameter $u = \widetilde{\omega}/\widetilde{\Delta}$ has been calculated as

$$u = \frac{\omega}{\Delta} + \frac{c\Gamma\omega}{\pi N(0)}[\omega^2 + \Gamma^2 + \varepsilon_d^2 + 2\Gamma\omega(1+u^2)^{-\frac{1}{2}}]^{-1}\,. \tag{18.18}$$

The order parameter satisfies the self-consistent equation

$$\Delta = 2\pi g N(0) T \sum_{\omega} \frac{1}{\sqrt{1+u^2}} \tag{18.19}$$

and using the relations

$$\frac{1}{gN(0)} = \ln\frac{2\gamma\omega_D}{\pi T_{c0}} \tag{18.20a}$$

$$\ln\frac{2\gamma\omega_D}{\pi T} = 2\pi T \sum_{\omega=0}^{\infty} \frac{1}{\omega} \tag{18.20b}$$

we get

$$\ln\frac{T}{T_{c0}} = 2\pi T \sum_{\omega=0}^{\infty}\left[\frac{1}{\Delta}\frac{1}{\sqrt{1+u^2}} - \frac{1}{\omega}\right]. \tag{18.21}$$

Near the critical temperature T_c, $\Delta \to 0$ and $u \to 1$, and from (18.18) we get

$$u = \frac{1}{\Delta}a_{-1} + a_1\Delta \tag{18.22}$$

with

$$a_{-1} = \omega\left[1 + \frac{b}{\varepsilon_d^2 + (\Gamma+\omega)^2}\right]$$
$$a_1 = \frac{b\Gamma}{[\varepsilon_d^2 + (\Gamma+\omega)^2 + b]^2} \tag{18.23}$$

where

$$b = \frac{x\Gamma}{\pi N(0)} \,. \tag{18.24}$$

Then Eq. (18.21) becomes

$$\ln \frac{T_{c0}}{T} = B_0(c,T) + \frac{1}{2} B_1(c,T)\left(\frac{\Delta}{2\pi T}\right)^2 + \dots \tag{18.25}$$

where

$$B_0 = 2\pi T \sum_{\omega=0}^{\infty} \left(\frac{1}{\omega} - \frac{1}{a_{-1}}\right) \tag{18.26a}$$

$$B_1 = (2\pi T)^3 \sum_{\omega=0}^{\infty} \left(2a_1 + \frac{1}{a_{-1}}\right)\left(\frac{1}{a_{-1}}\right)^2 . \tag{18.26b}$$

If we calculate the critical temperature T_c from (18.25) we get

$$\ln \frac{T_c}{T_{c0}} = B_0(c,T) \tag{18.27}$$

where

$$\begin{aligned} B_0 = & \frac{b}{\Gamma^2 + (\varepsilon_d^2 + b^2)^{\frac{1}{2}}} \left\{ \Psi\left(\frac{1}{2} + \frac{\Gamma + i\alpha}{2\pi T}\right) - \Psi\left(\frac{1}{2}\right) \right. \\ & \left. - \frac{\Gamma}{\alpha} \operatorname{Im} \Psi\left(\frac{1}{2} + \frac{\Gamma + i\alpha}{2\pi T}\right) \right\} \end{aligned} \tag{18.28}$$

and

$$\alpha^2 = \varepsilon_d^2 + b^2 \,.$$

For a large impurity concentration, the critical temperature can be obtained from (18.27)–(18.28) as

$$T_c \cong T_{c0} \exp\left[- \frac{b}{\varepsilon_d^2 + \Gamma^2} \ln \frac{4\gamma b^{\frac{1}{2}}}{2\pi T_{c0}} \right] \tag{18.29}$$

an equation which shows an exponential decreasing of T_c with the concentration of the impurities. An interesting feature of this approach is the nonexistence of a critical concentration.

These calculations have been performed in the simple case of $U = 0$. However, in the superconducting alloys containing the transition metal impurities, the Coulombic interaction cannot be considered, in all cases, very small and such a model is unrealistic.

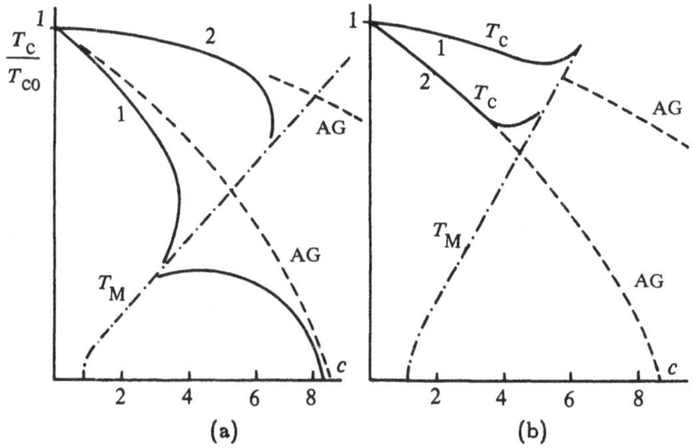

Fig. 17. The critical temperature T_c as function of the impurities concentration c: (a) in the presence of the ferromagnetic correlations, (b) in the presence of the antiferromagnetic correlations.

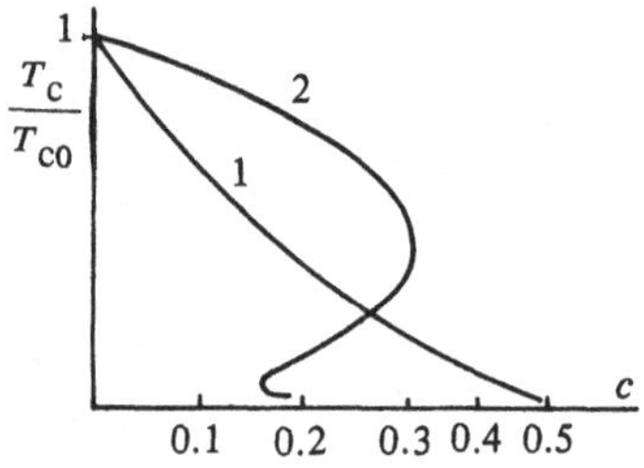

Fig. 18. The critical temperature T_c as function of the magnetic impurities in the Kondo regime: (1) the case $T_k \cong T_{c0}$, (2) the case $T_k \ll T_{c0}$.

In the following, we will consider the model proposed by Keiser[27] which will take the d-electrons correlation effect on the superconductivity.

In the strong coupling formalism, the Green function for s-electrons, respectively d-electrons are

$$\widehat{G}_s^{-1}(\mathbf{p};\omega) = \omega Z(\omega)\widehat{\tau}_0 - \varepsilon(\mathbf{p})\widehat{\tau}_3 - \Phi(\omega)\widehat{\tau}_1 \equiv \omega\widehat{\tau}_0 - \varepsilon(\mathbf{p})\widehat{\tau}_3 - \Sigma(\mathbf{p};\omega) \quad (18.30)$$

$$\widehat{G}_d(\omega) = \omega Z(\omega)\widehat{\tau}_0 - E\widehat{\tau}_3 - \Phi_d(\omega)\widehat{\tau}_1 \equiv \omega\widehat{\tau}_0 - E\widehat{\tau}_3 - \widehat{\Sigma}_d(\omega)\,. \quad (18.31)$$

The diagrams for $\widehat{\Sigma}(\mathbf{p};\omega)$ and $\widehat{\Sigma}(d;\omega)$ are

$\widehat{\Sigma}(\mathbf{p};\omega) \equiv$

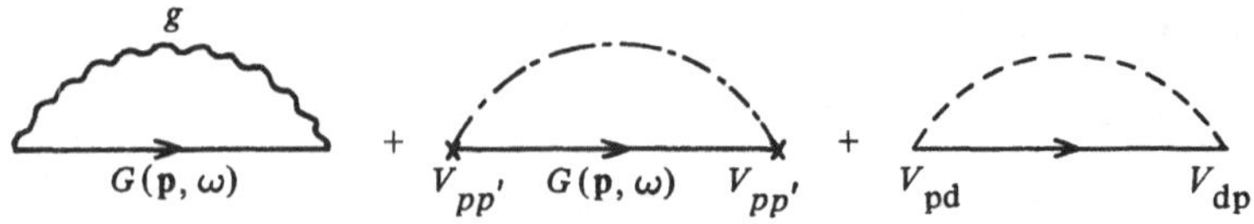

$\hat{\Sigma}(d;\omega) \equiv$ 

where

$$
\begin{aligned}
Z(\omega) &= 1 + c\frac{|V_{\rm pd}|^2 Z_{\rm d}(\omega)}{B(\omega)}\,;\\
Z_{\rm d}(\omega) &= 1 + \frac{|V_{\rm pd}|^2 \pi N(0) Z(\omega)}{[\Phi^2(\omega) - \omega^2 Z^2(\omega)]^{\frac{1}{2}}} + c\frac{|V_{pp'}|^2 \pi N(0) Z(\omega)}{[\Phi^2(\omega) - \omega^2 Z^2(\omega)]^{\frac{1}{2}}}\\
\Phi(\omega) &= \Delta + c\frac{|V_{\rm pd}|^2 \Phi_{\rm d}(\omega)}{B(\omega)}\\
\Phi_{\rm d}(\omega) &= \Delta_{\rm d} + |V_{\rm pd}|^2 \frac{\pi N(0)\Phi(\omega)}{B(\omega)} + c|V_{pp'}|^2 \frac{\pi N(0) Z(\omega)}{[\Phi^2(\omega) - \omega^2 Z^2(\omega)]^{\frac{1}{2}}}
\end{aligned}
\tag{18.32}
$$

where c is the impurity concentration, $N(0)$ the density of s-electrons at the Fermi level and

$$
B(\omega) = \Phi_{\rm d}^2(\omega) + E^2 - \omega^2 Z_{\rm d}^2(\omega)\,. \tag{18.33}
$$

The equations for Δ and $\Delta_{\rm d}$ are

$$
\Delta = g\pi T \sum_{\omega} \frac{\Phi(\omega)}{[\Phi^2(\omega) + \omega^2 Z^2(\omega)]^{\frac{1}{2}}} \tag{18.34}
$$

$$
\begin{aligned}
\Delta_{\rm d} &= -U\pi T \sum_{\omega} \frac{\Phi_{\rm d}(\omega)}{B(\omega)}\\
&= -U\pi T \sum_{\omega} \frac{\Delta_{\rm d}(\omega)}{B(\omega)} - U \sum_{\omega<\omega_{\rm D}} \frac{|V_{\rm pd}|^2 \pi N(0)\Phi(\omega)}{B(\omega)[\Phi^2(\omega) + \omega^2 Z^2(\omega)]^{\frac{1}{2}}}\,.
\end{aligned}
\tag{18.35}
$$

The order parameter Δ of the s-electrons is not identical to the gap $\Delta_{\rm g}(\omega)$ from the excitation spectrum.

The gap is defined by

$$
\Phi(\omega) = \Delta_{\rm g}(\omega) Z(\omega) \tag{18.36}
$$

and from (18.32) and (18.36) we get

$$
\Delta_{\rm g}(\omega) = \frac{\Delta(1 - dH(\omega))}{1 + H(\omega)} \tag{18.37}
$$

where

$$H(\omega) = \frac{c|V_{\rm pd}|^2}{B(\omega)} \tag{18.38}$$

$$d = -\Delta_{\rm d}/\Delta \; . \tag{18.39}$$

The density of states in excitation spectrum is calculated from

$$N(\omega) = \frac{1}{\pi}{\rm Im}\sum_{\mathbf{p}} G(\mathbf{p},\omega) = N(0){\rm Re}\left[\frac{\omega}{\sqrt{\omega^2-\Delta_{\rm g}^2(\omega)}}\right] . \tag{18.40}$$

The order parameter Δ is greatly reduced by the impurities, but the density of states $N(\omega)$ does not differ very much from the BCS shape. This is because $\Delta_{\rm g}(\omega)$ is, to a very good approximation, real and constant as we will see below.

If we consider $B(\omega)$ as

$$B(\omega) = E^2 + \Gamma^2 + \Delta_{\rm d}^2 - \omega^2 + 2\Gamma\frac{[\Delta_{\rm d}\Phi(\omega) - \omega^2 Z(\omega)]}{[\Phi^2(\omega) - \omega^2 Z^2(\omega)]^{\frac{1}{2}}} \tag{18.41}$$

for the case $\Gamma \gg \omega_{\rm D} \gg \Delta_{\rm g}(\omega)$ one obtains

$$B(\omega) = \Gamma^2 + E^2 \tag{18.42}$$

and the gap $\Delta_{\rm g}(\omega)$ is

$$\Delta_{\rm g}(\omega) = f\Delta$$

with

$$f = \frac{1-nd}{1+n} \qquad n = \frac{cN_{\rm d}(0)}{N(0)} \tag{18.43}$$

an equation which shows that $\Delta_{\rm g}(\omega)$ is a real constant. In the usual theory of superconductivity, the order parameter is identical to the gap. In the case of dilute alloys, there are differences and the gap $\Delta_{\rm g}$ is reduced below the order parameter Δ of the pure superconductor. This is because the resonance and non-resonance terms cancel out in the expression for $\Delta_{\rm g}(\omega)$ in contrast with the case of the magnetic states when these terms have opposite signs because of inexistence of the time-invariance of the Hamiltonian. These terms give rise to a large energy dependence of $\Delta_{\rm g}(\omega)$ and a large imaginary part leading to gapless superconductivity.

The denominator $(1+n)$ from f defined by (16.43) is a dilution effect due to d-electrons participating in superconductivity via the resonance scattering. The factor $(1-nd)$ from f represents the opposition to the electron pairing attraction of the Coulomb repulsion U between opposite spins of d-electrons.

With these results, one can calculate the order parameters Δ and Δ_d. From Eq. (18.35), Δ_d can be written as

$$\Delta_d = -U\Delta_d \pi T \sum_{\omega} \frac{1}{B(\omega)} - U\Gamma\pi T \sum_{\omega<\omega_D} \frac{\Phi(\omega)}{B(\omega)} [\Phi^2(\omega)+\omega^2 Z^2(\omega)]^{\frac{1}{2}} \quad (18.44)$$

and from the general relation (18.44) if $\Gamma \gg \Delta_g$ one obtains

$$B(\omega) = (\Gamma + |\omega|)^2 + E^2 \ . \quad (18.45)$$

Since $\Gamma \gg \omega_D$ the second term is important, and this approximation is opposite to $\Gamma \ll \omega_D$ performed by Rato and Blandin. The second term in Eq. (18.44) can be expressed using the equation for Δ and

$$\Delta_d \left[1 + U\pi T \sum_{\omega} \frac{1}{(\Gamma + |\omega|)^2 + E_d^2}\right] = -\frac{\Gamma U \Delta}{\pi g(\Gamma^2 + E_d^2)} \quad (18.46)$$

and since $\Gamma \gg 2\pi T$, the sum from the first term can be written as

$$\pi T \sum_{\omega} [(\Gamma + |\omega|)^2 + E_d^2]^{-1} \doteq \frac{1}{\pi E_d} \tanh^{-1} \frac{E_d}{\Gamma}$$

and finally $d = -\Delta_d/\Delta$ becomes

$$d = \frac{N_d(0)}{N(0)g} U_{\text{eff}} \quad (18.47)$$

where U_{eff} is given by (18.9). This equation shows clearly the physical meaning of d. The parameter d is temperature independent as well as concentration independent because we neglected the interaction between impurities. From the equation for Δ, we get

$$\Delta = g\pi T \sum_{\omega} \frac{\Delta_g}{[\Delta_g^2 + \omega^2]^{\frac{1}{2}}} \quad (18.48)$$

and at $T = 0$, this equation becomes

$$\Delta_g = f\Delta \ .$$

Using these two equations, one obtains an equation for Δ_g as

$$\Delta_g = fg\pi T \sum_{\omega<\omega_D} \frac{\Delta_g}{(\Delta_g^2 + \omega^2)^{\frac{1}{2}}} \tag{18.49}$$

which is in fact the usual BCS relation with $\Delta \longrightarrow \Delta_g$ and $g \longrightarrow fg$. The critical temperature will be obtained as

$$T_c = 1.14\omega_D \exp\left[-\frac{1}{fg}\right] \tag{18.50}$$

and using

$$\frac{1}{N(0)g} = \ln\frac{2e\gamma\omega_D}{\pi T_c}$$

the critical temperature for the alloy is given by

$$\ln\frac{T_c}{T_{c0}} = \frac{f-1}{fg} \tag{18.51}$$

one of the important result of this approach of dilute alloys. As a final result which is new in the theory of dilute alloys, we have to remark that if the concentration of the impurities reduces $\mathrm{Re}\,\Delta_g(0)$ to zero, both $\mathrm{Re}\,\Delta_g(\omega)$ and $\mathrm{Im}\,\Delta_g(\omega)$ are zero and the *superconductivity is quenched*, instead of gapless superconductivity. Another important feature of this model is the reduction of the effect of impurities on superconductivity due to the dependence of $N_d(0)$ on E_d which is essentially dependent on the number of electrons from the state "d".

We have to mention that the absence of the gapless superconductivity obtained by Keiser is also obtained in the *approximation* $\Gamma \gg \omega_D > \Delta_g(\omega)$ and $\omega \ll \omega_D$, which is opposite to the Ratto-Blandin approximation.

The factor f represents a weakening of the superconducting pairing, in contrast to the pair breaking which is produced from spin-flip scattering from paramagnetic impurities and which leads to gapless superconductivity.

19. Kondo Effect in Superconductors

In the Abrikoson-Gor'kov theory of dilute alloys, the scattering of electrons on impurity atoms has been considered in the first Born approximation. The amplitude of scattering, in this case, does not depend on the electron energy and temperature, and the critical temperature decreases with the concentration. At a critical concentration c_0, the superconducting state is completely suppressed.

In the normal metals containing paramagnetic impurities, the next perturbation term with respect to the s-d(f) interaction will give an amplitude of scattering which diverges logarithmically near the Fermi energy. This is known as *Kondo effect* in the normal metals and leads to logarithmic corrections in resistivity of the dilute alloys.

For the normal metals, it was shown that the behaviour of the amplitude scattering on a paramagnetic impurity depends on the sign of the s-d exchange integral. For $J < 0$ (antiferromagnetic coupling), the behaviour changes at the temperature

$$T_k \cong \varepsilon_0 \exp\left[-\frac{1}{JN(0)}\right]$$

called *Kondo temperature*. At this temperature, the amplitude scattering appears as being divergent but recently Wilson *et al.*[28] showed that the Kondo problem can be solved numerically using the Renormalization Group Theory (RNG), a method used in Quantum Electrodynamics and Critical Phenomena.

In the superconducting alloys, the binding energy of the Kondo state is of the order of T_k and the energy of the Cooper pairs is of the order of T_{c0}. Then if $T_k \simeq T_{c0}$, the properties of the alloys are drastically changed.

A great number of papers have been published concerning the Kondo effect in superconductors. We will start with the result obtained by Müller-Hartman and Zittartz[29] which are based on the Nagaoka[30] and Hamann[31] methods from the Kondo effect in normal metals. These results are equivalent to those obtained by Ludwig and Zukermann[32] using the effective electron-electron interaction due to impurities. The simplest model for the Kondo problem in superconductors is a single impurity interacting with superconducting electrons. This simple model can be described by the Hamiltonian

$$\mathcal{H} = \mathcal{H}_0 + \mathcal{H}_i \tag{19.1}$$

$$\mathcal{H}_0 = \sum_{\mathbf{p},\alpha} \varepsilon(\mathbf{p}) a^\dagger_{\mathbf{p}\alpha} a_{\mathbf{p}\alpha} - \Delta \sum_{\mathbf{p}} (a^\dagger_{\mathbf{p}\uparrow} a_{-\mathbf{p}\downarrow} + h.c.) \tag{19.1a}$$

$$\mathcal{H}_i = -\frac{J}{2N} \sum_{\mathbf{p},\mathbf{p}'} \sum_{\alpha,\beta} \mathbf{S} a^\dagger_{\mathbf{p}\alpha} \sigma_{\alpha\beta} a_{\mathbf{p}\beta} \; . \tag{19.1b}$$

The system has no translational invariance and the Green function for electrons will be specified by $\mathbf{p}$ and $\mathbf{p}'$.

In this simplified model, the Kondo effect appears from the multiple scattering of the conduction electrons on the magnetic impurity. The Green function for the pure superconductor is

$$\widehat{G}^0(\mathbf{p}, z) = \begin{bmatrix} z - \varepsilon(\mathbf{p}) & \Delta \\ \Delta & z + \varepsilon(\mathbf{p}) \end{bmatrix} \tag{19.2}$$

and the Green function $G(\mathbf{p}, \mathbf{p}'; z)$ for the alloy can be calculated from the equation

$$\widehat{G}(\mathbf{p}, \mathbf{p}'; z) = \widehat{G}^0(\mathbf{p}, \mathbf{p}'; z)\delta_{\mathbf{p},\mathbf{p}'} + \widehat{G}^0(\mathbf{p}; z)\frac{J}{N}\widehat{G}^0(\mathbf{p}'; z) \tag{19.3}$$

where the t-matrix is a solution of a singular integral equation obtained by Hamann.[31]

Before giving the results from Refs. 29 and 36, we will show the relation of these approximations with the other results presented in Sec. 16.

In the second Born approximation

$$\widehat{t}(z) = \frac{S(S+1)}{4}\widehat{F}(z) \tag{19.4}$$

where

$$\widehat{F}(z) = \frac{N(0)J}{2}\int_{-\infty}^{\infty} d\varepsilon \frac{\rho(\varepsilon)}{z^2 - \varepsilon^2 - \Delta^2}\begin{bmatrix} z + \varepsilon & \Delta \\ -\Delta & Z - \varepsilon \end{bmatrix} \tag{19.5}$$

$N(0)$ being the density of states of the conduction electrons at the Fermi surface, $\rho(\varepsilon)$ the energy dependent density of states. In this approximation, one gets the Abrikosov-Gor'kov result. In the classical limit, $J \longrightarrow 0, S \longrightarrow \infty$, JS = constant

$$\widehat{t}(z) = \left[1 - \left(\frac{S}{2}\widehat{F}(z)\right)^2\right]^{-1}\left(\frac{S}{2}\right)^2\widehat{F}(z) \tag{19.6}$$

and the Shiba result[5] can be re-obtained.

There is an important difference between these two simple results. No matter how small the coupling constant JS is, the scattering amplitude in the Shiba theory exhibits two poles within the energy gap. Similar bound states will appear in the solution of the Hamann integral equation, and these bound states have a central importance in the Kondo effect theory.

In the Abrikosov-Gor'kov theory, we get only the gapless superconductivity. We see then, that the approximation is essential in the Kondo problem.

The t-matrix has been obtained from the equation[30]

$$\left[1-\frac{S(S+1)}{4}\left(F_2^{(1)}(z)\right)\right]^2+\mathcal{F}_\omega\left\{\frac{F_2^{(1)}(\omega)-F_2^{(1)}(z)}{z-\omega}(1+\hat{t}(\omega))\right.$$
$$\left.\times\left(F_2^{(1)}(\omega)-F_2^{(1)}(z)\right)\right\}\hat{t}(z)$$
$$=\frac{S(S+1)}{4}F_2^{(0)}(z)+\mathcal{F}_\omega\left\{\frac{F_2^{(1)}(\omega)-F_2^{(1)}(z)}{z-\omega}\hat{t}(\omega)\right\} \tag{19.7}$$

where $\mathcal{F}_\omega$ denotes the principal value of a sum

$$\mathcal{F}_\omega\{A(\omega)\}=T\sum_\omega A(\omega)\,. \tag{19.8}$$

From this equation we can show that if $\rho(\varepsilon)=\rho(-\varepsilon)$, the t-matrix has the form

$$\hat{t}(z)=\begin{bmatrix} t_1(z) & t_2(z)\\ t_2(z) & t_1(z)\end{bmatrix} \tag{19.9}$$

with $t_1(z)=-t_2(z)$ and $t_2(z)=t_2(-z)$.
If we introduce the new matrix

$$\hat{t}(z)=t_1(z)-t_2(z) \tag{19.10}$$

the solution of Eq. (19.7) has been given by Müller-Hartman and Zittartz[29] as

$$\gamma\hat{t}(z)=\frac{1}{2\pi i}\sqrt{\frac{z^2-\Delta^2}{z+\Delta}}\left[1-\frac{X(z)}{\Phi(z)}\right]\,. \tag{19.11}$$

If $z\simeq\Delta$, the functions $X(z)$ and $\Phi(z)$ can be calculated as

$$X(z)=\gamma\left(\tau+\frac{a}{y-1}\right),\quad \Phi(z)=|\gamma|\sqrt{\tau^2+\pi^2S(S+1)} \tag{19.12}$$

where

$$y=z/\Delta\,,\qquad \gamma=\frac{JN(0)}{2}\,,\qquad \tau=\frac{1}{\gamma}+\frac{1}{g}=\ln\frac{T_k}{T}\,,$$
$$a=\tau-\operatorname{sign}\gamma\sqrt{\tau^2+\pi^2S(S+1)}\,,\quad y=\operatorname{sign}\gamma\frac{\tau}{\sqrt{\tau^2+\pi^2S(S+1)}}\,. \tag{19.13}$$

The density of the one-particle excitations can be calculated as

$$N(\omega) = -\frac{2}{\pi}\frac{1}{N}\text{Im}\sum_{\mathbf{p}} G^{11}(\mathbf{p};\omega + i\delta) \tag{19.14}$$

and using Eq. (19.3) we get

$$N(\omega) = N_0(\omega) + \frac{1}{N}n(\omega)$$

where $N_0(\omega)$ is the density of the one-particle excitations for a pure superconductor, and for

$$n(\omega) = \frac{a}{2}[\delta(\omega - \omega_0) + \delta(\omega + \omega_0)] - \frac{1}{2}[\delta(\omega - \Delta) + \delta(\omega + \Delta)] \tag{19.15}$$

where $\omega_0 = y_0\Delta$.

The presence of the magnetic impurity changes the energy of the superconducting state, and ω_0 from the additional density of states $n(\omega)$ can be considered as the level of impurity. The relation $\omega_0 = y_0\Delta$ shows that ω_0 is sensitive to the sign of the electron-impurity interaction. Another important result will be transparent if we calculate the matrix elements t_1 and t_2 as

$$\gamma t_1(t) = \frac{\sqrt{y^2-1}}{2\pi i}\frac{y(y-y_0)}{y^2-y}, \quad \gamma t_2 = \frac{\sqrt{y^2-1}}{2\pi i}\frac{y_0(y-y_0)}{y^2-y_0} \tag{19.16}$$

which for high impurity concentrations, give the same result as for the classical spin.

The bound state ω_0 is energy dependent and is determined by the magnitude of the s-d exchange. For $\gamma < 0$, this bound state can attain the middle of the superconducting gap if $T_k \sim T_c$. This case is one of the most complicated from the Kondo effect in superconductors.

The equation for the critical temperature T_c can be obtained from the gap equation and has the general form

$$\ln\frac{T_c}{T_{c0}} = \lim\frac{c\gamma}{N(0)\Delta}\int d\varepsilon N(\varepsilon)\pi T_c\sum_{\omega}(\hat{G}^0(\mathbf{p};\omega)\hat{t}(\omega)\hat{G}^0(\mathbf{p};\omega)) \tag{19.17}$$

an equation which can be transformed using (19.10) as

$$\ln\frac{T_c}{T_{c0}} = -\lim\frac{2\pi\gamma c T_c}{N(0)\Delta}\sum_{\omega}\frac{1}{\omega}\text{Re}\left\{t_2(\omega) - \frac{i\Delta}{\omega}t_1(\omega)\right\} \tag{19.18}$$

where $c = \frac{1}{N}$, N being the number of impurities.
For a finite impurity concentration, the self-energy of the electrons is

$$\Sigma(z) = ct(z) \tag{19.19}$$

the new Green function will be

$$\hat{G}^{-1}(\mathbf{p};\omega) = \begin{bmatrix} \tilde{\omega} - \varepsilon(\mathbf{p}) & \tilde{\Delta}(\mathbf{p},\omega) \\ \tilde{\Delta}(\mathbf{p},\omega) & \tilde{\omega} + \varepsilon(\mathbf{p}) \end{bmatrix} \tag{19.20}$$

where

$$\tilde{\omega} = \omega - c\frac{J}{2}t_1(\tilde{\omega},\tilde{\Delta}), \quad \tilde{\Delta} = \Delta - c\frac{J}{2}t_2(\tilde{\omega},\tilde{\Delta}) . \tag{19.21}$$

The critical temperature in this case becomes

$$\ln\frac{T_c}{T_{c0}} = \Psi\left(\frac{1}{2}\right) - \Psi\left(\frac{1}{2} + \rho\right) \tag{19.22}$$

where the pair-breaking ρ defined as

$$\rho = \frac{c}{4\pi^2 N(0)T_c}(1 - y_0^2) \tag{19.23}$$

is energy-dependent, and for

$$1 - y_0^2 = \frac{\pi^2 S(S+1)}{\ln^2 T_k/T_c + \pi^2 S(S+1)} . \tag{19.24}$$

This is the main result obtained by Müller-Hartman and Zittartz and in Fig. 18 we showed the dependence of the critical temperature T_c on the concentration.

From this figure, we see that re-entrance phenomenon appears only for $T_k < T_{c0}$ and if $T_k > T_{c0}$, this effect may disappear because

$$1 - y_0^2 \cong \pi^2 S(S+1)N(0)\left(\frac{J}{2}\right)^2 \tag{19.25}$$

which gives for T_c the Abrikosov-Gor'kov result.

In the above theory, the repulsive interaction between impurities electrons has been completely neglected. An interesting result has been obtained by Matsuura[33] and Sakurai[34] who considered the repulsive interaction between the electrons with antiparallel spins resulting from the virtual polarization of impurities in singlet states. These results are based on the Yoshida and Yamada[35] results concerning the Kondo effect in the normal metals. The important results obtained by these authors are that at $T = 0$ (in this case, the Suhl-Nagaoka approximation cannot be applied), the magnetic susceptibility is finite. Before we start to analyse the importance of the repulsion interaction for the Kondo problem in superconductors, we shall reconsider some results from Kondo effect in the normal metals.

If $T \gg T_k$, the impurity behaves as a magnetic impurity with an effective interaction which is energy and temperature dependent.

In the opposite case: $T \ll T_k$, the impurity behaves as non-magnetic and $\chi_i^{-1} \sim T_k$. When the temperature and energy change, the property of impurities gradually changes from one regime to the other, then we can see that the critical boundary obtained in the Hartree-Fock, is due to the approximation.

In the framework of these conclusions, we can discuss the impurity state in a superconductor.

In such a discussion, we will consider the following cases:

1. $T_{c0} \ll T_k \ll \omega_D$. In this situation, the electrons with energy $T_k < |\omega| < \omega_D$ cannot participate in superconductivity due to the pair-breaking effect and the cut-off ω_D is replaced by T_k. For the electrons with energy $|\omega| < T_k$, the electron-electron interaction is reduced by the repulsion ($U_{rep} \sim T_k N^2(0)$) the critical temperature T_c being

$$T_c \sim T_k \exp\left[-\frac{1}{gN(0)} - \frac{c}{T_k N(0)}\right] \tag{19.26}$$

and the critical concentration is

$$c_0 = T_k N^2(0) g$$

for the one-impurity approximation.

Let us consider the opposite case when the Kondo temperature is very low.

2. $T_k \ll T_{c0}$, but the critical temperature of the alloy satisfies $T_c \gg T_k$ the main effect is the pair-breaking one, the theory of Müller-Hartman and Zittartz[29] can be applied.

Indeed, in this case the energy dependence in the pair-breaking parameter plays only a secondary role here. At a finite concentration, the critical temperature T_c of the alloy is reduced and $T_c \cong T_k$, the interaction between electrons with energy $|\omega| < T_k$ being repulsive. For the temperature $T \simeq T_k$, the separation between pair-breaking and repulsion has no clear meaning, and is necessary a kind of interpolation which seems to be possible because there is no sharp transition in this region.

The critical temperature T_c as function of the concentration for a superconducting alloy can be calculated using the equation

$$\Delta(T) = gQ(T)\Delta \tag{19.27}$$

where $Q(T)$ is the two-particle Green function and can be considered as

$$Q(T) = \pi T \sum_{\omega} \gamma(\omega) G(\mathbf{p},\omega) G(\mathbf{p},-\omega) \tag{19.28}$$

where

$$G(\mathbf{p},\omega) = \frac{1}{i\omega - \varepsilon(\mathbf{p}) - \Sigma(\mathbf{p};\omega)} \tag{19.29}$$

with $\gamma(\omega)$ as the vertex correction. The electron-hole symmetry satisfies

$$\gamma(\omega) = \gamma(-\omega)\,, \quad \Sigma(\omega) = \Sigma(-\omega)$$

and if one performs the integral over ε (for a constant density of states) Eq. (19.28) becomes

$$Q(T) = N(0)\pi T \sum_{\omega} \frac{\gamma(\omega)}{|\omega| + \Sigma(\omega)}\,. \tag{19.30}$$

For a low concentration of impurities, we can perform the average on the impurities and the self-energy becomes proportional to the concentration. The vertex function can be written as

$$\begin{aligned} \gamma(\omega) &= 1 + \pi T \sum_{\omega,\mathbf{p}} \Gamma_{\uparrow\downarrow}(\omega,\omega') G(\mathbf{p},\omega) G(-\mathbf{p},-\omega)\gamma(\omega') \\ &= 1 + \pi N(0) \sum_{\omega'} \Gamma_{\uparrow\downarrow}(\omega,\omega') \frac{\gamma(\omega')}{|\omega'| + |\Sigma(\omega')|}\,. \end{aligned} \tag{19.31}$$

The self-energy of the conduction electrons can be written as

$$\Sigma(\omega) = c|V_{\mathrm{pd}}|^2 G(d,\omega) \tag{19.32}$$

where

$$G(d,\omega) = [i\omega + i\Gamma\,\mathrm{sign}\,\omega - \Sigma(d;\omega)]^{-1} \tag{19.33}$$

where Γ is the resonance width.

The self-energy $\Sigma(d;\omega)$ is due to the repulsive interaction U. The expressions of the self-energy and of the vertex function $\Gamma_{\uparrow\downarrow}(\omega,\omega')$ have been obtained by Yamada and Yoshida[35] for the Kondo problem in the normal dilute alloys and this result can be used because we are interested in the calculation of the critical temperature. The self-energy $\Sigma(d;\omega)$ is

$$\Sigma(d;\omega) = -(\chi_{\uparrow\uparrow} - 1)i\omega - \frac{i\,\mathrm{sign}\,\omega}{\Gamma}\chi^2_{\uparrow\downarrow}[(i\omega)^2 - (\pi\Gamma)^2] \tag{19.34}$$

where for a symmetric Anderson model we have

$$\chi_{\uparrow\uparrow} = \frac{\pi\Gamma}{4T_k} \,.$$

If the second term from (19.33) can be neglected, the Green function for the d-electrons becomes

$$G(d,\omega) = -\frac{i}{\Gamma}\frac{\mathrm{sign}\,\omega}{1 + (\pi\omega/4T_k)} \,. \tag{19.35}$$

Using the vertex function for the d-electrons $\Gamma_d(\omega,\omega')$, the vertex function $\Gamma_{\uparrow\downarrow}$ can be expressed as

$$\Gamma_{\uparrow\downarrow} = c|V|^4\left[\frac{1}{T}|G(d,\omega)|^2 - |G(d,\omega)|^2\Gamma_d(\omega,\omega')|G(d,\omega)|^2\right] \tag{19.36}$$

and in the low temperature limit

$$\Gamma_d(0,0) \cong \frac{\pi^2\Gamma^2}{4T_k} \,. \tag{19.37}$$

With these results, one can treat the two limits discussed above.

In the approximation $|\omega'| \gg T_k$, Eq. (19.36) reduces to

$$\Gamma_{\uparrow\downarrow} = -\frac{3c|V_{pd}|^2}{16T}\left[\frac{1}{3}\delta_{\omega,\omega'} + \frac{2}{3}\delta_{\omega,-\omega'}\right] \tag{19.38}$$

which is correct for $U \gg \Gamma$ and gives the Abrikosov-Gor'kov result.

In the most divergent approximation, the Kondo effect can be considered taking the scattering as

$$\tau(\omega) = \frac{1}{N(0)}\left[\ln\frac{|\omega|}{T_k}\right]^{-1} \tag{19.39}$$

which is valid only for $\omega \gg T_k$.

If we compare Eqs. (19.36) and (19.38) for $\Gamma_{\uparrow\downarrow}$, one can see that $\Gamma_d(\omega,\omega')$ should have a singularity at $|\omega+\omega'|\lesssim T_k$ when $|\omega|,|\omega'|>T_k$. When (19.36) is substituted in (19.27), the second term of the vertex $\Gamma_{\uparrow\downarrow}(\omega,\omega')$ has to be a slowly varying function of ω' compared to $\Gamma_\uparrow(\omega,\omega')$ for $|\omega|\gg T_k$.

In this case Matsuura *et al.*[36] considered for $|\omega|\gg T_k$ that

$$\Gamma_{\uparrow\downarrow}=c\Big[\frac{1}{T}\Gamma_1(\omega)\delta_{\omega,\omega'}-\frac{1}{4T_k N^2(0)}f(\omega)f(\omega')\Big] \qquad (19.40)$$

where

$$f(\omega)=\Big[1+\frac{\pi|\omega|}{4T_k}\Big]^{-2}.$$

The term Γ_1 is given for the case $\omega\ll T_k$ by

$$\Gamma_1=V^4|G(d,\omega)|^2=\frac{1}{\pi N(0)}f(\omega)\,. \qquad (19.41)$$

In the regime $\omega\gg T_k$ in $\Sigma(d;\omega)$, the spin-flip contribution (19.39) will be considered and

$$\Gamma_1\cong-\frac{3}{16}\frac{1}{N^2(0)}\Big[\ln\frac{|\omega|}{T}\Big]^{-2}. \qquad (19.42)$$

In order to calculate the critical temperature T_c, we define the pair-breaking parameter

$$c\alpha(\omega)=|\Sigma(\omega)|-c\pi N(0)\Gamma_1(\omega)\,. \qquad (19.43)$$

In the region $\omega\ll T_k$ one gets

$$\alpha(\omega)=\frac{|\omega|}{4N(0)T_k} \qquad (19.44a)$$

and for $\omega\gg T_k$

$$\alpha(\omega)=\frac{3\pi}{8N(0)}\Big[\ln\frac{|\omega|}{T_k}\Big]^{-2}. \qquad (19.44b)$$

In the intermediate region, the interpolation procedure supposes to consider a more realistic expression for $\alpha(\omega)$. Matsuura *et al.*[36] have chosen the expression

$$\alpha(\omega)=\begin{cases}\frac{1}{\pi N(0)}\Big[\Big(\frac{\pi|\omega|}{4T_k}\Big)-\frac{1}{2}\Big(\frac{\pi|\omega|}{4T_k}\Big)^2\Big]; & \frac{\pi|\omega|}{4T_k}\ll 1\\ \frac{1}{2\pi N(0)}\frac{\frac{3}{4}\pi^2}{\ln|\pi\omega/4T_k|+\frac{3}{4}\pi^2}; & \frac{\pi|\omega|}{4T_k}\gg 1\end{cases} \qquad (19.45)$$

for the simple case $S = \frac{1}{2}$.
This expression in the high frequency limit is similar to the Müller-Hartman-Zittartz pair-breaking parameter. For $\omega \sim T_k$, all these calculations are irrelevant because the separation of the impurity effects in two parts loses its meaning.

The critical temperature for a "Kondo alloy" will be obtained from (19.27) as

$$gQ(T_c) \simeq 1 \tag{19.46}$$

which becomes

$$\ln\frac{T_c}{T_{c0}} = 2\pi T_c \sum_\omega \left[\frac{1}{\omega + c\alpha(\omega)} - \frac{1}{\omega}\right] - \frac{c}{4T_k N(0)} \frac{\Phi_1(T_c, c)}{[1 + \Phi_2(T_c, c)A]} \tag{19.47}$$

where

$$A = \frac{c}{4T_k N(0)}, \quad \Phi_k(T_c, c) = 2\pi T_c \sum_\omega \frac{f^k(\omega)}{\omega + c\alpha(\omega)} . \tag{19.48}$$

For the case $T_c \gg T_k, \Phi_1$ and Φ_2 can be neglected and

$$\ln\frac{T_c}{T_{c0}} = 2\pi T_c \sum_\omega \left[\frac{1}{\omega} - \frac{1}{\omega + c\alpha(\omega)}\right] . \tag{19.49}$$

In the opposite case $T_c < T_k$, one may obtain the critical concentration c_0 for which $T_c = 0$. This expression is quite complicated and is given by a transcedental equation. For $T_k \ll T_{c0}$, this equation becomes simple

$$\int_{T_k}^\infty d\omega \left[\frac{1}{\omega} - \frac{1}{\omega + c_0\alpha(\omega)}\right] = \ln\frac{T_{c0}}{T_k} . \tag{19.50}$$

The lower bound of the critical concentration can be obtained from this equation if $\alpha(\omega \sim T_k^{-1}) = N(0)^{-1}$. The upper bound can be calculated using $\alpha(\omega \gg T_k) = J^2N(0)/4$. These values are

$$c_0 = \begin{cases} \frac{\pi^2}{\gamma} N(0)T_{c0}, & T_k \ll T_{c0} \\ (4N(0)T_k)/\ln(2\gamma T_k/\pi T_{c0}), & T_k \gg T_{c0} \end{cases} \tag{19.51a}$$

and we see that the upper bound is attained in the limit $T_k \to 0$, and in this case $c_0 \sim T_{c0}N(0)$ if T_k/T_{c0} is not too small. In the first case considered ($\omega_D \gg T_k \gg T_{c0}$), we get

$$c_0 \cong \frac{4T_{c0}N(0)}{\ln T_k/T_{c0}} \tag{19.51b}$$

an expression which is comparable to (19.51a).

An equivalent treatment of this problem, using the Anderson Hamiltonian has been given by Sakurai,[34] who stressed that in the previous analysis, the temperature dependence of the scattering amplitude has been completely neglected.

The theory presented in Ref. 34 is a Fermi-liquid theory which describes the low energy excitations and predicts the third transition temperature.

The problem of the spin interactions in a superconductor is still unsolved and the concentration dependence of the critical temperature is not enough to elucidate the dominant mechanism in the superconducting alloys.

The calculation of the other parameters and the study of the magnetic field can give important information concerning the validity of any model and on its theoretical description.

20. Localization and Superconductivity

The superconducting critical temperature T_c depends rather weakly on disorder in general. Anderson[1] showed that the non-magnetic impurities do not change the critical temperature. On the other hand, the electron-phonon interaction is of short range and we do not expect this interaction to be changed for $l \gg q_D^{-1}$. The analysis of Anderson, Abrikosov and Gor'kov[16–19] does not consider the localizing effect of strong disorder and the interference between interaction and disorder. Before considering this effect, let us briefly discuss, the basic problems from the electron-impurities interaction taking into consideration the basic ideas from the theory of the Anderson localization.[37] The new understanding is based on advances in the studies of the two aspects of systems consisting of electrons which contain a high concentration of non-magnetic impurities randomly distributed. This problem is equivalent to that of the description of the state of an electron in a random potential. The nature of the new wave function, as well as the interaction between electrons in the presence of a random potential are the main problems which have to be solved. The wave function may be drastically altered if the randomness is sufficiently strong. More exactly, this wave function can be associated to a localized state or to an extended state with the mean free path "l". The interaction between electrons can be so strongly affected that the traditional Fermi-liquid approach becomes unappropriate instead of a diffusive regime.

The first problems in which localization has been considered in "dirty" system were the transport problems in systems with different dimensions. If the system consisting of electrons and impurities has a large enough mean free path: $p_0 l \geq 1$, we consider that in the system there is a *weak disorder.* For $p_0 l \cong 1$, the localization effects become important and the localization correlation length becomes very large. In this case, the regime will be considered as having a *strong disorder.* If in the first case, the conductivity is given by the standard Drude formula ($\sigma = ne^2\tau_{tr}/m$) in the second one the conductivity is

$$\sigma(\omega) = \begin{cases} \dfrac{g_c}{L} \quad ; \quad \omega < \omega_c & (20.1a) \\[2ex] \dfrac{g_c}{L}\left(\dfrac{\omega}{\omega_c}\right)^{\frac{1}{3}} ; \quad \omega > \omega_c & (20.1b) \end{cases}$$

where $\omega_c = 1/eL^3$ and g_c is the critical conductance of the system. We see that in this case the system does not present an Ohmic behaviour and the conductance $g(L)$ depends on the scale size. The scaling-idea, as well as the Renormalization Group Method have been successfully applied in this regime by different authors.[38-40] However, we shall stress the importance of one of these papers from another point of view namely Abrahams *et al.*[38] showed using the scale theory the existence of an expansion parameter (dependent on the scattering time τ) defined by

$$\lambda = \frac{\hbar}{2\varepsilon_0 \pi \tau} \quad (\hbar = 1) \tag{20.2}$$

which gives us the possibility to treat the randomness from the metallic limit using the perturbation theory. This is the main point in the description of the interacting electronic systems and at the present time a theory, which presents many common points with the standard one, has been developed using the perturbation theory, propagators and the approximation known as "the average on the impurities". A summation of the particle-particle and particle-hole diagrams in the presence of a random impurity has given the first interesting result: both processes are described by a propagator which will be called *randomon* and has the analytical form

$$R = \frac{1}{2\pi\tau^2}\frac{1}{[Dq^2 + |\omega|]} \tag{20.3}$$

where D is the diffusion coefficient.

With the propagator (20.3) one may start to calculate the lowest correction in the electronic self-energy. In this case even the Hartree diagrams will give an important contribution to the processes from the electronic system.[41] In order to treat the effect of random impurities on superconductivity, we have to mention that the Coulomb interaction V_c can be recalculated in the presence of impurities. In the Random Phase Approximation, we get

$$U(\mathbf{q};\omega) = V^c(q)\Big/[1 + V^c(q)\Pi(\mathbf{q};\omega)] \tag{20.4}$$

where the polarization operator is

$$\Pi(\mathbf{q};\omega) = \frac{2N(0)Dq^2}{Dq^2 + \omega} \tag{20.5}$$

an equation which is valid for $|q| < q_0$ where $q_0 = 1/\sqrt{D\tau}$. The vertex correction is also affected by impurities and was calculated as

$$\Gamma(\mathbf{q};\omega) = \begin{cases} \tau^{-1}(D|q|^2 + |\omega|)^{-1}; & \omega(\omega - \omega') < 0 \\ 1 \end{cases} \tag{20.6}$$

where $Dq^2 \ll \tau^{-1}$ and $\omega \ll \tau^{-1}$.

The first theoretical attempt to describe the problem of the influence of disorder-induced changes on T_c are due to Keck and Schmid.[42] They used the Eliashberg equation, neglecting the Coulomb interaction and the obtained results appear to be reasonable if the materials have a resistivity $\rho < 10\,\mu\Omega$ cm. Later, Maekawa and Fukuyama[43] studied the corrections to both the pair propagator and the repulsive interaction in the weak disorder regime. For a two dimensional superconductor with $(\varepsilon_0\tau)^{-1} \ll 1$, the critical temperature has been obtained as

$$\ln\frac{T_c}{T_{c0}} = -C\rho\left(\ln\frac{1}{T_c\varepsilon_0}\right)^2 \tag{20.7}$$

and in three dimensional superconductor for $(\omega_D\tau)^{-1} \ll 1$

$$\ln\frac{T_c}{T_{c0}} = -C\rho^2 g^* \tag{20.8}$$

where C is a constant and g^* is given by

$$g^* = g\frac{1 + \mu\ln(\varepsilon_0/\omega_D)}{1 + \mu\ln\tau\varepsilon_0} \tag{20.9}$$

where μ is the Coulomb pseudopotential and g the BCS interaction. The strong disorder regime has been analysed first by Anderson *et al.*[44] and they showed that the strong scale dependent diffusion enhances the Coulomb repulsion and T_c decreases. Leavens[45] used the same Coulomb kernel

$$K_c(\omega, \omega') = \mu C(\omega - \omega')\theta(\varepsilon_0 - |\omega'|)\theta(\varepsilon_0 - |\omega|) \tag{20.10}$$

to study the importance of the frequency dependence $Z(\omega)$ in the Eliashberg equation. Taking

$$C(\omega) = \begin{cases} 1 + \pi(3/2p_0 l)^2 \ln \alpha\,; & \omega < (\tau\alpha)^{-1} \\ 1 - \pi(3/2p_0 l)^2 \ln |\omega|\alpha\,; & (\alpha\tau)^{-1} < \omega < \tau^{-1} \\ 1\,; & (\tau\omega)^{-1} < 1 \end{cases} \tag{20.11}$$

Crisan[46] obtained an analytical result and showed that the decreasing of the critical temperature can be obtained from the Eliashberg equations and is proportional to ρ^2 in the weak coupling approximation.

In the weakly localized regime, the strong spin fluctuations enhance the critical temperature. Ebisawa[47] calculated this effect for a three dimensional superconductor and obtained

$$\ln \frac{T_c}{T_{c0}} = -\frac{C}{\eta}\lambda^2 \tag{20.12}$$

where

$$\eta = 1 - N(0)\langle V^c(\mathbf{k} - \mathbf{k}'; 0)\rangle_{S(\varepsilon_0)}$$

and the constant C is given by

$$\begin{matrix} \dfrac{g\sqrt{3}\,\pi^3}{4} V^c\,; & \eta < \tau T_c < 1 \\ \dfrac{g\sqrt{3}\,\pi}{3} V^c \ln^2 \dfrac{1}{\pi\tau T_c}\,; & \tau T_c < \eta < 1\,. \end{matrix} \tag{20.13}$$

For a two-dimensional superconductor, a similar formula has been obtained but the decreasing of the temperature is proportional to λ ($\lambda \propto \rho$).

A correct description has to consider all Hartree-Fock contributions to the self-energy and to get a set of Eliashberg equations which will describe all effects. Using the exact eigenstates formalism, Belitz[48] obtained a set of equations which are similar to the Eliashberg equations containing the electron-electron correlation functions. We start with the Gor'kov equations written in the exact eigenstate basis as

$$\begin{aligned} &[i\omega - E_n - S_n(\omega)]G_n(\omega) + W_n(\omega)F_n^+(\omega) = 1 \\ &[i\omega - E_n - S_n(-\omega)]F_n^+(\omega) - W_n(\omega)G_n(\omega) = 0 \end{aligned} \tag{20.14}$$

where S_n and W_n are corrections to the phonon and Coulomb self-energy taking into account the Hartree contribution. The solutions of these equations are

$$G(\varepsilon,\omega)=\frac{i\omega Z(\omega)+[\varepsilon+Y(\varepsilon,\omega)]}{[i\omega Z(\varepsilon,\omega)]^2-[\varepsilon+Y(\varepsilon,\omega)]^2-W^2(\varepsilon,\omega)} \tag{20.15}$$

$$F^{+}(\varepsilon,\omega)=\frac{-W(\varepsilon,\omega)}{[i\omega Z(\varepsilon,\omega)]^2-[\varepsilon+Y(\varepsilon,\omega)]^2-W^2(\varepsilon,\omega)} \tag{20.16}$$

where we performed the average using

$$f(\varepsilon)=\frac{1}{N(0)}\langle\sum_n \delta(\varepsilon-E_n)\rangle\ .$$

With these results, the self-energy of the system can be calculated. Taking into consideration the phonon contribution and the Coulomb contribution, the Eliashberg equations have been obtained as

$$\begin{aligned}W(\varepsilon,\omega)&=\int d\varepsilon'\int d\nu\alpha^2(\nu)F^{F}(\varepsilon-\varepsilon',\nu)\\&\times\left\{n(\nu)[F(\varepsilon',\omega+\nu)+F(\varepsilon',\omega+\nu)]\right.\\&\left.+\int\frac{dx}{\pi}\mathrm{Im}\,F(\varepsilon',x)\left[-\frac{f(x)}{x-\omega-\nu}+\frac{1-f(x)}{x-\omega-\nu}\right]\right\}\\&+\int d\varepsilon' V^{c,W}(\varepsilon-\varepsilon')\int\frac{dx}{\pi}f(x)\mathrm{Im}\,F(\varepsilon',x)\end{aligned} \tag{20.16a}$$

$$\begin{aligned}\times\omega[1-Z(\varepsilon,\omega)]&=\int d\varepsilon'\int d\nu\alpha^2(\nu)F^{F}(\varepsilon'-\varepsilon,\nu)\\&\times\{n(\nu')[G_{-}(\varepsilon',\omega+\nu)+G_{+}(\varepsilon',\omega-\varepsilon)]\}\\&+\int\frac{dx}{\pi}\mathrm{Im}\,G(\varepsilon',x)\left[\frac{f(x)}{(x-\nu)^2-\omega^2}+\frac{1-f(x)}{(x+\nu)^2-\omega^2}\right]\end{aligned} \tag{20.16b}$$

and an equation for $Y(\varepsilon,\omega)$ which is

$$Y(\varepsilon,\omega) = \int d\varepsilon' \int d\nu \alpha^2(\nu) F^{\mathrm{F}}(\varepsilon-\varepsilon',\nu)\{n(\nu)[G_+(\varepsilon',\omega+\nu') + G_-(\varepsilon',\omega-\nu)]\} + \int \frac{dx}{\pi} \mathrm{Im}\, G(\varepsilon',x)\Big[f(x)\frac{x-\nu}{(x-\nu)^2+\omega^2} + (1-f(x))\frac{x+\nu}{(x+\nu)^2-\omega^2}\Big] + \int d\varepsilon' \int d\nu \Big[\frac{4}{\nu}\alpha^2(\nu)F^{\mathrm{H}}(\varepsilon-\varepsilon',\nu) + V^{\mathrm{c,Y}}(\varepsilon-\varepsilon')\delta(\nu)\Big] \int \frac{dx}{\pi} f(x) \mathrm{Im}\, G(\varepsilon',x)\,. \tag{20.16c}$$

These equations have been obtained taking the self-energy

$$S(\varepsilon,i\omega) = i\omega[1-Z(\omega)] + Y(\varepsilon,i\omega)$$

and $G_\pm = \frac{1}{2}[G(\varepsilon,i\omega+i0) \pm G(\varepsilon,-i\omega-i0)]$.
The phonon contributions in (20.16a)-(20.16c) have the forms

$$\alpha^2(\nu)F^{\mathrm{F}}(\varepsilon,\nu) = \sum_{i,\mathrm{q}} R_i^{\mathrm{F}}(\mathrm{q},\varepsilon) B_i(\mathrm{q},\nu) \tag{20.17}$$

$$\alpha^2(\nu)F^{\mathrm{H}}(\varepsilon,\nu) = \sum_{i,\mathrm{q}} R_i^{\mathrm{H}}(\mathrm{q},\varepsilon) B(\mathrm{q},\nu) \tag{20.18}$$

$B(\mathrm{q},\varepsilon)$ being the spectral function of the phononic Green function. The term (20.18) is due to the Hartree contribution. The Coulomb kernels are defined as

$$V^{\mathrm{c,W}}(\varepsilon) = \sum_{\mathrm{q}} R_{\mathrm{c}}^{\mathrm{F}}(\mathrm{q},\varepsilon) V^{\mathrm{c}}(\mathrm{q}) \tag{20.19a}$$

$$V^{\mathrm{c,V}}(\varepsilon) = \sum_{\mathrm{q}} (R_{\mathrm{c}}^{\mathrm{F}}(\mathrm{q},\varepsilon) - 2R_{\mathrm{c}}^{\mathrm{H}}(\mathrm{q},\varepsilon) V^{\mathrm{c}}(\mathrm{q}))\,. \tag{20.19b}$$

In the relations (20.17)–(20.19), the functions $R(\mathrm{q},\varepsilon)$ are the vertex corrections calculated using the Kubo formalism.

Now we shall show that Eqs. (20.16) give the results obtained in the previous theories. Let us consider the weak disorder limit. In this limit, the scale for frequency variation of R and U is given by the Fermi energy. In this case, if we consider Y as frequency independent, the first two contributions from (20.16c) disappear and the last term gives a constant which renormalizes the chemical potential. In the same approximation, (20.19a) gives an energy independent Coulomb potential and the phonon contribution is independent of energy. In this case, one obtains the standard strong coupling-equations. In the "jelium model", the vertex corrections can be calculated, and only the longitudinal phonons (in the limit $q \to 0$) give the contribution:

$$\alpha^2(\nu)F^{\rm F}(\varepsilon,\nu) = \alpha^2(\nu)F(\nu) = g_{\rm L}^2 N(0) \int_0^{2p_0} \frac{q\,dq}{2p_0^2} B_{\rm L}(q,\nu) \tag{20.20}$$

where $g_{\rm L}$ is the electron-phonon coupling constant. Neglecting the Coulomb interaction, the correlations functions given by the stress operators of the longitudinal and transverse phonons gives the Keck and Schmid result. The Anderson *et al.*[44] result can be re-obtained taking

$$V^{\rm c,W} = \mu + \frac{1}{\pi N(0)} \sum_{\bf q} \frac{Dq^2 g(q)}{\varepsilon^2 + D^2 q^4} \tag{20.21}$$

where D is the diffusion constant and $g(q)$ is the q-dependent compressibility. Anderson used a scale dependent diffusion constant (near the Anderson transition) and obtained a strong decrease of $T_{\rm c}$.

Before calculating the critical temperature $T_{\rm c}$ from Eqs. (20.16), we mention that the new aspects of this theory are: the energy dependence of the self-energies and the occurrence of the additional renormalization function Y.

The first feature appears because with the increasing of the disorder, the Kubo functions which enter the different vertices develop frequency depend on a scale smaller than the Fermi energy, and in this way the kernels become energy dependent.

The second characteristic is the existence of Y which is in fact a part of the normal self-energy which is an even function of frequency. Even for strong disorder, the main contribution to Y is frequency independent. In fact, in BCS such a contribution exists but tends to suppress the superconductivity. The function Y has been interpreted[49] as Altshuler-Aronov[40] corrections to the Green functions. In fact we see that there is a phonon contribution of Hartree type to this one-particle renormalization which has been obtained using the BCS model by Fukuyama *et al.*,[41] but it was neglected. The Y-contribution is important for systems with a strong electron-phonon coupling, the high-$T_{\rm c}$ superconductors are good candidates and the strong-coupling superconductors should be analysed taking into consideration this contribution.

The critical temperature T_c can be calculated[50] making the standard simplifications in (20.16) concerning the energy independence of W and Z. However, we have the energy dependence in Y which will be developed using the basic approximation for $\delta Y = Y - Y^0$ as

$$\delta Y(\varepsilon, \omega) = \delta Y(0, \omega) - \varepsilon Y'(\omega) \tag{20.22}$$

where Y^0 is the contribution for zero disorder (in the clean limit of a jelium model), $Y(0,\omega)$ is zero in our approximation, and $Y'(\omega) = d\delta Y(\varepsilon,\omega)/d\varepsilon\big|_{\varepsilon=0}$.

Using (20.22), the Green functions (20.15)–(20.16) become

$$G(\varepsilon, \omega) = \frac{1}{1+Y'} \frac{\omega \bar{Z}(\omega)\,\mathrm{sign}\,\omega}{[\omega^2 \bar{Z}(\omega) - \varepsilon^2 - \bar{W}^2(\omega)]^{\frac{1}{2}}} \tag{20.23}$$

$$F(\varepsilon, \omega) = \frac{1}{1+Y'} \frac{\bar{W}(\omega)\,\mathrm{sign}\,\omega}{[\omega^2 \bar{Z}^2(\omega) - \varepsilon^2 - \bar{W}^2(\omega)]^{\frac{1}{2}}} \tag{20.24}$$

where $\bar{Z}(\omega) = Z(\omega)/(1+Y')$ and $\bar{W}(\omega) = W(\omega)/(1+Y')$. In Eqs. (20.16), the integrals over the energy can be performed and if we define $-i\pi g(\omega) = \int d\varepsilon G(\varepsilon,\omega)$ and $i\pi f(\omega) = \int d\varepsilon F(\varepsilon,\omega)$, Eqs. (20.23)–(20.24) become

$$g(\omega) = \frac{1}{1+Y'} \frac{\omega \bar{Z}(\omega)\,\mathrm{sign}\,\omega}{[\omega^2 \bar{Z}^2(\omega) - \bar{W}^2(\omega)]^{\frac{1}{2}}} \tag{20.25}$$

$$f(\omega) = \frac{1}{1+Y'} \frac{\bar{W}(\omega)\,\mathrm{sign}\,\omega}{[\omega^2 \bar{Z}^2(\omega) - \bar{W}^2(\omega)]^{\frac{1}{2}}} \tag{20.26}$$

and because $\mathrm{Re}\, g(\omega)$ is an even function on frequency it is easy to show that $Y(0,\omega) = 0$. With these results, Eqs. (20.16) have been transformed as

$$W(\omega) = -V^{\mathrm{c,W}} \int_0^\infty dx \tanh\frac{x}{2T} \mathrm{Re}\, f(x) + \int d\nu \alpha^2(\nu) F^{\mathrm{H}}(\nu) \int dx \mathrm{Re}\, f(x) \left[\frac{f(x)}{x-\omega-\nu} + \frac{1-f(x)}{x-\omega+\nu}\right] \tag{20.27}$$

$$[1 - Z(\omega)]\omega = -\int d\nu \alpha^2(\nu) F^{\mathrm{H}}(\nu) \int dx \mathrm{Re}\, g(x) \left[\frac{f(x)}{x-\omega-\nu} + \frac{1-f(x)}{x-\omega+\nu}\right]. \tag{20.28}$$

The new function Y' has been determined as

$$Y' = \int d\nu \Big[\frac{4}{\nu}\alpha^2(\nu)F^{\mathrm{H}}(\nu) + \delta V^{\mathrm{c,Y}}\delta(\nu)\Big] \int \frac{dx}{4\pi} f(x) \times \lim_{\varepsilon\to 0}\Big[\mathrm{Im}\, G(\varepsilon, x) - \mathrm{Im}\, G(-\varepsilon, x)\Big] \qquad (20.29)$$

where $\delta V^{\mathrm{c,Y}}$ denotes the quantity (20.19b) with the respective clean-limit contribution substractions.

With these results, we will calculate T_c defining the gap $\Delta(\omega) = W(\omega)/Z(\omega) = \bar{W}(\omega)/\bar{Z}(\omega)$ and taking $W \to 0$ for $T \to T_c$. After some algebra, (20.29) becomes

$$Y' = \delta V^{\mathrm{c,Y}} 4 \int \frac{d\nu}{\nu}\alpha^2(\nu)F^{\mathrm{H}}(\nu) \qquad (20.30)$$

and the critical temperature T_c has been obtained as

$$T_c = \frac{\omega_{\mathrm{D}}}{1.45}\exp\Big[-\frac{1.04(1+\tilde{\lambda}+Y')}{\lambda - \mu^*(1+0.62\tilde{\lambda}/(1+Y'))}\Big] \qquad (20.31)$$

where

$$\tilde{\lambda} = 2\int \frac{d\nu}{\nu}\alpha^2(\nu)F^{\mathrm{F}}(\nu) \qquad (20.32)$$

and

$$\mu^* = V^{\mathrm{c,W}}\Big[1 + \frac{V^{\mathrm{c,W}}}{1+Y'}\ln\frac{E_{\mathrm{B}}}{\omega_0}\Big] \qquad (20.33)$$

where E_{B} is of the order of the bandwidth and ω_0 is of the order of the Debye frequency.

Equation (20.31) has been approximated using the relation

$$V^{\mathrm{c,W}} = \mu\Big[1 + (1+\varsigma)\frac{6}{\pi}\rho/\rho_{\mathrm{M}}\Big] \qquad (20.34)$$

where ρ is the resistivity which is a measure of the disorder and $\rho_{\mathrm{M}} = 3\pi^2/p_0$ is the Mott resistivity, and ς has been calculated as function of the Fermi wave vector k_0 and the screening parameter χ as

$$\varsigma = \frac{2p_0}{\chi}\tan^{-1}\Big(\frac{2p_0}{\chi}\Big)\ln\Big[1 - \Big(\frac{2p_0}{\chi}\Big)^2\Big].$$

Using now

$$2\int \frac{d\nu}{\nu}\delta\alpha^2(\nu)F^{\mathrm{F}}(\nu) = \lambda\frac{6}{\pi}\frac{\rho}{\rho_{\mathrm{M}}} \qquad (20.35)$$

$$2\int \frac{d\nu}{\nu}\delta\alpha^2(\nu)F^{\mathrm{H}}(\nu) = 2\lambda\frac{6}{\pi}\frac{\rho}{\rho_{\mathrm{M}}} \qquad (20.36)$$

we obtain from (20.31) T_c as

$$T_c = \frac{\omega_D}{1.45} \exp\left[-\frac{1.04[1+\lambda f_1(\hat{\rho}) + f_2(\hat{\rho})]}{\lambda f_1(\hat{\rho}) - \mu^*(\rho)[1+1.062 f_1(\hat{\rho})/(1+f_2(\hat{\rho}))]}\right] \tag{20.37}$$

where $\hat{\rho} = \rho/\rho_M$ and

$$\mu^*(\rho) = \mu f_2(\rho)\left[1 + \frac{\mu f_3(\rho)}{1+f_2(\rho)} \ln \frac{E_B}{\omega_0}\right] \tag{20.38}$$

with

$$\begin{aligned} f_1(x) &= 1 + \frac{6}{\pi}x \\ f_2(x) &= [\mu(\varsigma - 3) + 4\lambda]\frac{6}{\pi}x \\ f_3(x) &= [1 + (1+\varsigma)]\frac{6}{\pi}x\,. \end{aligned} \tag{20.39}$$

If we neglect the renormalization of λ due to disorder, taking $f_1 = 1$ and $f_3 = 1$ we get

$$T_c = \frac{\omega_D}{1.45} \exp\left[\frac{-1}{(\lambda - \mu^*)(1-f_2)}\right] \tag{20.40}$$

a result obtained in Ref. 49.
The renormalization $\mu = \mu f_3$ has been obtained by Anderson *et al.*[44] and is given by the anomalous diffusion near the Anderson transition.

If we do not neglect f_1, Eq. (20.37) will contain the renormalized Cooper-propagator and the diffusion pole is stronger than the stress renormalization.

An important experimental result which has to be explained is the variation from positive to negative values of the coefficient $\alpha = \frac{1}{\rho}\left(\frac{d\rho}{dT}\right)$. The same behaviour has been obtained for $dT_c/d\rho$ as function of the critical temperature T_c^0 of a clean superconductor. In the framework of the above theory, this behaviour can be obtained for a weak disorder regime in a jelium model. The Coulomb contribution U_c has been recalculated[51] as

$$V^{c,W} = \mu + \delta\mu \tag{20.41}$$

where the pseudopotential μ is

$$\mu = \frac{1}{2}y^2 \ln\left(1 + \frac{1}{y^2}\right)$$

and the correction $\delta\mu$ is

$$\delta\mu = \frac{1}{2}\widehat{\rho}y\Big\{\pi\tan^{-1}\frac{1}{y} - \tan^{-1}\frac{1}{y} + \frac{1}{1+y^2} - 2\Big\} \tag{20.42}$$

with $y = x/2\rho_0$.

The Coulomb contribution in Y' has been obtained as

$$Y_c' = \widehat{\rho}y\Big\{\Big(\frac{\pi}{2} - 2\pi^2\Big)\tan^{-1}\frac{1}{y} + \Big(4\pi - \frac{1}{2}\Big)\tan\frac{1}{y} + \frac{1/2}{1+y} - 1\Big\}\,. \tag{20.43}$$

The phonon contribution has been evaluated by the calculation of the electron-phonon vertex and the electron-phonon contribution in Y'.

Using a simple Debye spectrum for phonons, the parameter $\widetilde{\lambda} = \lambda + \delta\lambda$ and Y' have been calculated as

$$\widetilde{\lambda} \equiv \lambda + \delta\lambda = \lambda + \widehat{\rho}\lambda\Big[n\Big(\frac{C_L}{C_T}\Big)^2 - 8 + \frac{\pi^2}{2}\Big]\Big/x_D\pi \tag{20.44}$$

and

$$Y' = \widehat{\rho}\bigg\{y\Big[\frac{3}{2}\Big(\tan^{-1}\frac{1}{y}\Big)^2 - \frac{\pi}{2}\tan^{-1}\frac{1}{y} + \frac{1/2}{1+y^2} - 1\Big] - \frac{1024}{g}x_D\lambda\bigg\} \tag{20.45}$$

where λ is the electron-phonon coupling for a clean system and $x_D = q_D/2p_0$, q_D being the wave vector associated with the Debye energy ω_D, C_L, C_T being the sound velocities. With these results, the critical temperature T_c is

$$\begin{aligned}\frac{T_c - T_c^0}{T_c^0} = -\frac{1.04(1+\lambda)}{[\lambda - \mu^*(1+0.62\lambda)]^2}\bigg\{&\frac{\delta\lambda + Y'}{1+\lambda}[\lambda - \mu^*(1+0.62\lambda)] \\ &+ \mu^*(1+0.62\mu)\Big[\frac{\delta\mu^*}{\mu^*} + \frac{0.62}{1+0.62\lambda}\Big(\frac{\delta\lambda}{\lambda} - Y'\Big) - \delta\lambda\Big]\bigg\}\end{aligned} \tag{20.46}$$

where

$$\mu^* = \mu\Big/\Big[1 + \mu\ln\frac{E_B}{\omega_D}\Big]$$

and the correction $\delta\lambda$ is given by (20.43). The correction $\delta\mu^*/\mu^*$ has been calculated as

$$\frac{\delta\mu^*}{\mu^*} = \Big[\frac{\delta\mu}{\mu} + \mu Y'\ln\frac{E_B}{\omega_D}\Big]\Big/\Big[1 + \mu\ln\frac{E_B}{\omega_D}\Big]\,. \tag{20.47}$$

If we neglect the Coulomb repulsion, the derivative $dT_c/d\rho$ behaves like

$$\frac{1}{\rho}\frac{T_c - T_c^0}{T_c^0} = \frac{1.04}{\lambda}(a - b\lambda) \tag{20.48}$$

where

$$a = \delta\lambda/\widehat{\rho}\lambda\,, \quad b = Y'(y=0)/\widehat{\rho} = 1024x_D/g\ .$$

For small λ, the variation $(T_c - T_c^0)/(T_c^0\rho) > 0$ is positive and proportional to $1/\lambda$ and for large λ this function is negative and saturates at $-1.04b$. Since the electron-phonon coupling constant is a measure of T_c, we expect a positive correction with a strong $-T_c$ dependence at small T_c, and a negative correction with a weak $-T_c$ dependence at large T_c. This result is quite general and in fact it occurs from the competition between attractive and repulsive electron-electron interaction.

We may conclude this paragraph with the general result that for a dirty superconductor, the BCS and the strong-coupling approximations give essentially different results. Then, if one of these mechanisms will be adopted for the high temperature superconductivity, the study of the disorder due to the impurities will be important because in these materials the electron-phonon coupling constant seems to be very important and strong.

Appendix

We start by considering the motion of a single electron in an array of potentials $U(\mathbf{r} - \mathbf{R}_i)$ located at positions $\mathbf{R}_i$ which are randomly placed. These potentials are due to magnetic or non-magnetic impurities and in the following, we will present the procedure of performing the "average on the impurities".

In order to calculate the physical quantities from the theory of the dilute alloys, it is convenient to use the second-quantization in which we can treat the many-electron states. If the electrons of energy $\varepsilon(\mathbf{p})$ interact with the impurities, the Hamiltonian of the system is

$$\mathcal{H} = \sum_{\mathbf{p}} \varepsilon(\mathbf{p}) c_{\mathbf{p}}^\dagger c_{\mathbf{p}} + \sum_{\mathbf{q}} U(\mathbf{q})\rho(\mathbf{q}) \sum_{\mathbf{p}} c_{\mathbf{p}+\mathbf{q}}^\dagger c_{\mathbf{p}} \tag{A.1}$$

and in fact the difference between (A.1) and the single-impurity Hamiltonian appears in the factor

$$\rho_{\mathbf{q}} = \sum_j \exp(-i\,\mathbf{q}\cdot\mathbf{R}_j) \tag{A.2}$$

which is the Fourier transform of the density function $\sum_i \delta(\mathbf{r}-\mathbf{R}_i)$ for the scattering centers, and $U(\mathbf{q})$ is the Fourier transform of the potential $U(\mathbf{r})$.

We calculate for the beginning, the Green function for the electrons using the equation of motion method. In this way, the off-diagonal Green function is

$$\left[i\frac{\partial}{\partial t}-\varepsilon(\mathbf{p})\right]F(\mathbf{p},\mathbf{p}';t)=\delta_{\mathbf{p},\mathbf{p}'}\delta(t)+\sum_{\mathbf{q}}U(\mathbf{q})\rho(\mathbf{q})F(\mathbf{p}+\mathbf{q},\mathbf{p};t) \quad (A.3)$$

where we see that in the right-hand side of Eq. (A.3) a term which contains the randomly varying term $\rho_{\mathbf{q}}$ appears. The solution of (A.3) can be obtained taking the Fourier transform of (A.3) and iterating to produce a series solution to the problem:

$$\begin{aligned}F(\mathbf{p},\mathbf{p}')&=G^0(\mathbf{p})\delta_{\mathbf{p},\mathbf{p}'}+G^0(\mathbf{p})U(\mathbf{p}-\mathbf{p}')\rho(\mathbf{p}-\mathbf{p}')G^0(\mathbf{p}')\\&\quad+\sum_{\mathbf{q}}G^0(\mathbf{p})U(\mathbf{q})\rho(\mathbf{q})G^0(\mathbf{p}+\mathbf{q})U(\mathbf{p}-\mathbf{q})\rho(\mathbf{p}-\mathbf{q}-\mathbf{p}')G^0(\mathbf{p}')\,.\end{aligned} \quad (A.4)$$

In the special case of the one-electron Green funcion $G(\mathbf{p})=F(\mathbf{p},\mathbf{p})$, (A.4) becomes

$$\begin{aligned}G(\mathbf{p})&=G^0(\mathbf{p})+G^0(\mathbf{p})[U(\mathbf{q})\rho(\mathbf{q})]_{\rho=0}G^0(\mathbf{p})\\&\quad+\sum_{\mathbf{q}}G^0(\mathbf{p})U(\mathbf{q})\rho(\mathbf{q})G^0(\mathbf{p}+\mathbf{q})U(-\mathbf{q})\rho(-\mathbf{q})G^0(\mathbf{p})\end{aligned} \quad (A.5)$$

and a general term of such series will be represented by the diagram

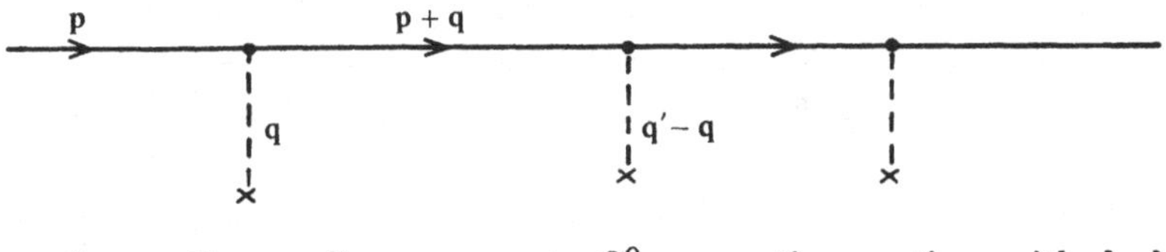

where the continuous line represents G^0 connecting vertices with dashed lines which represent $U(\mathbf{q})\rho(\mathbf{q})$.

If in our system the finite number of N impurities are in the positions $R_1, R_2, \ldots, R_N$, then the Green function is a functional of this set of position vectors.

$$G=G[\mathbf{R}_1,\mathbf{R}_2,\ldots,\mathbf{R}_N] \quad (A.6)$$

and the ensemble average is defined as

$$\bar{G}=\prod_{i=1}^{N}\frac{1}{V}\int d^3R_i G[R_1,\ldots,R_N] \quad (A.7)$$

where V is the volume of the system. In this ensemble we assume that each impurity placed in a site $\mathbf{R}_i$ is uncorrelated with each other. This averaging procedure may now be applied, term by term to Eq. (A.5). We consider the term of the general form $\overline{\rho(q_1)\rho(q_2)\ldots\rho(q_n)}$. Using (A.7), one finds

$$\bar{\rho}(\mathbf{q}) = \frac{N}{V}\int d^3\mathbf{R}\exp(-i\,\mathbf{q}\cdot\mathbf{R}) = N\delta_{\mathbf{q},0}\ . \tag{A.8}$$

The second-order term

$$\rho(\mathbf{q}_1)\rho(\mathbf{q}_2) = \sum_{i,j}\overline{e^{-i\,\mathbf{q}_1\cdot\mathbf{R}_1}e^{-i\,\mathbf{q}_2\cdot\mathbf{R}_2}} \tag{A.9}$$

can be transformed as

$$\overline{\rho(\mathbf{q}_1)\rho(\mathbf{q}_2)} = N^2\delta_{\mathbf{q}_1,0}\delta_{\mathbf{q}_2,0} + N\delta_{\mathbf{q}_1+\mathbf{q}_2,0}\ . \tag{A.10}$$

For the calculation of $G(\mathbf{p})$ we need (A.10) written as

$$\overline{\rho(\mathbf{q})\rho(-\mathbf{q})} = N^{-2}\delta_{\mathbf{q},0} + N \tag{A.11}$$

and we have to mention that it is very important for the occurrence in this average of a second order Born scattering term given by the single atom. This term which is proportional to N occurs without restrictions on the momentum change q in the intermediate state. With these results, we can consider Eq. (A.5). The first-order term (in the interaction U) represented by the diagram

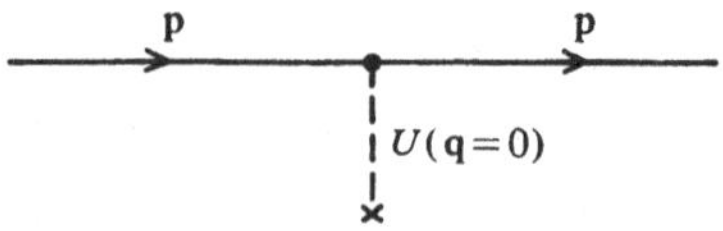

which gives the contribution

$$G^0(\mathbf{p})NU(\mathbf{q}=0)G^0(\mathbf{p}) \tag{A.12}$$

in the expression for $G(\mathbf{p})$.

The second-order term may be represented by the diagrams

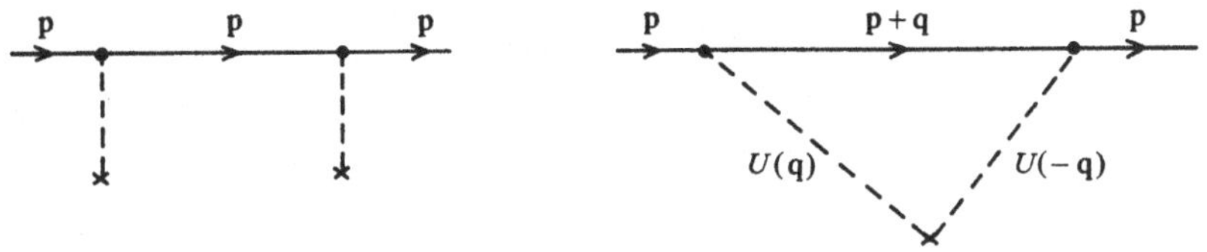

which gives the contribution

$$G^0(\mathbf{p})NU(\mathbf{q}=0)G^0(\mathbf{p})NG(\mathbf{p})U(\mathbf{q}=0)$$
$$+G^0(\mathbf{p})\sum_{\mathbf{q}} NG^0(\mathbf{p}+\mathbf{q})U(\mathbf{q})U(-\mathbf{q})G^0(\mathbf{p})\ . \tag{A.13}$$

In these diagrams, the single dashed lines each ending in a cross may be considered as independent scattering (in the lowest Born approximation) from two different impurities, while the pair of dashed lines ending in a single cross corresponds to a second Born approximation scattering from the same impurity.

The general term in the series is obtained by including all possible repeated scatterings on a single impurity. If we consider only the diagrams which cannot be divided into two subdiagrams connected by a line representing $G^0(\mathbf{p})$, which will be called irreducible diagrams, we can carry out the summation in Eq. (A.5). All the irreducible diagrams will be included in the self-energy

$$\hat{\Sigma}(\mathbf{p};\varepsilon)=\sum_{i=0}^{\infty}\hat{\Sigma}^i(\mathbf{p};\varepsilon) \tag{A.14}$$

and the sum in (A.5) will be performed collecting the terms of the form $G^0(\mathbf{p})\Sigma^i G^0(\mathbf{p})$. In this way, we get

$$G^{-1}(\mathbf{p},\varepsilon)=G^{0^{-1}}(\mathbf{p},\varepsilon)-\hat{\Sigma}(\mathbf{p};\varepsilon)\ . \tag{A.15}$$

The approximation performed by considering only the irreducible diagrams is in fact equivalent to the classification of the diagrams according to their dependence on the impurities concentration $c=N/V$.

The first order diagram $\Sigma^{(1)}$ is

$$\Sigma^{(1)}=c\int dxU(x)$$

and leads to a shift in the energy of the electron.

The second order correction is

$$\Sigma^{(2)} = N \sum_{\mathbf{p}'} |U(\mathbf{p}-\mathbf{p}')|^2 G^0(\mathbf{p}', \varepsilon) \tag{A.16}$$

and it has been applied in the problems from this chapter.

References

1. P. W. Anderson, *I. Phys. Chem. Solids* **11** (1959) 26.
2. A. A. Abrikosov and L. P. Gor'kov, *Zh. Eksp. Teor. Fiz.* **39** (1962) 1781 (*Soviet Phys. JETP* **12** (1961) 1243).
3. S. Skalski, O. Betbeder-Matibet and P. R. Weiss, *Phys. Rev.* **136A** (1964) 1500.
4. V. Ambegaokar and A. Griffin, *Phys. Rev.* **137A** (1965) 1151.
5. H. Shiba, *Prog. Theor. Phys.* **40** (1968) 435.
6. L. P. Gor'kov and A. I. Rusinov, *Zh. Eksp. Teor. Fiz.* **46** (1964) 1363 (*Soviet Phys. JETP* **19** (1964) 922).
7. J. Keller and K. Benda, *J. Low Temp. Phys.* **2** (1970) 241.
8. W. A. Smith and G. Vertoghen, *Phys. Lett.* **43A** (1973) 455.
9. M. Crisan and H. Jones, *J. Low Temp. Phys.* **18** (1975) 297.
10. P. G. de Gennes and G. Sarma, *Solid State Commun.* **4** (1966) 449.
11. A. M. Toxen, P. Kwok and R. J. Gambino, *Phys. Rev. Lett.* **21** (1968) 792.
12. D. Rainer, *Z. Physik* **252** (1972) 174.
13. P. Entel and W. Klose, *J. Low. Temp. Phys.* **17** (1974) 529.
14. J. Keller in: *Proc. Int'l Conf. Ternary Superconductors*, eds. G. K. Shenoy, B. D. Dunlop and F. Y. Fraden (North-Holland, Amsterdam, 1981).
15. P. Fulde and J. Keller in *Superconductivity in Ternary Compounds II*, p. 249, eds. M. B. Maple and O. Fischer (Springer Verlag, Berlin, Heidelberg, New York, 1982).
16. C. M. Soukonlis and G. S. Grest, *Phys. Rev.* **B21** (1980) 5119.
17. M. Crisan, M. Gulacsi and Zs. Gulacsi, *Proc. of LT-17* p. 106, Karlsruhe (1984) eds. U. Ekern, A. Schmid, W. Weber and H. Wühlal *Physica* B + C.
18. A. A. Abrikosov, *Adv. Phys.* **29** (1980) 869.
19. K. H. Bennemann, *Phys. Rev. Lett.* **17** (1966) 438.
20. P. W. Anderson, *Phys. Rev.* **124** (1961) 41.
21. K. Machida and F. Shibata, *Prog. Theor. Phys.* **47** (1972) 1817.
22. C. F. Rato and A. Blandin, *Phys. Rev.* **156** (1967) 513.
23. K. Takanaka and F. Takano, *Prog. Theor. Phys.* **36** (1966) 1086.
24. M. J. Zuckermann, *Phys. Rev.* **140A** (1965) 889.
25. J. R. Schrieffer and D. C. Mattis, *Phys.* **140A** (1965) 1412.
26. Y. Okabe and A. D. Nagi, *Phys. Rev.* **B28** (1983) 2455.
27. A. B. Keiser, *J. Phys.* **C3** (1970) 410.

28. K. G. Wilson, *Rev. Mod. Phys.* **47** (1975) 773; Krishna-murthy, K. G. Wilson and J. W. Wilkins, *Phys. Rev.* **B21** (1980) 1003.
29. E. Müller-Hartman and J. Zittartz, *Phys. Rev. Lett.* **26** (1971) 428. *Z. Physik* **232** (1970) 11.
30. Y. Nagaoka, *Phys. Rev.* **A138** (1965) 1112.
31. D. R. Hamann, *Phys. Rev.* **158** (1967) 570.
32. A. Ludwig and M. J. Zuckermann, *J. Phys.* **F1** (1971) 516.
33. T. Matsuura, *Prog. Theor. Phys.* **57** (1977) 1823.
34. A. Sakurai, *Phys. Rev.* **B17** (1978) 1195.
35. K. Yoshida and K. Yamada, *Prog. Theor. Phys.* **53** (1975) 1286.
36. T. Matsuura, S. Ichinose and Y. Nagaoka, *Prog. Theor. Phys.* **57** (1977) 713.
37. P. W. Anderson, *Phys. Rev.* **109** (1958) 1492.
38. E. Abrahams, P. W. Anderson, D. C. Licciarello and T. V. Ramakrishnan, *Phys. Rev. Lett.* **42** (1979) 673.
39. F. Wegner, *Phys. Reports* **67** (1980) 15.
40. B. L. Altshuler and A. G. Aronov, *Solid State Commun.* **30** (1979) 115.
41. H. Fukuyama, *Prog. Theor. Phys. Supp.* **34** (1985) 47; H. Fukuyama, H. Ebysawa and S. Maekawa, *J. Phys. Soc. Jpn.* **53** (1984) 1919.
42. S. Maekawa and H. Fukuyama, *J. Phys. Soc. Jpn.* **51** (1982) 1380.
43. P. W. Anderson, K. A. Muttalib and T. V. Ramakrishnan, *Phys. Rev.* **828** (1983) 117.
44. C. R. Leavens, *Phys. Rev.* **B31** (1985) 6072.
45. M. Crisan, *Phys. Lett.* **A124** (1987) 195.
46. H. Ebisawa, H. Fukuyama and S. Maekawa, *J. Phys. Soc. Jpn.* **54** (1985) 4735.
47. D. Belitz, *Phys. Rev.* **35** (1987) 1636.
48. D. Belitz, *J. Phys.* **F15** (1985) 2315.
49. D. Belitz, *Phys. Rev.* **35** (1987) 1651.
50. D. Belitz, *Phys. Rev.* **36** (1987) 47.

IV
SUPERCONDUCTORS IN A MAGNETIC FIELD

In the previous chapters we have presented the phenomenological as well as the microscopic theory of the superconducting state neglecting the influence of the magnetic field, on the electrons spins. Various properties of the superconductors can be explained by calculating different parameters from the microscopic theory. This procedure is justified as long as the magnetic field of interest is small as in the case of linear response approximation. In the framework of this approximation, the Meissner effect has been explained for the type I superconductors.

However, for type II superconductors with χ large, the upper critical field H_{c2} can be very high and the contribution of the electron susceptibility (Pauli paramagnetism) cannot be neglected in the thermodynamic balance.

The transition curve is also influenced by the magnetic field, which gives rise to a triple point in the T-H plane, if the order parameter is constant. If one considers that the order parameter is spatially dependent as $\Delta(\mathbf{r}) = \Delta(0)\exp(i\mathbf{q}\cdot\mathbf{r})$, the state with $q_0 \neq 0$ corresponds to a "depaired" superconducting state in which the Cooper pairs have a single non-vanishing center of mass momentum. This order parameter gives a permanent current which is compensated by an opposite current due to unpaired electrons. Fulde and Ferrell studied the depaired state at $T = 0$ (called now Fulde-Ferrell state) and the order of the transition between the BCS state and Fulde-Ferrell state is also an important feature of the problem.

However, the superconductivity can be destroyed by the magnetic field and one of the most important problems is to calculate the critical fields: the lower critical field H_{c1} and the upper critical field H_{c2}, for the type II superconductors.

The transition from the normal state to the mixed state (the two states are separated by the transition line $H_{c2}(T)$) is of the second order and that from the mixed state to the Meissner state (the two states are separated by $H_{c1}(T)$) is of the first order.

We are going to discuss the general theory of a superconductor in a magnetic field and to calculate, using the microscopic theory, the critical fields $H_{c1}(T)$ and $H_{c2}(T)$. These results will be applied for the magnetic superconductors, as well as for the new class of superconducting materials (like the heavy fermion superconductors) which present a non-phononic mechanism for the electron-electron attraction.

21. Paramagnetic Effects in Superconductors

a) The superconducting state in the presence of magnetic field

The critical field $H_c(T)$ called the thermodynamical field for the transition between the normal to superconducting state is given by the difference of free energies:

$$F_s(T) - F_n(T) = \frac{H_c^2(T)}{8\pi} . \tag{21.1}$$

For the normal phase, in the presence of the magnetic field H, the energy is

$$F_n(T,H) = F_n(T) - \frac{1}{2}\chi_n(T)H^2 \tag{21.2}$$

and for the superconducting phase, this relation becomes

$$F_s(T,H) = F_s(T) - \frac{1}{2}\chi_s(T)H^2 \tag{21.3}$$

where χ_n and χ_s are the magnetic susceptibilities of the normal and superconducting states, respectively. From (21.1)–(21.3), we obtain the critical paramagnetic field $H_p(T)$ at which the transition from the superconducting into normal state takes place due to the paramagnetic effects. This field is given by

$$H_p(T) = \frac{H_c(T)}{\sqrt{4\pi(\chi_n - \chi_s)}} . \tag{21.4}$$

At $T = 0$, this magnitude may be expressed in terms of the microscopic properties of the metal. Indeed (21.1) becomes

$$\frac{H_c^2(0)}{8\pi} = N(0)\frac{\Delta_0^2(0)}{2}\,, \qquad \chi_n(0) = 2\mu_0^2 N(0) \tag{21.5}$$

where μ_0 is the Bohr magneton. If one takes $\chi_s = 0$ at $T = 0$, we get

$$\mu_0 H_p(0) = \frac{\Delta_0(0)}{2} \tag{21.6}$$

where $\Delta_0(0)$ is the gap of the superconductor in the absence of the magnetic field at $T = 0$.
This value determines the field limit above which the superconductivity cannot exist, and the field is called Clogston-Chandrasekhar[1,2] limit.

The existence of such a limit is connected with the fact that $\chi_s(T)$ decreases with the temperature and goes to zero at $T = 0$ for the electrons with opposite spins which form the Cooper pairs. To polarize the superconductor, it is necessary to break the pairs, i.e., the field which should be applied is such that the Zeeman energy $\mu_0 H$ is of the order of the gap.

The microscopic theory of the behaviour of a superconductor in a magnetic field can be developed using the Gor'kov equations (9.9) written in an external field. These equations are

$$\begin{aligned} &[i\omega - \varepsilon(\mathbf{p}) + \mu_0 H]G_{\uparrow\uparrow}(\mathbf{p},\omega) - \Delta F^{\dagger}_{\downarrow\uparrow}(\mathbf{p},\omega) = 1 \\ &[i\omega + \varepsilon(\mathbf{p}) + \mu_0 H]F^{\dagger}_{\downarrow\uparrow}(\mathbf{p},\omega) + \Delta^{\dagger} G_{\uparrow\uparrow}(\mathbf{p},\omega) = 0\,. \end{aligned} \tag{21.7}$$

From these equations, we obtain the self-consistent equation for the gap as

$$1 = g\pi T \sum_{\omega} \int \frac{d^3\mathbf{p}}{(2\pi)^3} \frac{1}{\varepsilon^2(\mathbf{p}) + \Delta^2 - (i\omega + \mu_0 H)^2}\,. \tag{21.8}$$

For $T = 0$, this equation will be re-written as

$$1 = gN(0)\int_{-\infty}^{\infty} \frac{d\omega}{2\pi i} \int_{-\omega_D}^{\omega_D} \frac{d\varepsilon}{\varepsilon^2 + \Delta^2 - (\omega - \mu_0 H + i\delta)^2} \tag{21.9}$$

and after the usual algebra we get:

$$\begin{aligned} \ln\frac{\Delta(T)}{\Delta_0(0)} = \int_{-\infty}^{\infty} \frac{d\omega}{2\pi i} \int_{-\omega_D}^{\omega_D} d\varepsilon \Bigg[&\frac{1}{\varepsilon^2 + \Delta^2 - (\omega - \mu_0 H + i\delta)^2} \\ &- \frac{1}{\varepsilon^2 + \Delta^2 - (\omega - \mu_0 H - i\delta)^2} \Bigg] \end{aligned} \tag{21.10}$$

which can be written as

$$\ln \frac{\Delta(T)}{\Delta_0(0)} = \int_0^{\omega_D} \frac{d\varepsilon}{\sqrt{\varepsilon^2+\Delta^2}} \left[\theta(\sqrt{\varepsilon^2+\Delta^2} - \mu_0 H) - 1\right] \tag{21.11}$$

where

$$\theta(z) = \begin{cases} 1; & z > 0 \\ 0; & z < 0 \,. \end{cases}$$

If $\mu_0 H < \Delta$, the solution of Eq. (21.11) has the form

$$\Delta = \Delta_0(0) = 2\omega_D \exp[-1/N(0)g] \tag{21.12}$$

and for $\mu_0 H > \Delta$, Sarma[3] obtained the solution

$$\Delta^2 = 2\mu_0 H \Delta_0(0) - \Delta_0^2(0) \tag{21.13}$$

which is allowed in the interval

$$\Delta_0(0)/2 < \mu_0 H < \Delta_0 \,. \tag{21.14}$$

In order to obtain information about stability of the superconducting phases corresponding to the solutions (21.13) and (21.14), we calculate the free energy of these phases using the general relations

$$F_s - F_n = \int_0^{\Delta} \Delta^2 d\left(\frac{1}{g}\right) \,. \tag{21.15}$$

For the BCS solution (21.9), one obtains

$$F_s - F_n = -\frac{N(0)}{2}\Delta_0^2(0) \,. \tag{21.16}$$

For the second solution (21.13), we get

$$F_s - F_n = N(0)[\Delta_0^2(0)/2 - 2\mu_0 H \Delta_0^2(0) + 2(\mu_0 H)^2] \tag{21.17}$$

and in the interval $\Delta_0(0)/2 < \mu_0 H < \Delta_0(0)$, this energy is lower than the BCS energy, thus the phase is stable.

We have to mention that all this calculation is in the mean field approximation and the real stability seems to be questionable in the mean field approximation.

Let us consider Eq. (21.8) at $T \neq 0$. The summation over frequencies can be performed in the usual way and we obtain

$$1 = gN(0)\int_0^{\omega_D} \frac{d\varepsilon}{\sqrt{\varepsilon^2+\Delta^2}}\left[\tanh\frac{\sqrt{\varepsilon^2+\Delta^2}+\mu_0 H}{2T} - \tanh\frac{\sqrt{\varepsilon^2+\Delta^2}-\mu_0 H}{2T}\right] \tag{21.18}$$

an equation which can be written as

$$\ln\frac{\Delta(T,H)}{\Delta_0(0)} = -\int_0^{\infty}\frac{d\varepsilon}{\sqrt{\varepsilon^2+\Delta^2}}[f(E+\mu_0 H)+f(E-\mu_0 H)] \tag{21.19}$$

where $E^2 = \varepsilon^2+\Delta^2$ and $f(x)$ is the Fermi function.

In the low temperatures domain, $T \ll T_c$ for the weak fields $\mu_0 H \ll \Delta$, from (21.20) we get the solution

$$\ln\frac{\Delta(T,H)}{\Delta_0(0)} = 2\sum_{n=0}^{\infty}(-1)^n\cosh\frac{n\mu_0 H}{2T}K_0\left(\frac{n\Delta}{T}\right) \tag{21.20}$$

where $K_0(x)$ is the MacDonald function. The expression (21.20) can be approximated as

$$\ln\frac{\Delta(T,H)}{\Delta_0(0)} = -\sqrt{\frac{2\pi T}{\Delta}}\left(1-\frac{T}{8\Delta}\right)\exp\left[-\left(\frac{\Delta-\mu_0 H}{T}\right)\right], \mu_0 H \ll \Delta \tag{21.21}$$

and

$$\ln\frac{\Delta(T,H)}{\Delta_0(0)} = -\sqrt{\frac{2\pi T}{\Delta}}\sum_{n=1}^{\infty}\frac{(-1)^{n+1}}{\sqrt{n}}\exp\left[-\frac{n(\Delta-\mu_0 H)}{T}\right], \quad \mu_0 H \gg \Delta\,. \tag{21.22}$$

Near the critical temperature T_c, one expands $[\varepsilon^2+\Delta^2-(i\omega+\mu_0 H)^2]^{-1}$ in power series in terms of Δ which is assumed to be small in this temperature range, and performing the integral over ε (in the limit $\omega_D \to \infty$), we obtain

$$\ln\frac{T}{T_c} = \Psi\left(\frac{1}{2}\right) - \operatorname{Re}\Psi\left(\frac{1}{2}+i\rho\right) + \frac{1}{2}f_1(\rho)\left(\frac{\Delta}{2\pi T}\right)^2 + \frac{3}{8}f_2(\rho)\left(\frac{\Delta}{2\pi T}\right)^4 + \dots \tag{21.23}$$

where

$$\rho = \frac{\mu_0 H}{2\pi T}$$

$$\psi(z) = \sum_{n=0}^{\infty}\left[\frac{1}{n+1} - \frac{1}{n+z}\right]$$

$$f_1(z) = \mathrm{Re}\sum_{n=0}^{\infty}\frac{1}{n+\frac{1}{2}+z}, \qquad f_2(z) = \mathrm{Re}\sum_{n=0}^{\infty}\frac{1}{(n+\frac{1}{2}+z)^5}.$$

In the approximation of high fields ($\rho \gg 1$) from (21.23), the critical temperature can be approximated as

$$T_c^2 = \frac{6}{\pi^2}(\mu_0 H)^2 \ln\frac{2\mu_0 H}{\Delta_0(0)}.$$

If $\rho \ll 1$, Eq. (21.23) gives

$$T_c/T_{c0} = 1 - 7\zeta(3)(\mu_0 H/2\pi T_{c0})^2.$$

The free energy (21.17) can be calculated for the case of small field as

$$F_s - F_n = N(0)\left[\frac{(\pi T)^2}{3} + (\mu_0 H)^2 - \Delta^2/2\right].$$

Using now (21.17) and (21.23), the free energy is

$$F_s - F_n = N(0)\left[f_1(\rho)\frac{\Delta^4}{(4\pi T)^2} - f_2(\rho)\frac{\Delta^6}{4(2\pi T)^4} + \ldots\right]. \qquad (21.24)$$

In an external field, the phase transition changes its order in the point given by the equation

$$f_1(\rho) = 0 \qquad (21.25)$$

which gives $\rho_0 = 0.308$ and $T_c/T_{c0} = 0.566$.
These simple calculations show that:
— if $\rho < \rho_0$ and $f_1(\rho_0) > 0$, the transition is of the second order,
— if $\rho < \rho_0$ and $f_1(\rho_0) < 0$, the transition is of the first order.
The critical temperature $T_c/\Delta_0(0)$ as function of the external field H is given by Fig. 19.

These results have been obtained by Sarma[3] for a BCS superconductor. However, the magnetic field can be given by a "molecular field" created by magnetic impurities. In this case, the external field denoted by I will be

$$I = cSa$$

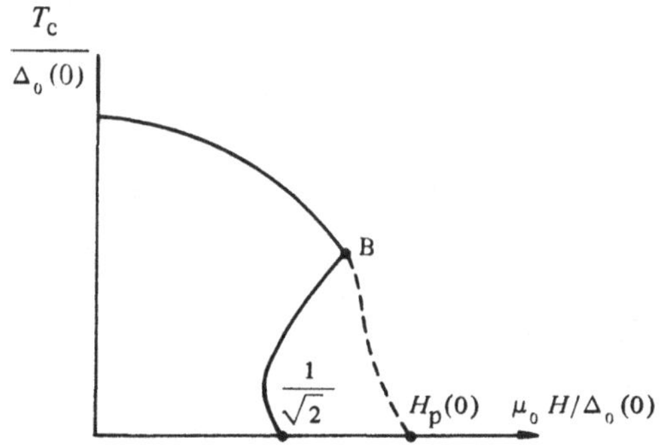

Fig. 19. The critical temperature T_c as function of the magnetic field H.

where S is the spin of an impurity and a is a constant. The temperature dependence of this molecular field can be described by the Brillouin function

$$I = caB_s[aMS/T]; \quad M = -\frac{\partial F(T,I)}{\partial I}. \tag{21.26}$$

Equation (21.18) can be transformed by $\mu_0 H \to I$ and a first order phase transition appears at the point $I_0 = \Delta_0/\sqrt{2}$. Larkin and Ovchinnikov[4] considered also the states in which the pairs have a momentum and showed that for any values of I which correspond to a certain range of impurity concentrations, these states are energetically more favourable. We can show that the normal state is unstable with respect to the formation of pairs with nonzero momentum.

The instability of this state can be obtained in the Random Phase Approximation as

$$1 = -g\int d\omega' \int \frac{d^3\mathbf{p}'}{(2\pi)^3}\mathrm{Tr}\left\{\hat{G}^0\left(\frac{\omega}{2}+\omega';\mathbf{p}+\frac{q}{2}\right)\hat{\sigma}_y\hat{G}\left(\frac{\omega}{2}-\omega';\frac{\mathbf{q}}{2}-\mathbf{p}\right)\right\}$$

where $\hat{G}^0$ is

$$\hat{G}^0(\mathbf{p};\omega) = [\omega - \varepsilon(\mathbf{p}) - I\hat{\sigma}_z + i\delta\,\mathrm{sign}\,\omega]^{-1}. \tag{21.27}$$

This condition is equivalent to

$$1 = \frac{N(0)g}{2}\Pi(q,I) \tag{21.28}$$

where the polarization is

$$\Pi = \ln\frac{4\omega_D^2}{v_0 q^2 - 4I^2} + 2 - \frac{2I}{v_0 q}\ln\frac{v_0 q + 2I}{v_0 q - 2I}.$$

For $q = 0$, (21.28) gives $I = \Delta_0(0)/2$, a point which is in the region of metastability of the normal state.

However, when $q \neq 0$, Eq. (21.28) has a solution for $I > \Delta_0(0)/\sqrt{2}$. The maximum of $\Pi(q)$ is given by

$$\frac{\partial \Pi(q)}{\partial q} = 0$$

and from this condition one gets

$$q_0 = \text{constant}\,\frac{I}{v_0}$$

which gives

$$I = I_0 = 0.775\Delta_0$$

and the free energy is

$$F(I_0) = -1.14\frac{\Delta_0^2(0)}{4}\,.$$

If the order parameter is considered now as

$$\hat{\Delta}(\mathbf{r}) = i\hat{\sigma}_y \sum_m \Delta_m \exp(i\,\mathbf{q}_m \cdot \mathbf{r}) \tag{21.29}$$

we have to choose the specific symmetry which gives a lowest free energy. The most favourable form for (21.29) appears to be

$$\Delta(\mathbf{r}) = \Delta \cos(\mathbf{q} \cdot \mathbf{r}) \tag{21.30}$$

and if one analyses the current in the linear response approximation, we have to consider also the vector potential A as spatially dependent.

The excitation spectrum has been obtained[4] for the one dimensional case and the calculations showed that this state called inhomogeneous state has excitations with small energy.

The influence of the impurities on this state has been treated by Aslamazov[5] using the Green function

$$G^{\pm}(\omega, \mathbf{p}) = [i\omega - \varepsilon(\mathbf{p}) \pm I + i\,\mathrm{sign}\,\omega/2\tau]^{-1} \tag{21.31}$$

and τ is the scattering time.

The critical temperature has been obtained from the equation

$$1+|g|\pi T\sum_{n=-\infty}^{\infty}\int\frac{d^3\mathbf{p}}{(2\pi)^3}\mathrm{Tr}\left\{K(\omega;\mathbf{p};\mathbf{p}-\mathbf{q})\right\}=0 \tag{21.32}$$

where $K(\omega;\mathbf{p};\mathbf{p}-\mathbf{q})$ is the Fourier transform of the average $\overline{\widehat{G}(\omega;\mathbf{r},\mathbf{s})\widehat{\sigma}_y\widehat{G}(-\omega;\mathbf{r},\mathbf{s})}$.
From (29.31) and (21.32), one gets

$$\ln\frac{T}{T_c}+\frac{1}{2}\left[\Psi\left(\frac{1}{2}+iz\right)-\Psi\left(\frac{1}{2}\right)\right]+\frac{\tau v_0^2q^2}{6\pi T}\Psi'\left(\frac{1}{2}+z\right)+h.c.=0 \tag{21.33}$$

where $z=I/2\pi T$, and T_{c0} is the critical temperature for a pure superconductor. In the limit $T\ll I$, from (21.33) we get

$$\ln\frac{2I}{\Delta_0(T)}+\frac{2\tau v_0^2q^2}{3}\left[\frac{\pi\exp(-I/T)}{T}-4\tau\ln\frac{1}{2\tau I}\right]+\frac{2\tau^2v_0^4q^4}{9I^2}=0 \tag{21.34}$$

where $\Delta_0(T)$ is the gap of the clean superconductor. This equation defines a temperature T below which in the system there are pairs with $q=0$. The equation which defines this temperature is

$$\pi\exp\left[-\frac{\Delta_0}{2T^*}\right]-4\tau T^*\ln\frac{1}{\tau\Delta_0}=0 \tag{21.35}$$

which gives

$$T^*=-\frac{\Delta_0}{2\ln\tau\Delta_0}\,. \tag{21.36}$$

For the temperatures above T^*, there is an instability point:

$$I=\frac{\Delta_0}{2}\left(1+\frac{a^2}{2}\right),\quad a=2\tau\Delta_0\ln\frac{1}{\tau\Delta_0}-\frac{\pi\Delta_0}{2T}\exp\left(-\frac{\Delta_0}{2T}\right) \tag{21.37}$$

and the momentum of the pairs is

$$q=\frac{1}{v_0}\left[\frac{3a\Delta_0}{4\tau}\right]^{\frac{1}{2}}\,. \tag{21.38}$$

This state may appear for the triplet states in the antiferromagnetic superconductors.

b) The Fulde-Ferrell state in superconductors

The superconducting state with an order parameter which oscillates in the space has been obtained in the previous model due to the action of an external magnetic field. This state has been studied by Fulde and Ferrell[6] and by Takada and Izuyama[7] for the case of a superconductor with $q \neq 0$.

The BCS Hamiltonian in the presence of the magnetic field is

$$\mathcal{H}_{\text{eff}} = \sum_{\mathbf{p},\alpha} \varepsilon_\alpha(\mathbf{p}) c^\dagger_{\mathbf{p}\alpha} c_{\mathbf{p}\alpha} + \sum_{\mathbf{p}} [\Delta_{\mathbf{q}} c^\dagger_{\mathbf{p}+\frac{\mathbf{q}}{2}\uparrow} c^\dagger_{-\mathbf{p}+\frac{\mathbf{q}}{2}\downarrow} + h.c.] \tag{21.39}$$

where

$$\varepsilon_\alpha(\mathbf{p}) = \varepsilon(\mathbf{p}) - \mu_0 h\alpha$$
$$\alpha = \begin{cases} 1 & \text{for the spin up} \\ -1 & \text{for the spin down} \end{cases}$$

h is the external magnetic field, and $\Delta_{\mathbf{q}}$ is defined as

$$\Delta_{\mathbf{q}} = -g \sum_{\mathbf{p}} \langle c_{\mathbf{p}+\frac{\mathbf{q}}{2}\uparrow} c_{-\mathbf{p}-\frac{\mathbf{q}}{2}\downarrow} \rangle \,. \tag{21.40}$$

The Green functions which can be obtained from the Gor'kov equations are

$$G_\alpha(\mathbf{p},\omega) = \frac{\omega + \varepsilon_{-\alpha}(-\alpha\mathbf{p} + \mathbf{q}/2)}{[\omega + E_\alpha(\mathbf{p})][\omega + E_{-\alpha}(\mathbf{p})]} \,;$$
$$F^\dagger_\alpha(\mathbf{p},\omega) = \frac{\Delta^\dagger}{[\omega + E_\alpha(\mathbf{p})][\omega + E_\alpha(\mathbf{p})]} \tag{21.41}$$

where

$$E_\alpha(\mathbf{p}) = \alpha\left[\frac{v_0 q}{2}x - h\right] + \sqrt{\varepsilon^2(\mathbf{p}) + \Delta^2_{\mathbf{q}}} \,. \tag{21.42}$$

From this equation, we see that

$$E_\uparrow(\mathbf{p}) + E_\downarrow(\mathbf{p}) = 2\sqrt{\varepsilon^2 + \Delta^2} \tag{21.43}$$

and these two types of solutions which describe this superconducting state.

If

$$\frac{v_0 q}{2} + h < \Delta_q$$

both $E_\uparrow$ and $E_\downarrow$ are always positive in the whole k space and the solution is a BCS one.

In the opposite case

$$\frac{v_0 q}{2} + h > \Delta_q \tag{21.44}$$

$E_\alpha(\mathbf{p})$ can be negative for some values of k as one can see from (21.42). The Cooper pairs with the negative energy must be destroyed, while these pairs remain for the states with $E_\uparrow(\mathbf{p})$ and $E_\downarrow(\mathbf{p}) < 0$.

In this case, the $\mathbf{p}$-space near the Fermi surface is divided into two regions: the pairing region and the depairing one. Concretely for $E_\uparrow(\mathbf{p}) > 0$ and $E_\downarrow(\mathbf{p}) < 0$, the electrons $(\mathbf{p} + \mathbf{q}/2; \uparrow)$ and $(-\mathbf{p} + \mathbf{q}/2; \downarrow)$ are paired. In the region $E_\alpha(\mathbf{p}) < 0$, the electron states $(\alpha\mathbf{p} + \frac{\mathbf{q}}{2}, \alpha)$ are totally blocked by the electrons with their unpaired spins, while the states $(-\alpha\mathbf{p} + \mathbf{q}/2, -\alpha)$ are unoccupied because $E_{-\alpha}(\mathbf{p})$ must be positive in this case.

This state is called the *Fulde-Ferrell (FF) state* and we have to analyse the conditions for the occurrence of such a state, its stability and the properties at the finite temperatures of a superconductor in this state.

First, we try to give some parameters which characterize the blocking effect specific for this state. If one considers the Fermi momenta $p_{0\uparrow}$ and $p_{0\downarrow}$ of the up-spin, respectively down-spin electrons and defines q as $q = \bar{q}(p_{0\uparrow} - p_{0\downarrow})$, the energy (21.42) can be written as

$$E_\alpha(\mathbf{p}) = \alpha(\bar{q}x - 1)h + \sqrt{\varepsilon^2(\mathbf{p}) + \Delta_q^2} \tag{21.45}$$

where

$$p_{0\uparrow} - p_{0\downarrow} = (h/\mu_0)p_0 \ .$$

The Fulde-Ferrell state will be examined using (21.45) for two cases, $0 < \bar{q} < 1$ and $\bar{q} > 1$.

In the first case, $E_\downarrow(\mathbf{p})$ is always positive and the blocking appears only for the up-spin electrons if the condition $h > \Delta_q/(\bar{q}+1)$ is satisfied. The blocked region is given by the values of x and $\uparrow$ satisfying

$$\begin{aligned} &-1 \le x \le \min(1, \Phi^-) \\ &-(\bar{q}+1)hx_1 \le \varepsilon(p) \le (\bar{q}+1)hx_1 \end{aligned} \tag{21.46}$$

where

$$\Phi^\pm(\varepsilon) = \frac{h \pm \sqrt{\varepsilon^2(p) + \Delta^2}}{\bar{q}h} \ , \qquad x_1 = \sqrt{1 - \frac{\Delta_q^2}{h^2(\bar{q}+1)^2}} \ . \tag{21.47}$$

In the second case ($\bar{q} > 1$), the blocking of the electrons with down spins is also possible and if $h > \Delta_q/(\bar{q}+1)$, the blocked region is

$$\begin{aligned} &-1 \le x \le \Phi^- \\ &-(\bar{q}+1)hx_1 \le \varepsilon(\mathbf{p}) \le (\mathbf{q}+1)hx_1 \ . \end{aligned} \tag{21.48}$$

The blocking of the down-spin electrons occurs in the case $h > \Delta_q/(\bar{q}-1)$ and its region is

$$\begin{gathered} \Phi^\dagger \le x \le 1 \\ -(\bar{q}-1)hx_2 \le \varepsilon(p) \le (\bar{q}-1)hx_2 \end{gathered} \tag{21.49}$$

where

$$x_2 = \sqrt{1-\Delta_q^2/h^2(\bar{q}-1)^2} - \begin{cases} 0; & 0 \le x_1 \le \frac{2\sqrt{\bar{q}}}{\bar{q}+1} \\ \sqrt{1-(\bar{q}+1)^2/(\bar{q}-1)^2}; & \frac{2\sqrt{\bar{q}}}{\bar{q}+1} \le x_1 \le 1\,. \end{cases} \tag{21.50}$$

The parameters x_1 and x_2 defined by (21.47) and (21.50) are the measure of the blocking and take values between zero and unity.

In the limit $x_1 = x_2 = 0$, we get the BCS solution and for $x_1 = x_2 = 1$, the solution corresponds to the normal state. The blocking induced the momentum in the Fulde-Ferrell ground state, but the direction of this momentum is opposite to that of q (the Cooper pair momentum) and these two momenta cancel each other so that the Bloch theorem of nonexistence of the mass current in the ground state is satisfied.

One of the most important results obtained in this theory is that the Fulde-Ferrell state can be a ground state if x_1 is close to unity. Another important problem is to obtain the lowest energy of the Fulde-Ferrell state. For a value of h, there is only one q which gives the lowest energy. Another important result is that the Fulde-Ferrell state is a ground state if its moment is zero. These results have been obtained in the Hartree-Fock approximation but seem to be quite relevant for this problem.

Let us discuss the thermodynamics of the Fulde-Ferrell state.

From (21.41), the gap can be written as

$$1 = -g\sum_{\mathbf{p},\omega} \frac{1}{[\omega - E_\alpha(\mathbf{p})][\omega - E_{-\alpha}(\mathbf{p})]} \tag{21.51}$$

and following the usual way we get

$$\begin{aligned} \ln\frac{T}{T_{c0}} = \frac{1}{2}\int_{-1}^{1} dx\Big[\Psi\Big(\frac{1}{2}\Big) - \Psi\Big(\frac{1}{2}+i\rho(x)\Big)\Big] \\ + \frac{1}{8}\Big(\frac{\Delta_q(T)}{2\pi T}\Big)^2 \int_{-1}^{1} dx \operatorname{Re}\Psi''\Big(\frac{1}{2}+i\rho(x)\Big) \end{aligned} \tag{21.52}$$

where $\psi(z)$ is the di-gamma function and

$$\rho(x) = \frac{4(\bar{q}x - 1)}{2\pi T} . \tag{21.53}$$

This equation can be transformed as

$$\ln \frac{T}{T_{c0}} = g_0(\bar{q}, h, T) + \left(\frac{\Delta_g}{2\pi T}\right)^2 g_1(\bar{q}, h, T) \tag{21.54}$$

where

$$g_0 = \sum_{n=0}^{\infty} \left[\frac{\pi T}{h\bar{q}}\left\{\tan^{-1}\frac{h(\bar{q}-1)}{2\pi T(n+\frac{1}{2})} + \tan^{-1}\frac{h(\bar{q}+1)}{2\pi T(n+\frac{1}{2})}\right\} - \frac{1}{n+\frac{1}{2}}\right] ,$$

$$g_1 = \frac{1}{8}\frac{2\pi T}{h\bar{q}}\mathrm{Im}\left[\Psi'\left(\frac{1}{2} + i\rho(1)\right) - \psi'\left(\frac{1}{2} + i\rho(-1)\right)\right] .$$

From this result, one may obtain the phase diagram given in Fig. 20. The transition from the normal phase in the BCS and FF phases is of the second order, but the transition from the BCS in the FF phase is of the first order. The linear response theory showed the existence of the Meissner effect in the Flude-Ferrell state, which is in fact a gapless superconducting state.

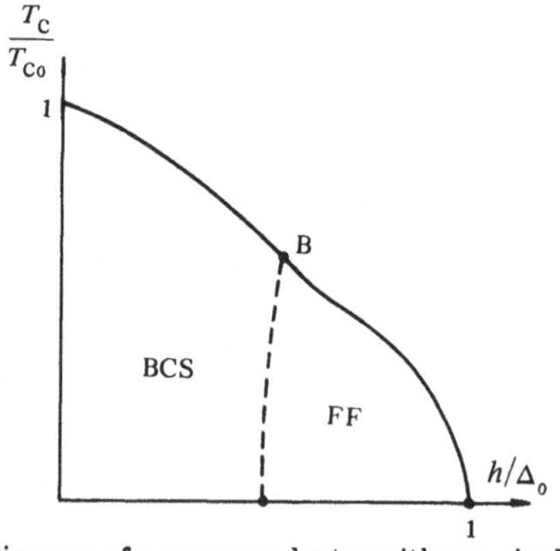

Fig. 20. The phase diagram of a superconductor with unpaired phase. The BCS is the usual states and FF state contains the blocked states.

This model is similar with that studied by Sarma but in Ref. 3, only the states with $\mathbf{q} = 0$ are considered. The energies of such states are always higher than those of the normal and BCS states.

22. Critical Fields of a Superconductor

In Chapter 1, we showed that the superconducting state is destroyed in an external magnetic field if this field is greater than a critical value which is temperature dependent. The important critical fields are the lower critical field H_{c1} and the upper critical field H_{c2}.

In the H-T plane below the $H_{c1}(T)$-line, the superconductor presents a complete Meissner effect, and above the $H_{c2}(T)$-line, the superconductivity is completely destroyed. Between these two lines, the superconductor is in the mixed state and has a vortex structure.

A qualitative discussion can be done using the free energy as function of the magnetic field H given in Fig. 21. The horizontal lines AC and DD′ represent the energy of the normal state, respectively of the superconducting state in the absence of the magnetic field H. The energy of a type I superconductor with a complete Meissner effect is represented by the parabolic line DEB where B denotes the thermodynamic critical field H_c. A type II superconductor with field penetration in the form of vortices has the energy showed by the line DEC. This curve is tangent to the AC line in C which is the upper critical field H_{c2}. The transition from the normal to the type II superconducting state is of the second order. The penetration of the magnetic field and the occurrence of the mixed state starts with a field H_{c1} which is known as the lower critical field.

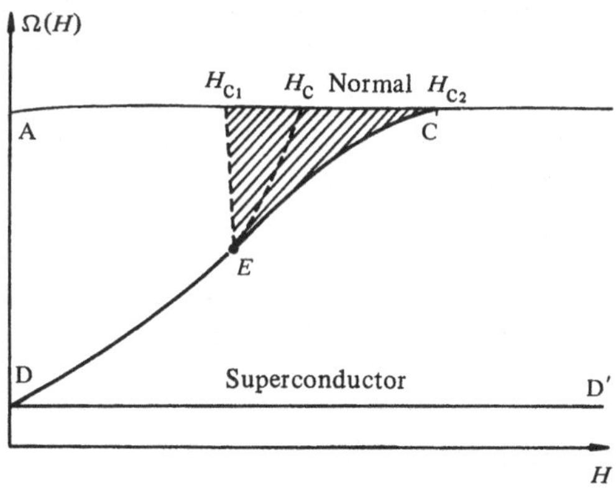

Fig. 21. The energy dependence of a superconductor as function of the magnetic field.

Both of these fields are temperature dependent and become zero at the critical temperature T_c.

The transition from the vortex-state in the complete Meissner state is a first order phase transition.

We will present the microscopic calculation of temperature dependence of these two critical fields.

a) The lower critical field

The Gibbs potential for the state with one vortex of the energy E_v is given by the equation

$$G = E_v - \frac{\Phi_0 H}{4\pi} \tag{22.1}$$

where $\Phi_0 = \pi/e$ is the quant of magnetic current. The instability of this state appears at $G = 0$ and gives the lower critical field as

$$H_{c1} = \frac{4\pi}{\Phi_0} E_v(T) \tag{22.2}$$

or if one considers in (22.1) also the contribution of the core of the vortex which has the energy

$$E_v = \frac{\Phi_0^2}{32\lambda^2} \tag{22.3}$$

the lower critical field will be given by

$$H_{c1}(T) = \frac{4\pi}{\Phi_0} E_v(T) + \frac{\Phi_0^2}{32\lambda^2} \,. \tag{22.4}$$

The temperature dependence of the vortex energy can be obtained from the general equation

$$E_v = F_H(\Delta) - F_0(\Delta) = \int d^3\mathbf{r} \int_0^g \frac{\Delta^2(\mathbf{r}) - \Delta^2(0)}{\Delta} d\left(\frac{1}{g}\right) \,. \tag{22.5}$$

From the equation of the order parameter

$$\frac{\Delta}{g} = \pi T \sum_\omega \int \frac{d^3\mathbf{p}}{(2\pi)^3} F(\mathbf{p}, \omega) \,, \tag{22.6}$$

one may calculate

$$d\left(\frac{1}{g}\right) = C(\Delta)\frac{d\Delta}{\Delta} \,, \qquad C(\Delta) = -\frac{mp_0}{2\pi}\pi T \Sigma[\omega^2 + \Delta^2]^{-\frac{3}{2}} \tag{22.7}$$

and from (22.5), the vortex energy becomes

$$E_v = 4\pi \int_0^\Delta d\Delta \int_0^\infty r dr \Delta_1(r) C(\Delta) \tag{22.8}$$

where we used

$$\Delta^2(\mathbf{r}) - \Delta^2(0) \cong 2\Delta(\mathbf{r})\Delta_1(\mathbf{r}) \ .$$

The order parameter $\Delta(\mathbf{r})$ has been expanded as

$$\Delta(\mathbf{r}) = \Delta(0) + \Delta_1(\mathbf{r}) \tag{22.9}$$

and $\Delta_1(\mathbf{r})$ can be calculated using an equation which is similar to (22.8) and is

$$\frac{\Delta_1(\mathbf{r})}{g} = \pi T \sum_{\omega} [\widehat{G}(\mathbf{r},\mathbf{r};\omega) - \widehat{G}^0(\mathbf{r},\mathbf{r}';\omega)] \ . \tag{22.10}$$

The Green function $\widehat{G}(\mathbf{r},\mathbf{r};\omega)$ is defined as (9.9) and from the Gor'kov equations in the presence of the electromagnetic field, one obtains

$$\widehat{G}(\mathbf{r},\mathbf{r}';\omega) = \widehat{G}^0(\mathbf{r},\mathbf{r}';\omega) + \frac{ie}{m}\int d^3\mathbf{r}_1 \widehat{G}^0(\mathbf{r},\mathbf{r}_1;\omega)\widehat{\sigma}_z\left(A(\mathbf{r}_1)\frac{\partial}{\partial r_1}\right) \times G(\mathbf{r}_1,\mathbf{r};\omega) - \int d^3\mathbf{r}_1 \widehat{G}^0(\mathbf{r},\mathbf{r}_1;\omega)\Delta_1\widehat{G}(\mathbf{r}_1,\mathbf{r};\omega) \tag{22.11}$$

where the electromagnetic vector potential $\mathbf{A}(\mathbf{r})$ is given as

$$A(\mathbf{r}) = \frac{1}{2|e|\lambda}K_1(r/\lambda) \ , \tag{22.12}$$

$K_1(x)$ being the first order Bessel function.
The correction $\Delta_1(\mathbf{r})$ contains only $A^2(\mathbf{r})$. Indeed, it is not difficult to show that if one considers the gauge $\nabla \cdot \mathbf{A}(\mathbf{r}) = 0$, the linear terms in $\Delta_1(\mathbf{r})$ are zero and the first correction $\Delta_1(\mathbf{r})$ is proportional to the square root of the vector potential. The Fourier transform of the first order correction of (22.11) contains the terms

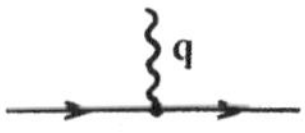

$$= -\frac{e}{m}\widehat{G}^0(\mathbf{p};\omega)\widehat{\sigma}_z(\mathbf{p}\cdot\mathbf{A}(\mathbf{q}))\widehat{G}^0(\mathbf{p}+\mathbf{q};\omega)$$

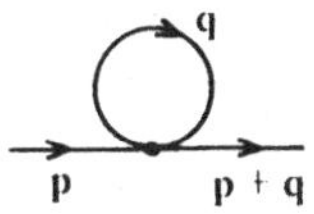

$$= \widehat{G}^0(\mathbf{p};\omega)\Delta_1(\mathbf{q})\widehat{G}^0(\mathbf{p}+\mathbf{q};\omega)$$

and the second order correction $G^{(2)}(\mathbf{p},\mathbf{q};\omega)$ is given by the diagram

$\widehat{G}^{(2)}(\mathbf{p}, \mathbf{p}-\mathbf{q}; \omega) =$ 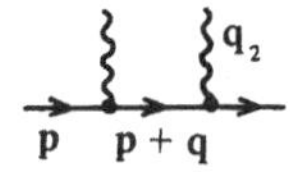 $2\pi\delta(q + q_1 + q_2) +$ 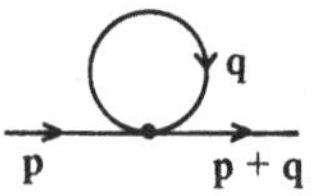

Using these results, the Fourier transform of (22.10) gives

$$\left[|g|^{-1} - \pi T \sum_{\omega} \int \frac{d^3\mathbf{p}}{(2\pi)^3} \Big(G(\mathbf{p}, \omega) G(\mathbf{p}+\mathbf{q}, \omega) - F(\mathbf{p}, \omega) F(\mathbf{p}+\mathbf{q}, \omega) \Big) \right] \Delta_1(q)$$
$$= \left(\frac{e}{m}\right) \pi T \sum_{\omega} \int \frac{d^3\mathbf{p}}{(2\pi)^3} \iint \frac{dq_1 dq_2}{2\pi} (FFF + GGG - GFG^{\dagger})$$
$$\times (\mathbf{p} \cdot \mathbf{A}(\mathbf{q}_1))(\mathbf{p} \cdot \mathbf{A}(\mathbf{q}_2)) \delta(q + q_1 + q_2) \,. \tag{22.13}$$

This equation can be written as

$$L(q)\Delta_1(q) = \iint \frac{dq_1 dq_2}{2\pi} L_2(q, q_1, q_2) A(q_1) A(q_2) \tag{22.14}$$

where

$$L(q) = \pi T \sum_{\omega} [\omega^2 + \Delta^2]^{-\frac{1}{2}} \left\{ 1 - \frac{2\omega^2}{v_0 q(\omega^2 + \Delta^2)^{\frac{1}{2}}} \tan^{-1} \frac{v_0 q}{2(\omega^2 + \Delta^2)} \right\} \tag{22.15}$$

and for $q + q_1 + q_2 = 0$

$$L(q) = \frac{\Delta}{2} (ev_0)^2 \pi T \sum_{\omega} (\omega^2 + \Delta^2)^{-\frac{1}{2}} \int_{-1}^{1} d\mu (1 - \mu^2)$$
$$\times \left\{ [2i(\omega^2 + \Delta^2)^2 - qv_0\mu]^{-1} [2i(\omega^2 + \Delta^2)^{\frac{1}{2}} + v_0 q_2 \mu]^{-1} + q_1 \rightleftarrows q_2 \left(1 - \frac{\Delta^2}{\omega^2}\right) \right.$$
$$- \frac{\Delta^2}{\omega^2 + \Delta^2} [2i(\omega^2 + \Delta^2)^{\frac{1}{2}} + v_0 q\mu]^{-1}$$
$$\left. \times [2i(\omega^2 + \Delta^2)^{\frac{1}{2}} - v_0 q_2 \mu]^{-1} \right\} . \tag{22.16}$$

For a London superconductor $\lambda \gg \xi_0$, Eqs. (22.15) and (22.16) become

$$L(q) = \pi T \sum_{\omega} \frac{\Delta^2}{(\omega^2 + \Delta^2)^{\frac{3}{2}}} \tag{22.17}$$

$$L_2(q) = -\frac{\Delta}{3} (ev_0)^2 \pi T \sum_{\omega} \frac{1}{(\omega^2 + \Delta^2)^{\frac{3}{2}}} \,. \tag{22.18}$$

Equation (22.14) can be approximated as

$$L(0)\Delta_1(\mathbf{r}) = -L_2(0)A^2(\mathbf{r}) \tag{22.19}$$

and one can consider the impurities taking $(\Delta\omega) \to (\eta_{tr}\Delta, \eta_{tr}\omega)$. In this case, from (22.20) the first order correction is

$$\Delta_1(\mathbf{r}) = -\frac{(ev_0 A(\mathbf{r}))^2}{3\Delta} \frac{\sum\limits_\omega \eta_{tr}^{-1}(\omega^2+\Delta^2)^{-\frac{3}{2}}}{\sum\limits_\omega (\omega^2+\Delta^2)^{-\frac{3}{2}}} . \tag{22.20}$$

From (22.7) and (22.20), one obtains

$$\Delta_1(\mathbf{r})C(\Delta) = \frac{mp_0}{2\pi}\left(\frac{eA(\mathbf{r})}{3}\right)^2 \sum_\omega \eta_{tr}^{-1}(\omega^2+\Delta^2)^{-\frac{3}{2}} \tag{22.21}$$

and using (22.2),(22.8) and (22.20), the lower critical field becomes

$$\begin{aligned} H_{c1} &= \frac{4\pi}{\Phi_0}\left[\frac{\pi^2\sigma(T)}{e^2}\left(\Delta(T)\tanh\frac{\Delta(T)}{2T}\right)\ln\chi(T)\right] , \\ \sigma(T) &= \frac{ne^2\tau_{tr}}{m} \end{aligned} \tag{22.22}$$

equation which can be written as

$$H_{c1} = H_c \frac{\ln\chi(T)}{\chi(T)\sqrt{2}} \tag{22.23}$$

where $\chi(T)$ was defined by (4.18).

b) The upper critical field

The upper critical field H_{c2} as function of temperature has been calculated first by Helfand and Werthammer,[8] Werthammer *et al.*,[9] Maki,[10] Caroli *et al.*[13] in the framework of the BCS theory. In the presence of magnetic field, the Gor'kov equation for the order parameter becomes

$$\Delta(\mathbf{r})\ln\frac{T}{T_c} = \int d^3\mathbf{r}'\Delta(\mathbf{r}')X(\mathbf{r},\mathbf{r}') \tag{22.24}$$

where $X(\mathbf{r},\mathbf{r}')$ is the kernel which depends on the magnetic field. The largest value of the magnetic field H for which the integral Eq. (22.24) has a solution, defines the upper critical field H_{c2} which will be temperature dependent. The kernel can be defined as

$$X(\mathbf{r},\mathbf{r}') = \sum_{n=-\infty}^{\infty} \left[\frac{1}{|2n+1|}\delta(\mathbf{r}-\mathbf{r}') - S(\mathbf{r},\mathbf{r}';\omega)\right] \tag{22.25}$$

where $S(\mathbf{r},\mathbf{r}';\omega)$ can be defined by an integral equation

$$S(\mathbf{r},\mathbf{r}';\omega) = S_0(\mathbf{r},\mathbf{r}';\omega) + (2\pi T\tau)^{-1}\int d^3\mathbf{r}'' S_0(\mathbf{r},\mathbf{r}'';\omega)S(\mathbf{r}'',\mathbf{r}';\omega) \tag{22.26}$$

and the kernel $S_0(\mathbf{r},\mathbf{r}';\omega)$ (see Appendix) is

$$S_0(\mathbf{r},\mathbf{r}';\omega) = \frac{T}{2v_0|\mathbf{r}-\mathbf{r}'|}\exp\left[-\left(\frac{2|\omega|}{v_0}+\frac{1}{l}\right)|\mathbf{r}-\mathbf{r}'| + 2ie\int_{\mathbf{r}'}^{\mathbf{r}} d\mathbf{s}A(\mathbf{s})\right] . \tag{22.27}$$

The integral Eq. (22.26) will be analysed taking X as defined by (22.25) as an operator and re-writing (22.24) as

$$\Delta \ln \frac{T}{T_c} = \hat{X}\Delta \tag{22.28}$$

where

$$\hat{X} = \sum_{n=-\infty}^{\infty}\left[\frac{1}{|2n+1|}\hat{1} - \hat{S}(\omega)\right] . \tag{22.29}$$

Equation (22.28) may be considered as an equation of the form:

$$\hat{S}(\omega) = \hat{S}_0(\omega) + (2\pi\tau)^{-1}\hat{S}_0(\omega)\hat{S}(\omega) \tag{22.30}$$

where the operator $\hat{S}_0(\omega)$ is hermitian with real and orthogonal eigenfunctions φ_ω which are associated to the eigenvalues $s(\omega)$.
Equation (22.30) can be written as

$$\hat{S}_0(\omega)\varphi_\omega = s(\omega)[1+(2\pi T\tau)^{-1}s(\omega)]^{-1}\varphi_\omega \tag{22.31}$$

where $s(\omega)$ is the eigenvalue obtained from the equation

$$\hat{S}_0(\omega)\varphi_\omega = s(\omega)\varphi_\omega . \tag{22.32}$$

From (22.29), we get

$$X^\dagger\varphi = \sum_{n=-\infty}^{\infty}\left[\frac{1}{2n+1} - \frac{s(\omega)}{1-(2\pi T\tau)^{-1}s(\omega)}\right]$$

which is equivalent to

$$(X\Delta, \varphi_\omega) = (\Delta\varphi_\omega)\ln\frac{T}{T_c}$$

a relation which can be written as

$$(\Delta, \varphi_\omega)\Big[\ln\frac{T_c}{T} - \sum_{n=-\infty}^{\infty}\Big(\frac{1}{2n+1} - \frac{s(\omega)}{1-(2\pi T\tau)^{-1}s(\omega)}\Big)\Big] = 0\,. \qquad (22.33)$$

Then the eigenfunction φ_ω becomes

$$\varphi_\omega = \text{constant}\ \Delta^* \qquad (22.34)$$

and

$$(\Delta, \varphi_\omega) \neq 0\,.$$

Equation (22.24) can be transformed as

$$\ln\frac{T_c}{T} = \sum_{n=-\infty}^{\infty}\Big[\frac{1}{2n+1} - \frac{s(\omega)}{1-(2\pi T\tau)^{-1}s(\omega)}\Big]\,. \qquad (22.35)$$

The eigenvalue (22.32) can be obtained solving (22.33) (see Appendix) and we get

$$s(\omega) = \frac{4T}{v_0(2|e|H)^{\frac{1}{2}}}\int_0^\infty dw\exp(-w^2)\tan^{-1}w\alpha(\omega) \qquad (22.36)$$

where

$$\alpha(\omega) = \frac{h^{\frac{1}{2}}}{|2n+1|t+\lambda}\,, \quad t = \frac{T}{T_c}\,, \quad \lambda = \frac{0.8\xi_0}{l}\,. \qquad (22.37)$$

and l is the mean free path of the electrons.
Equation (22.35) becomes

$$\ln\frac{1}{t} = \sum_{n=-\infty}^{\infty}\Big[\frac{1}{2n+1} - \frac{(t/h^{\frac{1}{2}})J(\alpha(\omega))}{1-(\rho/h^{\frac{1}{2}})J(\alpha(\omega))}\Big] \qquad (22.38)$$

where

$$\rho = \frac{1}{2\pi T_c\tau} \qquad (22.39)$$

and the integral $J(\alpha(\omega))$ is

$$J(\alpha(\omega)) = 2\int_0^\infty dw \exp(-w^2)\tan^{-1}\alpha(\omega)w \ . \tag{22.40}$$

In the clean limit $\lambda \to 0$ and from (22.35) and (22.37), one obtains

$$\ln\frac{1}{t} = \sum_{n=-\infty}^{\infty}\left\{\frac{1}{2n+1} - \frac{t}{h^{\frac{1}{2}}}\int\frac{d^3\rho}{\rho^2}\exp\left[-\frac{(2n+1)t\rho}{h^{\frac{1}{2}}} - \frac{\rho_\perp^2}{4}\right]\right\} \tag{22.41}$$

which is equivalent to

$$\ln\frac{1}{t} = \lim_{y\to 1}\left\{\ln\frac{2}{1-y} - \left[-\ln\frac{h^{\frac{1}{2}}(1-y)}{t} + \frac{1}{2}\ln\gamma - 1\right]\right\} \tag{22.42}$$

where y is a factor of convergence. Near the critical temperature $t \to 1, h \to 0$ and (22.40) can be expanded as

$$J(\alpha(\omega)) = \sum_{n=0}^{N}(-1)^n\frac{\alpha(\omega)}{2n+1}$$

and from (22.38), one obtains the equation for $h(t)$ as

$$\ln\frac{1}{t} = \sum_{n=-\infty}^{\infty}\left\{\frac{1}{2n+1} - \left[(2n+1) - \frac{h/3t}{(2n+1)t+\lambda}\right]\right\} \tag{22.43}$$

which in the dirty limit $(\lambda \to \infty, h \sim \lambda)$ gives

$$\ln\frac{1}{t} = \Psi\left[\frac{1}{2} + \frac{1}{2}\left(\frac{h}{3\lambda t}\right)\right] - \Psi\left(\frac{1}{2}\right) \tag{22.44}$$

a result obtained by Maki.[10]

This general method has been applied in order to calculate the influence of different kinds of impurities on the upper critical field $h(t)$. Werthamer *et al.*[9] considered the scattering on the non-magnetic impurities taking the spin-orbital scattering. Later, Fulde and Maki[14] and Fischer[15] considered the scattering on the magnetic impurities. The main point of the calculation is the approximation performed in the theory of dilute alloys, namely the average on the impurities. This approach has been used in the calculation of the eigenvalue $s(\omega)$ of the operator

$$\widehat{S}(\mathbf{r},\mathbf{r}';\omega) = \widehat{S}_0(\mathbf{r},\mathbf{r}';\omega) + \int d^3\mathbf{r}_1 d^3\mathbf{r}_2 \widehat{S}_0(\mathbf{r},\mathbf{r}';\omega)\frac{N(0)}{T[\delta(\omega)]^2} \times \langle\widehat{U}(\mathbf{r}_1,\mathbf{r}_2)\widehat{S}(\mathbf{r}_1,\mathbf{r}_2;\omega)\widetilde{U}(\mathbf{r}_1,\mathbf{r}_2)\rangle_c \tag{22.45}$$

where the operator $U(\mathbf{r}_1, \mathbf{r}_2)$ is the Fourier transform of

$$\widehat{U}(\mathbf{p}, \mathbf{p}') = V(\mathbf{p}, \mathbf{p}') + \frac{J(\mathbf{p}, \mathbf{p}')}{2}\boldsymbol{\sigma} \cdot \mathbf{S} + V_{s0}(\mathbf{p}, \mathbf{p}')i(\mathbf{n} \times \mathbf{n}')\boldsymbol{\sigma} \tag{22.46}$$

with $\mathbf{n} = \mathbf{p}/|\mathbf{p}|$.
The kernel $S_0(\mathbf{r}, \mathbf{r}'; \omega)$ from (22.45) is

$$\widehat{S}_0(\mathbf{r}, \mathbf{r}'; \omega) = \frac{T}{2v_0|\mathbf{r} - \mathbf{r}'|} \exp\left[- 2i\mu_0\mathbf{H} \cdot \boldsymbol{\sigma} \operatorname{sign} \omega + 2|\omega| + \tau^{-1} + \frac{|\mathbf{r} - \mathbf{r}'|}{v_0} + 2ie\int_{r'}^{r} d\mathbf{s} \cdot A(\mathbf{s})\right] . \tag{22.47}$$

If we neglect the magnetic scattering, the second term in (22.45) can be approximated as

$$\int d^3\mathbf{r}_1 \langle \widehat{U}(\mathbf{r}, \mathbf{r}_1)\widehat{S}(\omega)\widehat{\widehat{U}}(\mathbf{r}_1, \mathbf{r})\rangle = c[\delta(0)]^2 \left\{ [|V|^2 + \frac{1}{6}|V_{s0}|^2] S^{(1)}_{(\omega)} + \left[|J|^2 - \frac{1}{8}|V|^2 S^{(2)}_{(\omega)}\boldsymbol{\sigma} \cdot \mathbf{H} \operatorname{sign} \omega\right] \right\} \tag{22.48}$$

where

$$\widehat{S}(\omega) = \widehat{S}^{(1)}_{(\omega)} + \widehat{S}^{(2)}_{(\omega)}\boldsymbol{\sigma} \cdot \mathbf{H} \operatorname{sign} \omega$$
$$\widehat{S}_0(\omega) = \widehat{S}_0^{(1)}(\omega) + S_0^{(2)}(\omega)\boldsymbol{\sigma} \cdot \mathbf{H} \operatorname{sign} \omega .$$

In order to calculate the critical field, we take

$$\frac{1}{2}\operatorname{Tr} \widehat{S}(\omega) = S^{(1)}(\omega) = \left\{ \operatorname{Re}\left[I^{-1}(\omega) - \left(\tau^{-1} - \frac{4}{3}\tau_s^{-1}\right)^{-1} (2\pi T)^{-1} - \frac{1}{3}(2\pi T\tau_s)^{-1}\right]\right\}^{-1} \tag{22.49}$$

where τ^{-1} and τ_s^{-1} have been defined by (16.75a)–(16.75c) and

$$I(\alpha(\omega)) = \frac{2\pi T}{v_0(2|e|H)^{\frac{1}{2}}} 2\int_0^\infty dw \exp(-w^2) \tan^{-1} \frac{1 + izw}{1 - izw} \tag{22.50}$$

where

$$\alpha(\omega) = v_0(2|e|H)^{\frac{1}{2}}[2|\omega| + \tau^{-1} + 2i\mu H]^{-1} . \tag{22.51}$$

Following the same way as for the (22.35), one gets

$$\ln\frac{T_c}{T}=\sum_{n=-\infty}^{\infty}\left\{\frac{1}{2n+1}-\left[2n+1+h_1\frac{T_c}{T}+\frac{(\alpha h_1 T_c/T)^2}{|2n+1|+(h_1+\lambda_{s0})T_c/T}\right]\right\} \tag{22.52}$$

with the notations

$$h_1=\frac{2|e|H(v_0^2\tau)}{6\pi T_c}\,,\qquad \lambda_{s0}=(3\tau_{s0}T_c)^{-1}\,,$$
$$\tau_{s0}^{-1}=\frac{cN(0)}{2}\int d\Omega|V_{s0}|^2\,,\qquad \alpha=\frac{3}{mv_0^2\tau}\,.$$

From (22.52), the equation for the critical field can be written as

$$\ln\frac{1}{t}=\left(1+\frac{i\lambda_{s0}}{4\gamma}\right)\Psi\left[\frac{1}{2}+\frac{1}{2t}\left(h_1+\frac{1}{2}\lambda_{s0}+i\gamma\right)\right]$$
$$+\left(1-\frac{i\lambda_{s0}}{4\gamma}\right)\Psi\left[\frac{1}{2}+\frac{1}{2t}\left(h_1+\frac{1}{2}\lambda_{s0}-i\gamma\right)\right] \tag{22.53}$$

where

$$\gamma=\left[(\alpha h_1)^2-\frac{1}{2}\lambda_{s0}^2\right]^{\frac{1}{2}}\,.$$

The influence of the magnetic impurities $J\neq 0$ can be considered in the framework of the theory of dilute alloys and the critical field H_{c2} will be given by

$$\ln\frac{T}{T_c}+\frac{1}{2}\left[\left(1+\frac{b}{\sqrt{b^2-I^2}}\right)\Psi\left(\frac{1}{2}+\rho_-\right)\right.$$
$$\left.+\frac{1}{2}\left(1-\frac{b}{\sqrt{b^2-I^2}}\right)\Psi\left(\frac{1}{2}+\rho_-\right)\right]-\Psi\left(\frac{1}{2}\right)=0 \tag{22.54}$$

where

$$\rho_\pm=\frac{a\pm\sqrt{b^2-I^2}}{2\pi T}\,,$$
$$a=\frac{1}{\tau}\frac{\langle S_x^2\rangle+\langle S_y^2\rangle}{S^2}+\tau_{s0}^{-1}+\varsigma\Delta\,,$$
$$b=\tau_{s0}^{-1}-\tau^{-1}\frac{\langle S_x^2\rangle}{S^2}\,,$$
$$\varsigma=\frac{2}{3}\tau_{tr}(qv_0)^2\,,$$
$$\tau_{tr}^{-1}=cN(0)\int d\Omega(1-\cos\theta)|V(0)|^2\,,$$
$$I=\frac{cJ}{2}\langle S_z\rangle$$

and the scattering times τ, τ_s, τ_{s0} are defined by (16.75a)–(16.75c).

The case of superconductors with a strong paramagnetism has been studied by Maki,[12] Caroli *et al.*[13] and the important conclusion obtained by similar calculations is that the parameters $k_1(T)$ and $k_2(T)$ from the Abrikosov theory are identical at $T = 0$ for the dirty superconductors.

The influence of the different pair-breaking mechanisms can change the standard behaviour of the upper critical field as we can see from Fig. 22.

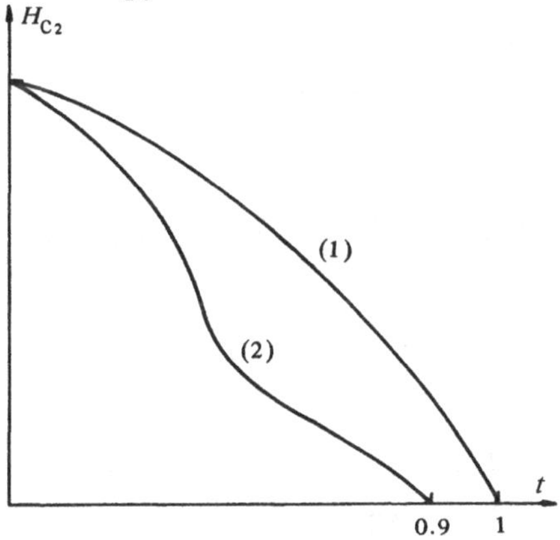

Fig. 22. Critical field H_{c2} as function of temperature t: (1) assuming $\lambda_{s0} = 0$, (2) $\lambda_{s0} = 1$ and $\lambda_m = 0.035$.

A detailed discussion of these effects has been presented by Fischer[15] in his investigations on the magnetic superconductors.

Another problem which has to be considered is the influence of an energy dependent density of states on the upper critical field $H_{c2}(T)$. This problem has been treated by Crisan[16] using the Eilenberger equations.

Appendix

1. *The kernel* $S(\mathbf{r}, \mathbf{r})$

The method for the calculation of the critical magnetic field has as basic idea the condition for divergence of the two-particle propagator

$$\Lambda(\mathbf{r}_1, \mathbf{r}_2; \omega) = g\pi T \sum_{\omega'} \int d^3\mathbf{r}' G(\mathbf{r}, \mathbf{r}_1; \omega) \Lambda(\mathbf{r}_1, \mathbf{r}_2; \omega') G(\mathbf{r}, \mathbf{r}_1; \omega - \omega') \,. \tag{A.1}$$

This equation can be re-written as

$$\Lambda(\mathbf{r}) = g\pi T \sum_{\omega'} \int d^3\mathbf{r}' S(\mathbf{r},\mathbf{r}';\omega')\Lambda(\mathbf{r}') \tag{A.2}$$

where

$$S(\mathbf{r},\mathbf{r}';\omega) = G(\mathbf{r},\mathbf{r}';\omega)G(\mathbf{r},\mathbf{r}';-\omega) \tag{A.3}$$

is the kernel of the integral equation (A.2) and can be calculated.

In the presence of the impurities, we perform the usual average on the impurities and (A.2) becomes

$$\bar{\Lambda}(\mathbf{r}) = g\pi T \sum_{\omega'} \int d^3\mathbf{r}' \bar{S}(\mathbf{r},\mathbf{r}';-\omega)\bar{\Lambda}(\mathbf{r}') \tag{A.4}$$

where

$$S(\mathbf{r},\mathbf{r}';\omega) = \overline{G(\mathbf{r},\mathbf{r}';\omega)G(\mathbf{r},\mathbf{r}';-\omega)} \,. \tag{A.5}$$

In order to calculate (A.5) we have to know the average G. The average G is given in terms of the Green function G_0, for the pure metal, by

$$\bar{G}(\mathbf{r},\mathbf{r}';\omega) = \bar{G}^0(\mathbf{r},\mathbf{r}';\omega) + \int d^3\mathbf{r}'' d^3\mathbf{r}''' G^0(\mathbf{r},\mathbf{r}''';\omega)\widehat{\Sigma}(\mathbf{r}'',\mathbf{r}''';\omega)\bar{G}(\mathbf{r}''',\mathbf{r}';\omega) \tag{A.6}$$

where

$$\begin{aligned}\widehat{\Sigma}(\mathbf{r}'',\mathbf{r}''';\omega) &= \overline{U(\mathbf{r}'')G(\mathbf{r}'',\mathbf{r}''';\omega)U(\mathbf{r}''')} \\ &= \Delta(\mathbf{r}''-\mathbf{r}''')\bar{G}(\mathbf{r}'',\mathbf{r}''';\omega)\end{aligned} \tag{A.7}$$

and

$$\Delta(\mathbf{r}''-\mathbf{r}''') = \sum_{\mathbf{q}} |U(\mathbf{q})|^2 \exp[i\mathbf{q}\cdot(\mathbf{r}-\mathbf{r}')] \,.$$

In the absence of the magnetic field, (A.6) is given by

$$\bar{G}(\mathbf{r},\mathbf{r}';\omega) = G^0(\mathbf{r},\mathbf{r};\omega)\exp\left[-\frac{1}{2l}|\mathbf{r}-\mathbf{r}'|\right] \tag{A.8}$$

where

$$l^{-1} = (v_0\tau)^{-1} = \frac{\pi N(0)}{v_0}\int d\Omega_{k,k'}|U(\mathbf{k}-\mathbf{k}')|^2 \,. \tag{A.9}$$

In order to see which is the effect of the magnetic field, we consider H in the x-direction and

$$\mathbf{A} = (0, A(x), 0)$$
$$H = \frac{\partial A(x)}{\partial x} .$$

The equations for the Green function are

$$\Big\{i\omega - \frac{1}{2m}[\nabla_x^2 + (\nabla_y + ieA(x))^2 + \nabla_z^2]\Big\} G^0(\mathbf{r},\mathbf{r}';\omega) = \delta(\mathbf{r}-\mathbf{r}') ,$$
$$\Big\{-i\omega - \frac{1}{2m}[\nabla_{x'} + (\nabla_{y'} - ieA(x))^2 + \nabla_{z'}^2]\Big\} G^0(\mathbf{r},\mathbf{r}';\omega) = \delta(\mathbf{r}-\mathbf{r}') . \tag{A.10}$$

A solution for Eqs. (A.10) is

$$\bar{G}(\mathbf{r},\mathbf{r}';\omega) = G^0(\mathbf{r},\mathbf{r}';\omega)\exp\Big[-\frac{1}{2l}|\mathbf{r}-\mathbf{r}'|\Big] \tag{A.11}$$

with

$$G^0(\mathbf{r},\mathbf{r}';\omega) = g_0(\mathbf{r}-\mathbf{r}';\omega)\exp[iA(y-y')] \tag{A.12}$$

where

$$g_0(\mathbf{r}) = \frac{m}{2\pi r}\exp\Big[-i(\operatorname{sign}\omega)p_0 r - \frac{|\omega|r}{v_0}\Big] . \tag{A.13}$$

Using, now in (A.6), the Green function (A.11), we obtain an equation for the kernel

$$\bar{S}(\mathbf{r},\mathbf{r}';\omega) = \bar{S}_0(\mathbf{r},\mathbf{r}';\omega) + (2\pi\tau N(0))^2\int d^3\mathbf{r}_1 S_0(\mathbf{r},\mathbf{r}_1;\omega)\bar{S}(\mathbf{r},\mathbf{r}';\omega) \tag{A.14}$$

and if we integrate over $\mathbf{r}$, Eq. (A.14) becomes

$$\int d^3\mathbf{r}\bar{S}(\mathbf{r},\mathbf{r}';\omega) = \int d^3\mathbf{r}S_0(\mathbf{r},\mathbf{r}';\omega)\Big[1-(2\pi N(0)\tau)^{-1}\int d^3\mathbf{r}S_0(\mathbf{r},\mathbf{r}';\omega)\Big] . \tag{A.15}$$

Using

$$\bar{G}(\mathbf{r},\omega) = \frac{m}{2\pi r}\exp\Big[\Big(\frac{ip_0}{\omega} - \frac{1}{v_0}\Big)|\omega|r - \frac{r}{2e}\Big]$$

we get for $S_0 = \bar{G}\,\bar{G}$, the expression

$$S_0(\mathbf{r};\omega) = \Big(\frac{m}{2\pi r}\Big)^2\exp\Big[\frac{2|\omega|r}{v_0} - \frac{r}{l}\Big] \tag{A.16}$$

a result which has been used in (22.27).

2. The calculation of the eigenvalue

Equation (22.32) will be written as

$$\frac{T}{2v_0}\int d^3\mathbf{r}'\frac{1}{|\mathbf{r}-\mathbf{r}'|^2}\exp\Big[-\Big(\frac{2|\omega|}{v_0}+\frac{1}{l}\Big)|\mathbf{r}-\mathbf{r}'| + 2ie\int_{r'}^{r} d\mathbf{s}\cdot\mathbf{A}(\mathbf{s})\Big]\varphi(\mathbf{r}';\omega) = s(\omega)\varphi(\mathbf{r};\omega) \qquad \text{(A.17)}$$

where

$$\varphi(\mathbf{r};\omega) = \exp[(\mathbf{r}-\mathbf{r}')\nabla]\varphi(\mathbf{r};\omega)\Big|_{\mathbf{r}''=\mathbf{r}} .$$

The integral over $\mathbf{s}$ can be calculated as

$$\int_{\mathbf{r}'}^{\mathbf{r}} d\mathbf{s}\cdot\mathbf{A}(\mathbf{s}) = -\int_{\mathbf{r}}^{\mathbf{r}'} d\mathbf{s}\cdot\mathbf{A}(\mathbf{s}) = \int_{\theta=0}^{1} d\theta(\mathbf{r}-\mathbf{r}')A[\mathbf{r}+\theta(\mathbf{r}'-\mathbf{r})] . \qquad \text{(A.18)}$$

Using now the Feynman identity

$$\exp\left\{\int_0^1 d\theta\exp(\theta P)Q\exp(-\theta P)\right\}\exp P = \exp[P+Q] ,$$

we get for (A.18)

$$\begin{aligned}
&\exp\left\{2ie\int_{\mathbf{r}'}^{\mathbf{r}} d\mathbf{s}\cdot\mathbf{A}(\mathbf{s})\right\}\varphi(\mathbf{r}';\omega)\\
&= \exp\Big[-\int_0^1 d\theta(\mathbf{r}-\mathbf{r}')2ieA[\mathbf{r}''+\theta(\mathbf{r}'-\mathbf{r}'')]\Big]\\
&\quad\times\exp[(\mathbf{r}-\mathbf{r}')\nabla]\varphi(\mathbf{r};\omega)\\
&= \exp[\mathbf{r}'-\mathbf{r}'']\Big|_{\mathbf{r}''=\mathbf{r}}
\end{aligned}$$

where we considered $P=0$ and

$$Q = (\mathbf{r}'-\mathbf{r}'')\cdot(\nabla-2ie\mathbf{A}) .$$

Using these results, (A.17) becomes

$$\frac{T}{2v_0}\int\frac{d^3\mathbf{r}'}{|\mathbf{r}'-\mathbf{r}''|^2}\exp\Big[-\Big(\frac{2|\omega|}{v_0}-\frac{1}{l}\Big)(\mathbf{r}'-\mathbf{r}'')\cdot(\nabla-2ie\mathbf{A})\Big]\varphi(\mathbf{r};\omega) = s(\omega)\varphi(\mathbf{r};\omega) . \qquad \text{(A.19)}$$

If we introduce the new variable

$$\boldsymbol{\rho} = (2|e|H)^{\frac{1}{2}}(\mathbf{r}' - \mathbf{r}'') ,$$

Eq. (A.19) becomes

$$\begin{aligned} \frac{T}{v_0} \int \frac{d^3\boldsymbol{\rho}}{\rho^2} \frac{2|e|H}{[2|e|H]^{\frac{3}{2}}} \exp\left[- \left(\frac{2|\omega|}{v_0} + \frac{1}{l} \right) \frac{\rho}{(2|e|H)^{\frac{1}{2}}} \right] \\ \times \exp[(2|e|H)^{-\frac{1}{2}} \boldsymbol{\rho} \cdot (\nabla - 2ie\mathbf{A})] = s(\omega)\varphi(\mathbf{r};\omega) \end{aligned} \qquad \text{(A.20)}$$

which can be transformed as

$$\begin{aligned} \frac{T}{2v_0(2|e|H)^{\frac{1}{2}}} \int \frac{d^3\rho}{\rho^2} \exp\left[- \frac{\rho}{\alpha(\omega)} \right] \\ \times \exp[(2|e|H)^{-\frac{1}{2}} \boldsymbol{\rho} \cdot (\nabla - 2ie\mathbf{A})]\varphi(\mathbf{r};\omega) = s(\omega)\varphi(\mathbf{r};\omega) . \end{aligned} \qquad \text{(A.21)}$$

We can introduce the operators

$$\mathcal{D}_1 = \mathbf{H} \cdot (\nabla - 2ie\mathbf{A}) \qquad \text{and} \qquad \mathcal{D}_2 = (\nabla - 2ie\mathbf{A}) \qquad \text{(A.22)}$$

which satisfy the commutation relation $[\mathcal{D}_1, \mathcal{D}_2]_- = 0$. The eigenfunction $\varphi(\mathbf{r},\omega)$ has to be an eigenfunction for these two operators and thus $\varphi(\mathbf{r};\omega)$ is independent of ω.

Now we introduce the operators:

$$\begin{aligned} a_\pm &= -\frac{i}{2(|e|H)^{\frac{1}{2}}}[(\nabla - 2ie\mathbf{A})_x \pm i(\nabla - 2ie\mathbf{A})_y] \\ a_0 &= (2|e|H)^{-\frac{1}{2}}(\nabla - 2ie\mathbf{A})_z \end{aligned} \qquad \text{(A.23)}$$

and taking $A = (-Hy, 0, 0)$, we can write the operators (A.22) using the operators (A.23) as

$$\mathcal{D}_1 = H(2|e|H)^{\frac{1}{2}} a_0 , \qquad \mathcal{D}_2 = -4|e|\left(a_- a_+ + \frac{1}{2} \right) + 2|e|Ha_0^2 \qquad \text{(A.24)}$$

with $a_0|0\rangle = 0$.

The operators a satisfy the commutation relations

$$[a_-, a_+]_- = 1 \qquad [a_\pm, a_0]_- = 0 .$$

In this formalism, we get

$$(2|e|H)^{-\frac{1}{2}}\rho(\nabla - 2ie\mathbf{A}) = \frac{\rho_\perp}{\sqrt{2}}(ica_- + ic^* a_+) + \rho_\| a_0 \tag{A.25}$$

where $c = e^{i\varphi}$.

Equation (A.20) can now be written as

$$s(\omega)|0\rangle = T(2v_0)^{-1}(2|e|H)^{-\frac{1}{2}} \int \frac{d^3\rho}{\rho^2} \exp\left[-\frac{\rho}{\alpha(\omega)}\right] \times \exp\left[\frac{\rho_\perp}{\sqrt{2}}(ica_- + ic^* a_+)\right] \tag{A.26}$$

and from this relation

$$s(\omega) = T(2v_0)^{-1}(2|e|H)^{-\frac{1}{2}} \int \frac{d^3\rho}{\rho^2} \exp\left[-\frac{\rho}{\alpha(\omega)}\right] \times \langle 0|\exp\left[\frac{\rho_\perp}{\sqrt{2}}(ica_- + ic^* a_+)\right]|0\rangle\,. \tag{A.27}$$

The mean value from (A.27) can be calculated using the identity

$$\exp(A+B) = (\exp A)(\exp B)\exp -\frac{1}{2}[A,B]_-$$

and we get

$$\begin{aligned}
&\langle 0|\exp\left[\frac{\rho_\perp}{\sqrt{2}}ica_- + \frac{\rho_\perp}{\sqrt{2}}ic^* a_+\right]\exp\left[\frac{\rho_\perp^2}{4}\right]|0\rangle \\
&= \exp\left[\frac{\rho_\perp^2}{4}\right]\sum_{n,m}\left(\frac{i\rho_\perp}{\sqrt{2}}\right)^{n+m}\frac{c^{n+m}}{n!m!}\langle 0|a_-^n a_+^m|0\rangle \\
&= \exp\left[-\frac{\rho_\perp^2}{4}\right]\sum_n\left(-\frac{\rho_\perp^2}{2}\right)^n\frac{|c|^{2n}}{n!} \\
&= \exp\left[-\frac{\rho_\perp^2}{4}\right]\sum_{n=0}^{\infty}\frac{(-\rho_\perp/2)^n}{n!}\,.
\end{aligned}$$

With this result (A.27) becomes

$$s(\omega) = T(2v_0)^{-1}(2|e|H)^{-\frac{1}{2}} \int \frac{d^3\rho}{\rho^2}\exp[-\rho/\alpha(\omega) - \rho_\perp^2/4] \tag{A.28}$$

and the Fourier transform of (A.28) is

$$s(\omega) = 4\pi T(v_0)^{-1}(2|e|H)^{-\frac{1}{2}} \int_0^\infty dw\,\exp(-w^2)\tan^{-1}\alpha(\omega)w \tag{A.29}$$

the result used in (22.35).

References

1. A. M. Clogston, *Phys. Rev. Lett.* **9** (1962) 266.
2. B. S. Chandrasekhar, *Appl. Phys. Lett.* **1** (1962) 7.
3. G. Sarma, *J. Phys. Chem. Solids* **24** (1962) 1029.
4. A. I. Larkin and Yu. Ovchinnikov, *Zh. Exp. Teor. Fiz.* **47** (1964) 1136 (*Sov. Phys. JETP* **20** (1965) 762).
5. L. G. Aslamazov, *Zh. Exp. Teor. Fiz.* **55** (1968) 2293.
6. P. Fulde and R. A. Ferrell, *Phys. Rev.* **135** (1964) 550.
7. S. Takada and T. Yzuyama, *Prog. Teor. Phys.* **41** (1969) 635, **43** (1970) 27.
8. E. Helfand and N. R. Werthamer, *Phys. Rev.* **147** (1966) 288.
9. N. R. Werthamer, E. Helfand and P. C. Hohenberg, *Phys. Rev.* **147** (1967) 295.
10. K. Maki, *Phys. Rev.* **141** (1966) 275.
11. K. Maki, *Phys. Rev.* **148** (1966) 362.
12. K. Maki, *Physics* **121** (1964) 1, 127.
13. C. Caroli, M. Cirot and P. G. de Gennes, *Solid State Commun.* **4** (1966) 17.
14. P. Fulde and K. Maki, *Phys. Rev.* **130A** (1965) 788.
15. O. Fischer, Helv. *Phys. Acta* **45** (1972) 330.
16. M. Crisan, *J. Low Temp. Phys.* **64** (1986) 233.

V
SUPERCONDUCTIVITY AND MAGNETIC ORDER

The first theoretical study on the interplay between superconductivity and magnetic order was published in 1957 by Ginzburg. He showed that coexistence between ferromagnetism and superconductivity is almost impossible. Later, Baltensperger and Strasler pointed out the possibility of the coexistence between superconductivity and antiferromagnetism. They showed that although the time reversal symmetry is broken by the antiferromagnetic order, the superconductivity may coexist with antiferromagnetism in a form of slightly modified pairing system.

The experimental attempts of the coexistence between superconductivity and magnetic order have been started by Matthias in 1959. The first researches have been done by alloying the usual superconductor with magnetic impurities. However, in this way the coexistence between superconductivity and magnetic order could not be obtained because the superconductivity is destroyed by paramagnetic impurities.

In 1977, Matthias and Maple in the USA, and Fischer in Switzerland discovered that the coexistence between superconductivity and magnetic order is possible in the ternary rare-earth compounds. These materials present different kinds of order: ferromagnetic, antiferromagnetic and spatial modulated spin order. The theory developed to explain the coexistence of superconductivity and magnetic order uses the results from the theory of the dilute alloys but the special features as the influence of the electromagnetic field gives rise to some new aspects.

23. Superconductivity and Ferromagnetism

Superconductivity and ferromagnetism are two states with long-range order that may exist in materials at low temperatures; they compete with one another and it is difficult to understand how they coexist. However, their coexistence has been demonstrated by the experimental results obtained on the ternary compounds of the rare-earths ($ErRh_4B_4$ and $HoMo_6S_8$)[1,2] and a simple model could be one with the rare-earth elements forming the ferromagnetic lattice, while the transition metal creates the superconducting state.

The Hamiltonian which describes such a model is

$$\begin{aligned} \mathcal{H} = & \sum_{\mathbf{p},\alpha} \varepsilon(\mathbf{p}) c^\dagger_{\mathbf{p}\alpha} c_{\mathbf{p}\alpha} - \Delta \sum_{\mathbf{p}} (c^\dagger_{\mathbf{p}\uparrow} c^\dagger_{-\mathbf{p}\downarrow} + c_{-\mathbf{p}\downarrow} c_{\mathbf{p}\uparrow}) \\ & - \frac{I}{2N} \sum_i \sum_{\substack{\mathbf{p},\mathbf{p}' \\ \alpha,\beta}} c^\dagger_{\mathbf{p}\alpha} \sigma_{\alpha\beta} \cdot c_{\mathbf{p}\beta} \mathbf{S}_i \\ & \times \exp[iR_i(\mathbf{p}-\mathbf{p}')] - \frac{1}{2} \sum_{i\neq j} J_{ij} \mathbf{S}_i \cdot \mathbf{S}_j \end{aligned} \tag{23.1}$$

where the first two terms describe the superconducting state, the third describes the interactions between electrons and spins (with the coupling constant I) and the last, the spin-spin interaction (with the coupling constant J) is supposed to be ferromagnetic.

Maekawa and Tachiki[3] calculated the critical temperature T_c of the ferromagnetic superconductor as

$$T_c = 1.14\omega_D \exp\left[-1\Big/\left(gN(0) - \frac{3C_0 I^2}{4p_0^2 a^2} \ln \frac{T_c - T_M + 4p_0^2 a^2}{T_c - T_M}\right)\right] \tag{23.2}$$

where g is the BCS electron-electron interaction and $N(0)$ the density of states. The spin-spin correlation has been considered as

$$\chi(q,T) = C_0/(T - T_M + a^2 q^2) \tag{23.3}$$

where T_M is the magnetic transition temperature and

$$C_0 = S(S+1)T_M \,.$$

If we consider the interaction between conduction electrons and the local moments in the Born approximation, we can use the results from the theory of dilute alloys to calculate all the physical quantities for the ferromagnetic superconductor. The only difference (which is expected to occur) is that for a ferromagnetic superconductor, we do not perform an "average on the impurities" but a summation on the regular lattice.

The Green functions G and G^0 will be considered of the form

$$\begin{aligned} \widehat{G}^{-1}(\mathbf{p};\omega) &= \widehat{G^0}^{-1}(\mathbf{p},\omega) - \widehat{\Sigma}(\mathbf{p};\omega) = i\omega - \varepsilon(\mathbf{p})\widehat{\tau}_3\widetilde{\Delta}\tau_1\widehat{\sigma}_y \\ \widehat{G^0}^{-1}(\mathbf{p},\omega) &= (i\omega - \varepsilon(\mathbf{p})\widehat{\tau}_3 - \Delta\widehat{\tau}_1\widehat{\sigma}_y)^{-1} \end{aligned} \tag{23.4}$$

where $\widehat{\Sigma}$ will be taken as

$$\Sigma(\mathbf{p};\omega) = -\frac{1}{2\tau(T)}\frac{(i\widetilde{\omega} - \widetilde{\Delta}\widehat{\tau}_1\widehat{\sigma}_y)}{(\widetilde{\omega}^2 + \widetilde{\Delta}^2)^{\frac{1}{2}}}\,\mathrm{sign}\;\omega\;. \tag{23.5}$$

Using for the spin-spin correlation the expression (23.3), the scattering time can be calculated as

$$\frac{1}{\tau(T)} = \frac{T}{\tau_s}f(T) \tag{23.6}$$

where τ_s has been defined by (16.32) and

$$f(T) = \frac{1}{(2ap_0)^2}\ln\frac{T - T_M + (2ap_0)^2}{T - T_M}\;. \tag{23.7}$$

Equation (23.4) gives

$$\frac{\omega}{\Delta} = u\left(1 - \frac{1}{\tau(T)}\frac{1}{\sqrt{1+u^2}}\right) \tag{23.8}$$

and the equation for the gap

$$\Delta = -g\pi T\sum_{\omega}\int\frac{d^3\mathbf{p}}{(2\pi)^3}\frac{1}{4}\mathrm{Tr}\,(\widehat{\tau}_1\widehat{\sigma}_y\widehat{G}(\mathbf{p};\omega)) \tag{23.9}$$

can be written in terms of u as

$$\Delta = gN(0)2\pi T\sum_{\omega}(1+u^2)^{-\frac{1}{2}}\;. \tag{23.10}$$

This equation can be linearized near the critical temperature T_c and we get

$$\ln\frac{T_c}{T_{c0}} = \Psi\left(\frac{1}{2}\right) - \Psi\left(\frac{1}{2} + \rho_c(T_c)\right) \tag{23.11}$$

where

$$\rho_c(T_c) = \frac{1}{2\pi T_c \tau(T_c)} . \tag{23.12}$$

This result has been obtained by Machida and Younger[4] and from Fig. 23, it is easy to see that there are always two temperatures which give solutions to this equation. The upper critical temperature T_{c1} which can be calculated as

$$T_{c1} = T_{c0} - \frac{\pi}{4}\frac{1}{\tau(T_c)} \tag{23.13}$$

can be transformed

$$\begin{aligned}\frac{T_{c1}}{T_{c0}} = 1 - \frac{\pi}{4}\frac{1}{\tau_{AG}(2ap_0)^2}\ln\frac{T_{c0}+(2ap_0)^2}{T_{c0}} \\ - \frac{\pi}{4}\frac{1}{T_{c0}+(2ap_0)^2}\frac{T_M}{T_{c0}}\end{aligned} \tag{23.14}$$

if $T_M \ll T_{c0}$. The second solution of (23.11) T_{c2} cannot be calculated analytically, but it is very near to T_M. The upper critical field H_{c2} has been calculated using the standard method presented in Ref. 22.

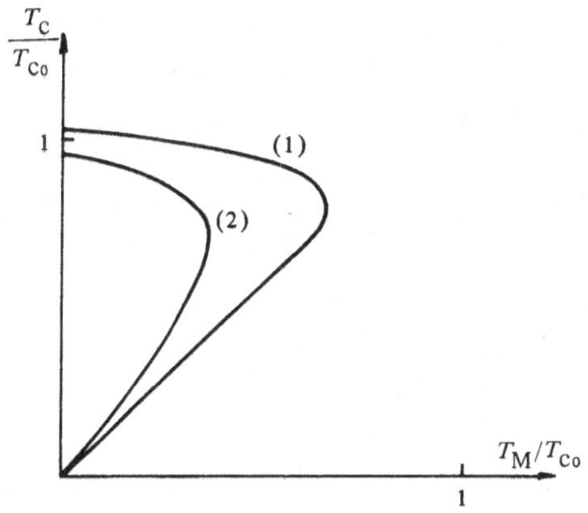

Fig. 23. The phase diagram for a ferromagnetic superconductor. The first line (1) corresponds to $(2p_0a)^2/T_{c0}=0.5$ and $N(0)(I/2)^2S(S+1)/T_{c0}=0.01$. The second line (2) is for the same value of the first parameter but the second parameter is 0.07.

The equation for the order parameter in the presence of the electromagnetic field A is

$$\Delta(x) = g\pi T\sum_\omega \int dx_1 Q_{\uparrow\downarrow}(x_1 - x,\omega)\left\{\exp\left[2e\int_x^{x_1} ds\cdot\mathbf{A}(s)\right]\Delta(x_1)\right\} \tag{23.15}$$

where

$$Q_{\uparrow\downarrow}(\mathbf{x};\omega) = G_{\downarrow}^{0}(\mathbf{x};\omega)G_{\uparrow}^{0}(\mathbf{x};-\omega) \tag{23.16}$$

and

$$G_{\alpha}^{0}(\mathbf{x},\omega) = -\frac{m}{2\pi|\mathbf{x}|}\exp\left\{ip_{0\alpha}|\mathbf{x}|\mathrm{sign}\omega - \frac{|\tilde{\omega}|}{v_{0\alpha}}|\mathbf{x}|\right\} \tag{23.17}$$

where

$$\tilde{\omega} = \omega + \frac{1}{2}(\tau_0^{-1} + \tau^{-1}(T)) . \tag{23.18}$$

In (23.18), τ is the scattering time for the non-magnetic impurities and $p_{0\alpha} = p_0 - \alpha h/v_0$.

The internal molecular field h acting on the conduction electrons contains a term which describes the interaction of the conduction electrons with the local moments and the Pauli paramagnetic term:

$$h = \frac{I\chi(T)}{2\mu_0 N} + \mu_0 B \tag{23.19}$$

where N is the number of the local spins per unit volume and

$$\chi(T) = [(\mu_0 g)^2 N/T]\langle S_{\mathbf{q}} S_{-\mathbf{q}}\rangle .$$

In the dirty limit $\tau^{-1} \gg T_c$, the upper critical field for a ferromagnetic superconductor is given by

$$\ln\frac{T}{T_{c0}} + \mathrm{Re}\,\Psi\left(\frac{1}{2} + \frac{1}{2\pi T\tau(T)} + \frac{ih}{2\pi T} + \frac{DeB}{2\pi T}\right) = \Psi\left(\frac{1}{2}\right) . \tag{23.20}$$

This equation can be generalized to include the effect of spin-orbit scattering. If we define the spin orbit scattering time by (16.75c), the equation for the upper critical field $H_{c2}(T)$ is

$$\ln\frac{T}{T_c} = \frac{1}{2}\left[\left(1 + \frac{b}{\sqrt{b^2 - h^2}}\right)\Psi\left(\frac{1}{2} + \rho_-\right) + \left(1 - \frac{b}{\sqrt{b^2 - h^2}}\right)\Psi\left(\frac{1}{2} + \rho_+\right)\right] - \Psi\left(\frac{1}{2}\right) \tag{23.21}$$

where

$$\rho_{\pm} = \frac{1}{2\pi T}\left[\frac{1}{\tau(T)} + \frac{1}{\tau_{s0}} + DeB \pm (b^2 - h^2)^{\frac{1}{2}}\right] \tag{23.22}$$

$b = 1/\tau_{s0}$, τ_{s0} being the spin-orbit scattering time.

This case becomes important in the high-field type II superconductors. We have to note that the molecular field acting on the conduction electrons is linear in I. If the exchange integral J is negative, the spin polarization counteracts the effect of Pauli paramagnetism and increases the upper critical field. This is the Jaccarino-Peter effect.[5]

From the phase diagram given in Fig. 24, we see that the system always exhibits a re-entrance phenomenon. In this model, there is no region where the two long-range orders coexist, but as we mentioned some ternary compounds present this coexistence. These results have been generalized[6] by the same authors taking the elastic and inelastic scattering of the electrons on the localized spins. These effects can be considered taking for the linearized equation of the order parameter the form

$$\Delta(T) = gQ(T)\Delta(T) \tag{23.23}$$

where

$$Q(T) = \pi T \sum_{\omega} \sum_{\mathbf{p}} \gamma(\omega) G(\mathbf{p}, \omega) G(\mathbf{p}, -\omega) \tag{23.24}$$

with the Green function

$$G^{-1}(\mathbf{p}, \omega) = i\omega - \varepsilon(\mathbf{p}) - \Sigma(\omega) \ . \tag{23.25}$$

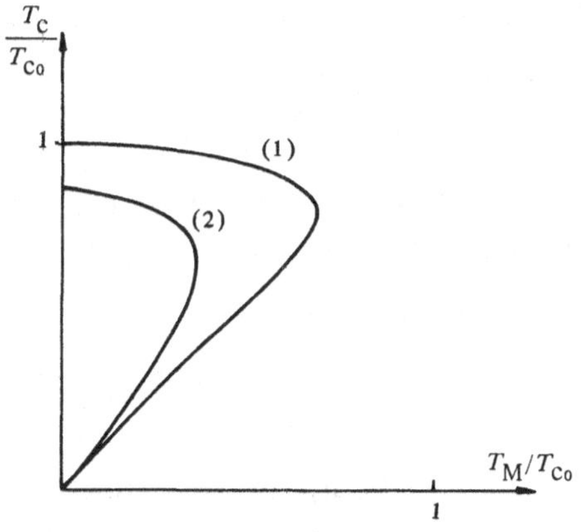

Fig. 24. The phase diagram of the ferromagnetic superconductor with dynamical spin-spin correlation. Line (1) is identical with the line (1) from Fig. 23 and corresponds to $\alpha = 1$, $\beta = 0$. Line (2) has been obtained for $\alpha = 0.5$, $\beta = 0.5$.

Performing the momentum integration in (23.24), we obtain

$$Q(T) = \pi T N(0) \sum_{\omega} \frac{\gamma(\omega)}{|\omega| + \Sigma(\omega)} \tag{23.26}$$

where $\gamma(\omega)$ will be calculated from the Dyson equation

$$\gamma(\omega) = 1 + \pi T \sum_{\mathbf{p}';\omega} \Gamma_{\uparrow\downarrow}(\mathbf{p},\omega;\mathbf{p}',\omega') G(\mathbf{p},\omega) G(-\mathbf{p}',-\omega')\gamma(\omega') \qquad (23.27)$$

where $\Gamma_{\uparrow\downarrow}$ is the irreducible four-point vertex given in Ref. 7. The self-energy $\Sigma(\omega)$ from (23.26) is given by

$$\hat{\Sigma}(\mathbf{p};\omega) = \left(\frac{I}{2}\right)^2 \pi T \sum_{\mathbf{p}',\omega'} G(\mathbf{p}',\omega)\chi(\mathbf{p}'-\mathbf{p};\omega-\omega') \qquad (23.28)$$

where $\chi(\mathbf{q},\omega)$ is the dynamical susceptibility of the localized spin system. Taking for the four-point vertex, the simple expression

$$\Gamma_{\uparrow\downarrow}(\mathbf{p},\omega;\mathbf{p}',\omega') = -\left(\frac{I}{2}\right)^2 \chi(\mathbf{p}-\mathbf{p}';\omega-\omega') \qquad (23.29)$$

and for the dynamical susceptibility $\chi(\mathbf{q};\omega)$ a diffusion-type equation.

$$\chi(\mathbf{q};\omega) = \chi(\mathbf{q}) \frac{D_0 q^2}{D_0 q^2 + |\omega|} \qquad (23.30)$$

where $\chi(\mathbf{q})$ is given by (23.3), we can calculate the scattering time.

In the limit $Dq^2 \ll |\omega|$ we reobtain the Abrikosov-Gor'kov result namely,

$$\frac{1}{\tau(T)} = \frac{Tf(T)}{\tau_s}$$

where $f(T)$ is given by (23.7) and $\tau_s^{-1} = 2\pi N(0)(I/2)^2 S(S+1)$.

In the inelastic case scattering, the frequencies with $\omega \neq \omega'$ are neglected and the critical temperature is given by

$$\frac{1}{gN(0)} = 2\pi T_c \sum_{\omega} (\omega + \rho_c)^{-1} \qquad (23.31)$$

which gives the upper and lower critical temperatures T_{c1} and T_{c2}. In the limit $D_0 q^2 \gg |\omega|$, we may neglect the frequency dependence in $\chi(\mathbf{q};\omega)$ and Eq. (23.28) gives

$$\Sigma(\omega) = \frac{1}{\tau(T)} \frac{-i\omega}{T} . \qquad (23.32)$$

Taking for (23.29), the expression

$$\Gamma_{\uparrow\downarrow}(\mathbf{p}\omega;\mathbf{p}'\omega') = -\left(\frac{I}{2}\right)^2 \chi(\mathbf{p}-\mathbf{p}')\delta_{\omega,\omega'} \qquad (23.33)$$

the critical temperature can be obtained as solution of the equation

$$\frac{1+\rho_c}{gN(0)-\rho_c} = 2\pi T_c \sum_{\omega} \frac{1}{\omega} . \tag{23.34}$$

We see that effective coupling which gives rise to the superconducting state is weakened by the repulsive paramagnon-mediated interaction. Equation (23.34) has two solutions T_{c1} and T_{c2}. The first critical temperature can be approximated for $(T_{c0} - T_{c1})/T_{c1} \ll 1$ and $\rho_{c1} \ll 1$ as

$$\frac{T_{c0}-T_{c1}}{T_{c1}} = \rho_{c1}\left[\frac{1}{(gN(0))^2} + \frac{1}{gN(0)}\right] . \tag{23.35}$$

The second critical temperature can be calculated only numerically. The dynamical character of $\chi(\mathbf{q};\omega)$ can be considered taking

$$\chi(\mathbf{q};\omega) = \chi(\mathbf{q})[\alpha + \beta\delta_{\omega,0}] \tag{23.36}$$

where α and β are adjustable parameters.

The critical temperature can be calculated from the standard relation $1 = gQ(T_c)$ where

$$Q(T) = \frac{N(0)\Phi(T)}{1+\alpha\rho\Phi(T)} \tag{23.37}$$

where

$$\Phi(T) = 2\pi T \sum_{\omega} \frac{1}{\omega(1+\alpha\rho)+\beta\rho} . \tag{23.38}$$

With these results, we also get two critical temperatures T_{c1} and T_{c2} and the first one T_{c1} can be approximated for $(T_{c0} - T_{c1})/T_{c0} \ll 1$ and $\rho_{c1} \ll 1$ as

$$\frac{T_{c0}-T_{c1}}{T_{c0}} = \rho_{c1}\left\{\alpha\left[\frac{1}{(gN(0))^2} + \frac{1}{gN(0)}\right] + 3\beta\varsigma(2)\right\} . \tag{23.39}$$

The upper critical field H_{c2} can be calculated in a similar way and the temperature dependence presents the same qualitative behaviour.

The most important result predicted by this model is the occurrence of the re-entrance effect. Indeed, the two critical temperatures T_{c1} and T_{c2} predicted in Refs. 4 and 6 have been discovered by different experiments.[1,2] Grest *et al.*[8] calculated the density of states $N_s(\omega)$ for a re-entrant superconductor, using the Eliashberg equations for the spin fluctuations. A reasonable agreement with the tunnelling experiments[9] has been obtained, but more accurate calculations taking the crystal field splitting have to be performed.

The re-entry condition for the ferromagnetic superconductor has been analyzed by Kuper *et al.*[10] in the framework of the Ginzburg-Landau Theory.

Another important feature of the ferromagnetic superconductor is the crossover from the second to a first order phase transition due to a magnetic field applied to the magnetic easy axis. The magnetization measurements showed that near the crossover temperature T^*, the magnetization $M(H,T)$ exhibits a downward quadrature $d^2M(H,T)/dH^2 > 0$ below the upper critical field H_{c2} in contrast with the ordinary type II superconductors where $d^2M(H,T)/dH^2 < 0$. The general theory given by Maki and Tsuneto[11] has been extended by Fujita *et al.*[12] for the ferromagnetic superconductors.

The Hamiltonian (23.1) has been written in the simple form

$$\mathcal{H}_{\text{eff}} = \sum_{\mathbf{p},\alpha} [\varepsilon(\mathbf{p}) - \mu_0 H] C^\dagger_{\mathbf{p}\alpha} C_{\mathbf{p}\alpha} + \Delta \sum_{\mathbf{p}} (C^\dagger_{\mathbf{p}\uparrow} C^\dagger_{-\mathbf{p}\downarrow} + h.c.) \tag{23.40}$$

where

$$H_{\text{eff}} = H_{\text{m}} + \frac{1}{N}\langle S_z \rangle \tag{23.41}$$

and

$$H_{\text{m}} = H + 4\pi M_{\text{f}}(T) \tag{23.42}$$

N_0 being the number of the magnetic ions per unit volume. The effective field (23.41) containing the contributions of the electron-spin and spin-spin interactions can be re-written as

$$H_{\text{eff}} = \alpha(T) H \tag{23.43}$$

where

$$\alpha(T) = 1 + \left(4\pi + \frac{I}{2\mu_0^2 N(0)}\right) \chi_f(T) \tag{23.44}$$

$\chi_f(T)$ being the magnetic susceptibility of the "f" moments. The order parameter can be calculated using the Hamiltonian (23.40) as

$$\ln \frac{T}{T_{c0}} = 2\pi T_{c0} \sum_{\omega} \left[\frac{1}{(\omega + i\rho)^2 + \Delta^2} - \frac{1}{\omega} \right] \tag{23.45}$$

where $\rho = \mu_0 H_{\text{eff}}$.

This equation can be used for the calculation of the Gibbs potential. In the second order approximation, we get

$$\Delta G = G_s - G_0 = N(0) \Big[\Delta^2 \ln \frac{T}{T_c} - \text{Re} \sum_{\omega} [(\omega + i\rho)^2 + \Delta^2]^{-\frac{1}{2}} + \text{Re} \sum_{\omega} [(\omega + i\rho)^{-1} + \Delta^2/\omega] \Big] \tag{23.46}$$

and the magnetization

$$\widetilde{M} = \frac{\partial \Delta G}{\partial H_{\text{eff}}}$$

becomes

$$\widetilde{M} = -\chi_{\text{p}} H_{\text{eff}} \,. \tag{23.47}$$

Equation (23.45) can be developed as

$$\ln \frac{T_{\text{c}}}{T} = f_0(\rho) + \Delta^2 f_1(\rho) - \frac{3}{8} f_2(\rho)\Delta^4 + \ldots \tag{23.48}$$

where

$$f_m = \text{Re}\, 2\pi T \sum_{\omega>0}^{\infty} (\omega + i\rho)^{-(2m+1)} \tag{23.49}$$

and from this equation we can calculate the crossover temperature. The coefficient of the fourth-order term in the Landau expression of the free energy changes the sign at $\rho = 0.308$ which indicates the crossover from a second order phase transition to a first order one at the temperature $T = T_{\text{c0}} = 0.556$.

The numerical calculations of the magnetization jump as function of temperature showed that this jump always appear at the transition point, below H_{c2} which has a convex quadrature.

Careful inspection of the experimental data[1,2] suggested that there might be a narrow coexistence region for temperatures just below T_{M}. The neutron scattering experiments near T_{M} performed for $HoMo_6S_8$ and $ErRh_4B_4$ appear to be very informative concerning the existence of a pure ferromagnetic phase. However, in both compounds the rare-earth magnetic moments are in a sinusoidally modulated state.

Thus, microscopic coexistence of superconductivity and magnetism occur in these ferromagnetic superconductors as well, but the interplay between these two phenomena transforms the ferromagnetic state with the superconducting regions into oscillatory magnetic state which will be described in Sec. 25.

24. Superconductivity and Antiferromagnetism

The coexistence between the superconducting state and the antiferromagnetic order of a localized spin system has been proved[12] in a large class of the rare-earths ternary alloys. These two phases are also competing but as in an antiferromagnet, the total magnetization is zero, the effect of the superconducting state being more complex.

The first problem which is important in an antiferromagnetic superconductor is the pairing mechanism. Privorotsky[13] considers the influence of the spin waves on the electron-electron interaction and showed that this interaction becomes attractive only for the pairs in the triplet state with zero total spin projection, the attraction being maximum for the electrons in the p-state. Baltensperger and Sträsler[14] reconsidered this problem taking also the electron-phonon contribution in the calculation of the effective electron-electron interaction. In this way, they showed that the singlet pairing is possible for the electronic time-reversed states, with a slight change in the electron-electron interaction which is weakened by the antiferromagnetic order.

A simple model, which seems to be appropriate for the ternary compounds, is based on the interaction between electrons and the sublattice magnetization $\langle S^z \rangle$ of the antiferromagnetic state. The Hamiltonian which describes such an interaction is

$$\mathcal{H}_{\text{e-S}} = -\sum_{\mathbf{p},\mathbf{Q}} H(\mathbf{Q})(c^{\dagger}_{\mathbf{p}\uparrow}c_{\mathbf{p}+\mathbf{Q}\uparrow} - c^{\dagger}_{\mathbf{p}\downarrow}c_{\mathbf{p}+\mathbf{Q}\downarrow}) \tag{24.1}$$

where the molecular field $H(\mathbf{Q})$ is

$$H(\mathbf{Q}) = \langle S^z \rangle \frac{I}{2} \sum_{\mathbf{p},\mathbf{Q}} \delta_{\mathbf{p},\mathbf{Q}}$$

and I is the "d-f" interaction.

In the absence of the interaction which breaks the time reversal symmetry, the total Hamiltonian commutes with the time-reversal operator T, i.e., $[\mathcal{H}T] = 0$. However, for an antiferromagnetic superconductor, the contribution (24.1) does not commute, i.e., $[\mathcal{H}_{e-S}, T] \neq 0$. However, if we consider the operator $Y = TR$ (R being the translation by a vector connecting the two sublattices), we have $[\mathcal{H}, Y] = 0$. This implies that if $\psi_{\mathbf{p}\alpha}(\mathbf{r})$ is an eigenfunction of the Hamiltonian $\mathcal{H}$ which describes the antiferromagnetic superconductor, then $Y\psi_{\mathbf{p}\alpha}(\mathbf{r})$ is also an eigenfunction with the same energy. The pairing in an antiferromagnetic superconductor appears between the states $\psi_{\mathbf{p}\alpha}(\mathbf{r})$ and $\exp(i\Phi)\psi_{\mathbf{p}\alpha}(\mathbf{r})$ where Φ is an arbitrary phase. Fulde and Keller[15] showed that the one-electron spectrum in the presence of the molecular field is

$$E(\mathrm{p}) = \frac{\varepsilon(\mathrm{p}) + \varepsilon(\mathrm{p}+\mathbf{Q})}{2} \pm \left[\frac{1}{2}\left(\varepsilon(\mathrm{p}) - \varepsilon(\mathrm{p}+\mathbf{Q})\right)^2 + H^2(Q)\right]^{\frac{1}{2}} \tag{24.2}$$

which has gaps at points $\mathbf{p} = \pm\mathbf{Q}/2$. The pairing of the electrons in an antiferromagnetic superconductor has been considered in different ways. Machida *et al.*[16] and Nass *et al.*[17] considered the Hamiltonian

$$\mathcal{H} = \sum_{\mathbf{p},\alpha} \varepsilon(\mathbf{p}) c^{\dagger}_{\mathbf{p}\alpha} c_{\mathbf{p}\alpha} + \sum_{\mathbf{p},\alpha} [\alpha H(Q) c^{\dagger}_{\mathbf{p}\alpha} c^{\dagger}_{\mathbf{p}+\mathbf{Q},\alpha} + h.c.] \\ + \Delta \sum_{\mathbf{p}} (c^{\dagger}_{-\mathbf{p}\downarrow} c^{\dagger}_{\mathbf{p}\uparrow} + h.c.) \tag{24.3}$$

where the order parameter Δ describes the pairing between the states $(\mathbf{p}\uparrow, -\mathbf{p}\downarrow)$. The pairing between the states has been neglected in this theory.

Zwicknagl and Fulde[18] used instead, energetically degenerate eigenstates $(n, \mathbf{p})$ of the Hamiltonian

$$\mathcal{H}_0 = \sum_{n} E_n(\mathbf{p}) c^{\dagger}_{n\mathbf{p}\alpha} c_{n\mathbf{p}\alpha}$$

and calculated the critical temperature T_c of an antiferromagnetic superconductor as

$$T_c = 1.13\omega_D \exp[-1/\eta] \tag{24.4}$$

where

$$\eta = g \sum_{i,\mathbf{p}} \delta[E_i(\mathbf{p}) - \varepsilon_0] \left[1 - \frac{4H^2(\mathbf{p})}{E_{+1}(\mathbf{p}) - E_{-1}(\mathbf{p})}\right] \tag{24.5}$$

and $E_{\pm 1}(\mathbf{p})$ corresponds to the two signs from (24.2). From (24.4), we see that the exchange field may change the effective coupling by a modification of the density of states and by a reduction of the electron-electron interaction due to its influence on the electronic wave function. The latter is more important and leads to a linear decrease of the critical temperature as function of the exchange field.

The model used for the ferromagnetic superconductor can be adopted for the antiferromagnetic superconductor with the mention that in (23.1) the last term gives rise to the antiferromagnetic order. In the Born approximation, Machida[19] calculated the scattering time as

$$\frac{1}{\tau(T)} = 2\pi N(0) \left(\frac{I}{2}\right)^2 \frac{1}{4} \int d\Omega_{\mathbf{q}} \chi(\mathbf{q}, T) \tag{24.6}$$

where the correlation function for the antiferromagnetic ordered spins has been defined by

$$\chi(\mathbf{q}; T) = \langle \mathbf{S}(\mathbf{q}) \cdot \mathbf{S}(-\mathbf{q}) \rangle\ . \tag{24.7}$$

The concrete expression of the scattering time τ given by (24.6) depends on the wave vector $\mathbf{Q}$ which characterizes the antiferromagnetic superlattice structure. If $Q \ll 2p_0$, the superconductivity is affected by the magnetic order because the conduction electrons are effectively scattered on the ordered magnetic ions. For $Q \simeq 2p_0$, the antiferromagnetic order does not affect the superconducting state drastically and the two states may coexist.

In the case $|\mathbf{Q}| \ll 2p_0$, the correlation of the spins given by (24.7) is

$$\chi_0(T) = \frac{S(S+1)T}{T+T_N} \tag{24.8}$$

where T_N is the Neel temperature of the antiferromagnet. Equation (24.6) becomes

$$\frac{1}{\tau(T)} = 2\pi N(0)\left(\frac{I}{2}\right)^2 \chi_0(T) \, . \tag{24.9}$$

The critical temperature T_c can be calculated in the same way as for a ferromagnetic superconductor and we get

$$\ln\frac{T_c}{T_{c0}} = \Psi\left(\frac{1}{2}\right) - \Psi\left(\frac{1}{2} + \rho(T_c)\right) \tag{24.10}$$

where

$$\rho(T_c) = \frac{1}{2\pi T_s} f(T_c) \tag{24.11}$$

$$f(T_c) = \begin{cases} \frac{1}{T+T_N} \, ; & T_c > T_N \\ \frac{1}{6T_N}\left(2 + \frac{\chi_{\parallel}(T_c)}{\chi_{\perp}(T_c)}\right) ; & T_c < T_N \end{cases} \tag{24.12}$$

$\chi_{\parallel}$ and $\chi_{\perp}$ being the parallel, respectively transverse components of $\chi(T)$ for an antiferromagnet. The critical temperature T_c as function of T_N is given in Fig. 25, and we see that a region of coexistence appears in this model.

The mechanism of the coexistence between the antiferromagnetic order and superconductivity is much more complicated than it appears in the simple model presented above. Indeed, the antiferromagnetic order in the spin system gives rise to a change in the density of the electronic states and in the electronic system there may appear the new condensed state: Charge-Density-Waves (CDW) and Spin-Density-Waves (SDW). The last ones give rise to the itinerant-electron antiferromagnetism if the triplet electronic states are considered.

Due to this particular feature of an antiferromagnetic superconductor, the non-magnetic impurities will have a strong effect on this state. The

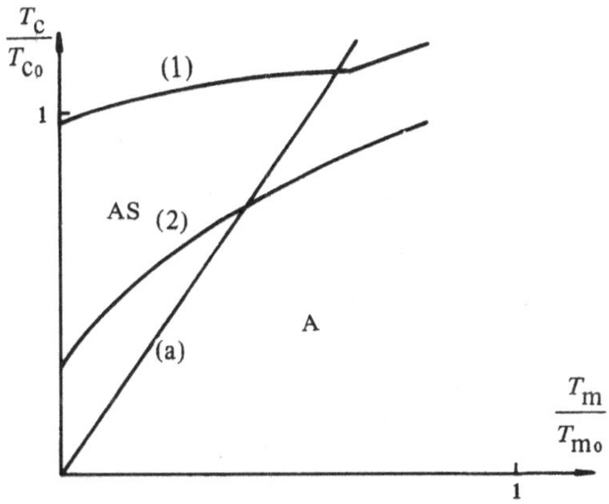

Fig. 25. The phase diagram for an antiferromagnetic superconductor. The lines correspond to $N(0)(I/2)^2 S\,(S+1)/T_{c0}$ as (1) 0.04 and (2) 0.13. Below the straight line (a) the system is only antiferromagnetic (A) and below the curves is an antiferromagnetic superconductor (AS).

Hamiltonian which describes the influence of the non-magnetic impurities on the antiferromagnetic superconductor is

$$\mathcal{H} = \mathcal{H}_{\rm BCS} - \sum_{\mathbf{p};\alpha,\beta} [H(Q) c^\dagger_{\mathbf{p}+\mathbf{Q},\alpha} \sigma_{\alpha\beta} c_{\mathbf{p}\beta} + h.c.] \\ - \frac{1}{N} \sum_{\mathbf{p},\mathbf{p}'} V(\mathbf{p}-\mathbf{p}') c^\dagger_{\mathbf{p}\alpha} c_{\mathbf{p}'\alpha'} \exp[i(\mathbf{p}-\mathbf{p}')\cdot \mathbf{R}_i] \qquad (24.13)$$

where V describes the scattering of the electrons of an antiferromagnetic superconductor on the non-magnetic impurities.

The strong-coupling formalism has been applied by Suzumura *et al.*[20] to calculate the critical temperature of a superconductor described by (24.13). The Green function

$$\widehat{G}^{-1} = \widehat{G}^{0^{-1}} - \widehat{\Sigma} \qquad (24.14)$$

with

$$\widehat{G}^{0^{-1}}(\mathbf{p},\omega) = [i\omega\widehat{1} - \varepsilon(\mathbf{p})\widehat{\tau}_3\widehat{\sigma}_z + \Delta\widehat{\tau}_2\widehat{\sigma}_y + H(Q)\widehat{\tau}_1\widehat{\sigma}_x] \qquad (24.15)$$

can be written as

$$\widehat{G}^{-1}(\mathbf{p},\omega) = [i\widetilde{\omega}\widehat{1} + i\widetilde{\Omega}\widehat{\tau}_1\widehat{\sigma}_x - \varepsilon(\mathbf{p})\tau_3\widehat{\sigma}_z + \widetilde{\Delta}\widehat{\tau}_2\widehat{\sigma}_y + \widetilde{H}\widehat{\tau}_1\widehat{\sigma}_x] \qquad (24.16)$$

where $\widetilde{\Omega}$ is a "pseudo gap" which has to be determined self-consistently.

From (24.14)–(24.16), we obtain

$$\widetilde{\omega} = \omega + \frac{1}{2\tau}\frac{\widetilde{\omega}_{\mp}}{\lambda_{\mp}}\,, \tag{24.17}$$

$$\widetilde{\Delta} = \Delta_{\pm}H(Q) + \frac{1}{2\tau}\frac{\widetilde{\Delta}_{\pm}}{\lambda_{\mp}}\,, \tag{24.18}$$

where

$$\lambda_{\mp} = (\widetilde{\omega}_{\pm}^2 + \widetilde{\Delta}_{\pm})^{\frac{1}{2}}\,, \tag{24.19a}$$

$$\frac{1}{\tau} = \pi N(0)c[V_1^2 + V_2^2]\,, \tag{24.19b}$$

$$\widetilde{\omega}_{\pm} = \omega \pm \widetilde{\Omega} \tag{24.19c}$$

and c is the concentration of the non-magnetic impurities. We considered that the potential $V = V_1 + V_2$, V_1 and V_2 correspond to the backward and forward scattering[20] are of the same order of magnitude. In order to calculate the critical temperature, we write the equation for the order parameter Δ. Following the standard way, we get

$$\Delta = gN(0)\pi T\sum_{\omega}\int_0^{\omega_D} d\varepsilon\left[\frac{\widetilde{\Delta}_+}{\varepsilon^2 + \lambda_+^2} + \frac{\widetilde{\Delta}_-}{\varepsilon^2 + \lambda_-^2}\right] \tag{24.20}$$

and if we introduce the notation

$$u_{\pm} = \frac{\widetilde{\omega}_{\pm}}{\widetilde{\Delta}_{\pm}}\,, \tag{24.21}$$

we get from (24.19)

$$\omega_{\pm} = u_{\pm}(\Delta + H(Q)) \pm \frac{u_+ - u_-}{2\tau(1 + u_{\mp}^2)^{\frac{1}{2}}} \tag{24.22}$$

and Eq. (24.20) becomes

$$\ln\frac{T_c}{T_{c0}} + E(\Delta) = 0 \tag{24.23}$$

where

$$E(\Delta) = \pi T\sum_{\omega}\left[\frac{1}{\omega}\left(\frac{1}{\sqrt{u_+^2 - 1}} - \frac{1}{\sqrt{u_-^2 - 1}}\right) - \frac{2}{\omega}\right]. \tag{24.24}$$

From (24.21), we get

$$u_\pm = u_0 \mp u_1\Delta + u_2\Delta^2 + \dots \tag{24.25}$$

where u_0 is the solution of the equation

$$\omega = u_0\Big[H(Q) - \frac{1}{2\tau}\frac{1}{(1+u_0^2)^{\frac{1}{2}}}\Big] . \tag{24.26}$$

The free energy can be calculated using the gap equation as

$$\frac{\delta F}{N(0)} = B_0(T)\Delta^2 + \frac{1}{4(2\pi T)^2}B_1(T)\Delta^4 + \dots \tag{24.27}$$

where $B_0(T)$ is

$$B_0(T) = \ln\frac{T}{T_{c0}} - 2\pi T\sum_{\omega>0}\Big[\frac{u_0}{H(Q)(1+u_0^2)^{\frac{1}{2}} - \tau_1^{-1}u_0^2} - \frac{1}{\omega}\Big] . \tag{24.28}$$

The expression for $B_1(T)$ is given in Ref. 20 and can be positive or negative.

The critical temperature T_c can be obtained from

$$B_0(T_c) = 0 \tag{24.29}$$

if $B_1(T_c) > 0$. For $B_1(T_c) < 0$, this equation will give a transition from the normal to the antiferromagnetic state but this transition is of the first order.

The critical temperature T_c can be obtained as

$$\frac{T_c}{T_{c0}} = 1 - \frac{\pi}{4}\frac{1}{T_{c0}}\tau_1 H(Q)\,; \quad \tau_1^{-1} \gg H^2(Q)/T_{c0} \tag{24.29a}$$

and

$$\begin{aligned}\ln\frac{T_c}{T_{c0}} &= 2\pi T_c\sum_\omega\Big\{\frac{\omega^2}{(\omega^2+H^2(Q))^{\frac{1}{2}}} - \frac{1}{\omega}\Big\}\\ &= \frac{1}{\tau_1}4\pi T_c H^2(Q)\sum_\omega\frac{\omega^2}{(\omega^2+H^2(Q))^3}\,;\\ &\quad \tau_1 < T_{c0}\,;\ H(Q)/\Delta(0)T < 0.5\,. \end{aligned}\tag{24.29b}$$

From these equations, we see that the pair breaking effect of the non-magnetic impurities appears only if $H(\mathbf{Q}) \neq 0$. If the scattering of the electrons on the non-magnetic impurities is strong, the superconducting state occurs when $\tau^{-1} \gg H(\mathbf{Q})$. In this limit, (24.29) (for $T \to 0$) becomes

$$\ln \frac{H(Q)}{\Delta(0)} = \int_0^\infty d\omega \Big[\frac{u_0 u_1}{H(Q)(1+u_0^2)^{\frac{3}{2}}} - \frac{1}{(\omega^2 + H^2(Q))^{\frac{1}{2}}} \Big] \tag{24.30}$$

which was solved (in the approximation $\tau_1^{-1} \gg H(Q)$) and one obtains

$$(\tau_1 \Delta(0))^{-1} = 2(H(Q)/\Delta(0))^2 - \frac{1}{2} \ln \frac{H(Q)}{\Delta(0)} + C \tag{24.31}$$

where $C \simeq 0.1$.

Going back to (24.23)–(24.26), we can calculate the order parameter for the superconducting gap at $T = 0$ as

$$\frac{\Delta(0)}{\Delta_0(0)} = 1 - \frac{\pi}{4} \frac{1}{\Delta_0} \tau_1^{-1} H(Q)\,; \quad \tau_1^{-1} \gg H^2(Q)/\Delta(0) \tag{24.31a}$$

and

$$\ln \frac{\Delta(0)}{\Delta_0(0)}$$
$$= 2\tau^{-1} H^2(Q) \int_0^\infty d\omega \frac{1}{[\omega^2 + (\Delta + H(Q))^2]^{\frac{3}{2}} [\omega^2 + (\Delta - H(Q))^2]^{\frac{3}{2}}}$$
$$\tau^{-1} \ll \Delta_0(0),\, H(Q)/\Delta_0(0) < 0.5\,. \tag{24.31b}$$

In the limit $\tau^{-1} \gg H(Q)$, (24.22) becomes

$$\frac{\omega}{\Delta} = u \Big[1 - \frac{\tau H^2(Q) \Delta}{(1+u^2)^{\frac{1}{2}}} \Big] \tag{24.32}$$

and an effective pair-breaking time

$$\tau_{\text{eff}}^{-1} = \tau H^2(Q) \tag{24.33}$$

can be defined. The superconducting state is completely destroyed when $\tau_{\text{eff}}^{-1} = \Delta_0/2$ which gives $\tau_1^{-1} = 2H^2(Q)/\Delta_0$. The gapless state occurs if $\Delta_0 \exp(-\pi/4) \ll \tau_{\text{eff}}^{-1}$, a condition which is equivalent to $2H^2(Q) < \Delta_0 \tau_1^{-1} < 2.2 H^2(Q)$ and $0 < \Delta/\Delta_0 < 0.46$.

A surprising result is the occurrence of the superconducting state for the large scattering time $1/\tau_1$ so that $\Delta_0 \tau_1^{-1} \gg 2H^2(Q)$. Another remarkable feature is the suppressing of the molecular field due to the antiferromagnetic spins by the non-magnetic impurities if the scattering is strong enough.

An important parameter which has been calculated for the antiferromagnetic superconductors is the upper critical field H_{c2}, which deviates from the usual behaviour (of dirty superconductors) near the antiferromagnetic critical temperature T_N. The first microscopic calculation of the upper critical field in an antiferromagnetic superconductor is due to Maekawa and Tachiki.[3] These authors neglected the molecular field $H(Q)$ and all spin-fluctuation-scattering processes have been considered as elastic; they could be combined with the phonons into an effective coupling constant. Machida *et al.*[16] included the perturbative effect of a non-zero $H(Q)$.

The calculation of $H_{c2}(T)$ for an antiferromagnetic superconductor implies the calculation of the kernel

$$K(\mathbf{r},\mathbf{r}';\omega) = G^{\uparrow}(\mathbf{r},\mathbf{r}';\omega)G^{\downarrow}(\mathbf{r},\mathbf{r}';-\omega) \tag{24.34}$$

which has the Fourier transform

$$\begin{aligned} &K(\mathbf{p},-\mathbf{p}';\omega) \\ &= \sum_{k_i} G^{\uparrow}(\mathbf{k}_1,\mathbf{k}_2;\omega)G^{\downarrow}(\mathbf{k}_3,\mathbf{k}_4;-\omega)\delta(\mathbf{k}_1+\mathbf{k}_3-\mathbf{p};0)\delta(\mathbf{k}_2+\mathbf{k}_4-\mathbf{p}';0) \end{aligned} \tag{24.35}$$

which will be considered only for $K(0,0;\omega)$ and $K(Q,0;\omega)$. The Green function

$$\begin{aligned} G_\alpha(\mathbf{p},\mathbf{p}';\omega) &= \frac{i\omega-\varepsilon(\mathbf{p}-\mathbf{Q})}{(i\omega-\varepsilon(\mathbf{p}))(i\omega-\varepsilon(\mathbf{p}-\mathbf{Q}))-H^2(Q)} \\ G_\alpha(\mathbf{p},\mathbf{p}-\mathbf{Q};\omega) &= \frac{-\alpha H(Q)}{(i\omega-\varepsilon(\mathbf{p}))(i\omega-\varepsilon(\mathbf{p}-\mathbf{Q}))-H^2(Q)} \end{aligned} \tag{24.36}$$

will be used to calculate the two kernels. In the approximation $H(\mathbf{Q}) \ll \varepsilon_0$, and for a three dimensional Fermi surface of the superconducting system, we get

$$K(0,0;\omega) = \frac{\pi N(0)}{|\omega|}\left[1-\frac{\pi H(Q)}{4\varepsilon_0}\right] \tag{24.37}$$

$$K(Q,0;\omega) = \frac{N(0)}{16\varepsilon_0}\left[1+O(T/\varepsilon_0)\right] \tag{24.38}$$

where in (24.38), a second order contribution in T/ε_0 has been neglected.

Using now the method presented in Sec. 22, we get

$$\alpha m(T) + \ln\frac{T}{T_c} - \Psi\left[\frac{1}{2} + \frac{DeB}{2\pi T}\right] - \Psi\left[\frac{1}{2}\right] = 0 \tag{24.39}$$

where

$$m = \frac{H(Q,T)}{H(Q,0)}, \qquad \alpha = \frac{1}{4}[S(S+1)]^{\frac{1}{2}}\frac{I}{\varepsilon_0 \tilde{g}N(0)} \tag{24.40}$$

and $\tilde{g}N(0) = gN(0) - m(T)$.

The scattering of the conduction electrons on the impurities can be considered in the same way as for a ferromagnetic superconductor, and we get

$$\alpha m(T) + \ln\frac{T}{T_c} + \frac{1}{2}\left\{\left[1 + \frac{b}{\sqrt{b^2 - h^2}}\Psi\left(\frac{1}{2} + \rho_-\right)\right]\right.$$
$$\left. + \left[1 + \frac{b}{\sqrt{b^2 - h^2}}\Psi\left(\frac{1}{2} + \rho_+\right)\right]\right\} = \Psi\left(\frac{1}{2}\right) \tag{24.41}$$

where b, h and and $\rho_\pm$ are given by (23.19) and (23.22).

These calculations have been improved by Ro and Levin[21] using the strong coupling theory formalism in order to consider the spin-fluctuation scattering. The importance of the spin-fluctuations scattering in the antiferromagnetic superconductor has been pointed out by Ramakrishnan and Varma.[22] Indeed, the approximation $H(\mathbf{Q}) \ll \varepsilon_0$ is very important for the calculation of H_{c2} in the antiferromagnetic superconductors. The relevant parameter for this treatment is $H(\mathbf{Q})/\omega_D, \omega_D$ is the Debye cut-off energy and the interplay between the effects of impurities and of the molecular field has to be treated in a more accurate way. Using, instead of (24.36), the Green functions

$$\tilde{G}(\mathbf{p},\mathbf{p};\omega) = \frac{\tilde{\omega} + \varepsilon(\mathbf{p})}{\tilde{\omega}^2 + \varepsilon^2(\mathbf{p}) + H^2(Q)};$$
$$\tilde{G}(\mathbf{p},\mathbf{p} \pm \mathbf{Q};\omega) = \frac{\tilde{H}(Q)}{\tilde{\omega}^2 + \varepsilon^2(\mathbf{p}) + \tilde{H}^2(Q)} \tag{24.42}$$

with

$$\tilde{\omega} = \omega + \pi T \sum_{\omega'} \int_0^{2p_0} \frac{q\,dq}{2p_0^2}\left\{D(\mathbf{q};\omega - \omega') + \chi^{z,z}(\mathbf{q};\omega - \omega')\right.$$
$$+ \frac{1}{2}[\chi^{+,-}(\mathbf{q};\omega - \omega') + \chi^{-,+}(\mathbf{q};\omega - \omega')]$$
$$\left. + \frac{1}{2}(\tau^{-1} - \tau_s^{-1})\frac{\omega'}{(\tilde{\omega}'^2 + \tilde{H}^2(Q))^{\frac{1}{2}}}\right\} \tag{24.43}$$

$$\widetilde{H}(Q) = H(Q) - \pi N(0)T\sum_{\omega'}\int_0^{2p_0}\frac{qdq}{2p_0^2}\Big\{D(\mathbf{q};\omega-\omega') + \chi^{z,z}(\mathbf{q};\omega-\omega')$$
$$+\frac{1}{2}[\chi^{+,-}(\mathbf{q};\omega-\omega') + \chi^{-,+}(\mathbf{q};\omega-\omega')]$$
$$+\frac{1}{2}(\tau^{-1}-\tau_s^{-1})\frac{\widetilde{\omega}'}{(\widetilde{\omega}'^2 + H^2(Q))^{\frac{1}{2}}}\Big\} \tag{24.44}$$

where $D(\mathbf{q};\Omega)$ is the phonon Green function and $\chi(\mathbf{q};\omega)$ the dynamical susceptibility of the localized spins.

With these results, we can calculate the kernel

$$K(\mathbf{p},\mathbf{p}';\omega) = \sum_{\mathbf{q}} g_{\text{eff}}(\mathbf{q};\omega-\omega')\sum_{p_1\dots p_4}\overline{G(\mathbf{p}_1,\mathbf{p}_2;\omega)G(\mathbf{p}_3,\mathbf{p}_4;-\omega)}$$
$$\times\,\delta(\mathbf{p}_1+\mathbf{p}_3;\mathbf{p})\delta(\mathbf{p}_2+\mathbf{p}_4;\mathbf{p}') \tag{24.45}$$

where $g_{\text{eff}} = D(\mathbf{q};\omega-\omega') - \chi^{z,z}(\mathbf{q};\omega-\omega') - \frac{1}{2}[\chi^{+,-}(\mathbf{q};\omega-\omega') + \chi^{-,+}(\mathbf{q};\omega-\omega')]$. The kernel (24.45) has been calculated[21] as

$$K(\pi,-\pi) \cong 2\pi N(0)\Big[1 - \frac{\widetilde{H}^2(Q)}{\widetilde{\omega}^2 + \widetilde{H}^2(Q)}\Big]\tan^{-1}\Big[\frac{v_0\pi}{2(\widetilde{\omega}^2 + \widetilde{H}^2(Q))^{\frac{1}{2}}}\Big] \tag{24.46}$$

where $\pi(x)$ has an eigenfunction

$$\pi^2(x)\Phi(x) = \alpha\Phi(x) \tag{24.47}$$

which is also an eigenfunction of the gap

$$\Delta(x,\omega)\Phi(x) = \Delta(\omega)\Phi(x)\ . \tag{24.48}$$

The eigenvalue α can be expressed by the upper critical field H_{c2} as

$$\alpha = |e|H_{c2} \tag{24.49}$$

and we can obtain a general equation for $H_{c2}(T)$ from (24.48).

In the weak coupling approximation and when the inelastic spin-fluctuations effects are neglected, the equation for H_{c2} has been obtained[21] as

$$\ln\frac{T}{T_{c0}} = \int_0^\infty d\omega\tanh\frac{\omega}{2T}\mathrm{Re}\Big\{\Big[\frac{\widetilde{H}(Q)}{T}(u^2-1)^{\frac{1}{2}} - \frac{ev_0H_{c2}}{6TH(Q)(u^2-1)}$$
$$-\frac{1}{T}(\tau^{-1}-\tau_s^{-1})\Big]^{-1} - \frac{1}{\omega}\Big\} \tag{24.50}$$

where $u = \tilde{\omega}/\tilde{\Delta}$, and from (24.43)–(24.44), one obtains

$$\begin{aligned}\frac{\omega}{H(Q)} &= u\left[1 - \frac{\tau_0^{-1}}{H(Q)}\frac{1}{(1-u^2)^{\frac{1}{2}}}\right] \\ \tau_0^{-1} &= \tau^{-1} + \frac{1}{3}\tau_s^{-1}\,.\end{aligned} \tag{24.51}$$

If the approximation $\tau_0^{-1} \gg H(Q)$, from (24.51), we get

$$u(\omega) = \frac{\omega}{H(Q)} \mp i\frac{\tau_0^{-1}}{H(Q)} \tag{24.52}$$

and

$$\begin{aligned}\tilde{\omega} &= \omega \mp (\tau^{-1} + \tau_0^{-1}) \\ \tilde{H}(Q) &= H(Q)\frac{1 \mp \frac{1}{2}(\tau^{-1} - \frac{1}{3}\tau_s^{-1})}{\omega \mp \frac{1}{2}(\tau^{-1} - \tau_s^{-1})}\,.\end{aligned} \tag{24.53}$$

In this approximation, the impurity scattering time enters only as a pure imaginary contribution to $\tilde{\omega}$, and since $|u|^2 \gg 1$ for $\omega \geq 0$, it follows that $H(Q)(u^2-1)^{\frac{1}{2}} \simeq \omega$ and the terms involving τ are cancelled in Eq. (24.50) and the non-magnetic impurities are not pair breaking in this case. However, we have to mention that this is not in contradiction with the general result obtained in the calculation of the critical temperature, but is strictly conditioned by the approximation $\tau^{-1} \gg H(Q)$. If the antiferromagnetic superconductor is sufficiently dirty, so that the normal state is gapless, the Anderson theorem will be applicable. We also have to mention that the impurities effects on the molecular field $H(Q)$ have been neglected, and a decrease of $H(Q)$ with the decreasing of τ^{-1} is expected which should reinforce all the explicit effects of τ.

In the limit $H(Q) \ll \tau^{-1}$, Eq. (24.50) has to be solved numerically. Taking for $H(Q)$, the temperature dependence

$$H(Q,T) = H(Q,0)[1 - (T/T_{\rm N})^2] \tag{24.54}$$

which is reasonable (because the molecular field is proportional to the sublattice magnetization), the critical field $H_{c2}(T)$ is obtained concretely as function of all parameters.

We show in Fig. 26a the results obtained by Machida *et al.*[19] and in Fig. 26b the results given by (24.50). We see that for any particular values of the parameters, the molecular field $H(Q)$ can destroy the superconductivity for a range of temperatures. If $H(Q)$ approaches the saturation, the superconductivity may reappear. If the system becomes dirty, the effect of the molecular field is decreased so that for such a system, the dip in H_{c2} at the Neel temperature T_N is not very strong. The difference may be done by the fact that in Ref. 19, the effects of impurities were not treated self-consistently and therefore $H(Q)$ was not renormalized. If the spin-spin and the electron-phonon interactions are strong, we have to adopt new approximations in the calculation of the upper critical field. The simple Einstein-like model has been adopted[21] for the spin waves and thus for $T \leq T_N$, the spin-waves energy is independent of the wave vector.

In this case, H_{c2} has been calculated only numerically and in the limit $\tau^{-1} \gg H(Q)$.

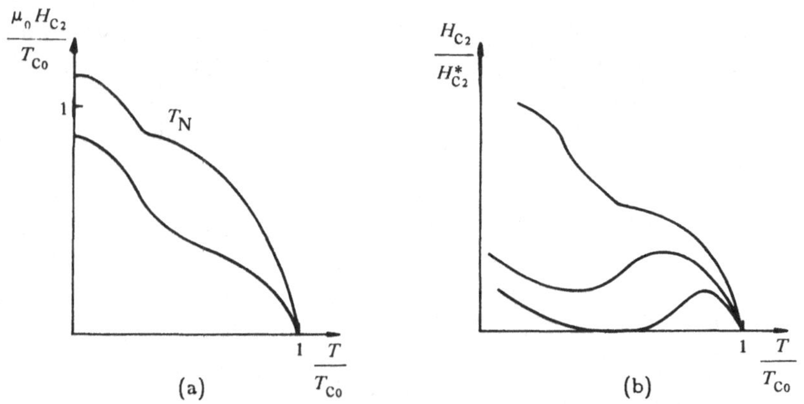

Figs. 26. The upper critical field of an antiferromagnetic superconductor: (a) obtained in Ref. 19 and (b) obtained in Ref. 21 for arbitrary-parameters of theory.

Even in this model,[21] which is in fact one dimensional, the effects due to the nonspherical Fermi surface, the electromagnetic effects, and the paramagnetic effects have been neglected. A different calculation of $H_{c2}(T)$ for the antiferromagnetic superconductors was given by Zwicknagl and Fulde,[23] who considered that the pairing in these superconductors appears between quasiparticle states. The $H_{c2}(T)$ was calculated in Ref. 23 in the strong-coupling formalism neglecting the spin fluctuations. This approximation was justified by the choosing of an appropriate phonon spectral function which is supposed to be modified by the magnetic quasiparticles.

Finally, we have to mention that the antiferromagnetic superconductor presents many other interesting features. Okabe and Nagi[24] showed that for a strong molecular field $H(Q)$, an "impurity band" can grow and in the dirty limit, the gapless superconductivity may appear under any conditions.

The superconducting state can change drastically the spin wave spectrum[25] of the antiferromagnetic spins. The fluctuations of the magnetization near T_N can also affect the behaviour of the superconducting state in the antiferromagnetic compounds.

The antiferromagnetic superconductors appear to be another important aspect of the high temperature superconductivity because the oxides which present a high T_c are antiferromagnets.

25. Magnetic Structures in Superconductors

A self consistent treatment of the coexistence between magnetic order and superconductivity has to consider the changes in the magnetic order due to the presence of the superconductivity. Long time before the occurrence of the microscopic models, Anderson and Suhl[26] suggested the occurrence of a magnetic complicated structure in the superconducting alloys, called "cryptoferromagnetism". This prediction was based on a simple analysis of the exchange energy of a system of localized spins which interact via conduction electrons. This Hamiltonian has the form

$$\mathcal{H}_{\rm exch} = -\sum_{i\neq j} J(\mathbf{R}_i - \mathbf{R}_j)\mathbf{S}_i \cdot \mathbf{S}_j = -\sum_{\mathbf{q}} |J(\mathbf{q})|\mathbf{S}(\mathbf{q}) \cdot \mathbf{S}(-\mathbf{q}) \tag{25.1}$$

where $J(\mathbf{q})$ is the Fourier transform of the exchange interaction, and is proportional to the magnetic susceptibility.
In the limit $ql \gg 1$ (l is the mean free path), the susceptibility $\chi_s(q)$ is

$$\frac{\chi_s(q)}{\chi_n} = 1 - \pi[2q\xi(T)]^{-1}\tan\frac{\Delta(T)}{2T} \tag{25.2}$$

where χ_n is the susceptibility of the metal in the normal state. The susceptibility $\chi_s(q)$ has a maximum at $T = 0$ for

$$Q_0 = [3\pi p_0^2 \xi_0^{-1}]^{-\frac{1}{3}} \ . \tag{25.3}$$

For $T \neq 0$, (25.2) has a maximum at

$$Q(T) = Q_0\left[\frac{\Delta(T)}{\Delta(0)}\tanh\frac{\Delta(T)}{2T}\right]^{\frac{1}{3}} \ . \tag{25.4}$$

The Ginzburg-Landau theory can be applied to study the interplay between superconductivity and magnetic order if we consider that $\Delta(\mathbf{r})$ and the magnetization density $M(T)$ are both small.

In this approximation, the free energy can be expanded in powers of the two other parameters. The total free energy

$$F = F_{\rm s}^{0}(\Delta) + F_{\rm s}^{0}(M) + F_{\rm s,m}^{0}(\Delta, M) \tag{25.5}$$

where

$$F_{\rm s}^{0}(\Delta) = \int d^3\mathbf{r}\left[\frac{\alpha_{\rm s}(T)}{2}|\Delta(\mathbf{r})|^2 + \frac{\beta_{\rm s}}{4}|\Delta(\mathbf{r})|^4 + \frac{\gamma_{\rm s}}{2}|\nabla\Delta(\mathbf{r})|^2\right] \tag{25.6a}$$

$$F_{\rm m}^{0}(M) = \int d^3\mathbf{r}\left[\frac{\alpha_{\rm m}(T)}{2}|M(\mathbf{r})|^2 + \frac{\beta_{\rm m}}{4}|M(\mathbf{r})|^4 + \frac{\gamma_{\rm m}}{2}|\Delta M(\mathbf{r})|^2\right]. \tag{25.6b}$$

The form of the coupling term $F_{\rm s,m}(\Delta, M)$ in (25.5) can be obtained if we consider that in the presence of a fixed $\Delta(\mathbf{r})$, the free energy functional of the spin system is

$$F_{\rm m}(M,\Delta) = F_{\rm m}^{0}(M) + F_{\rm s,m}(\Delta, M) \tag{25.6c}$$

and for a uniform order parameter $\Delta(\mathbf{r}) = \Delta$, $F_{\rm m}(M,\Delta)$ can be approximated as

$$F_{\rm m}(M,\Delta) = N\left[\frac{1}{2}\sum_{q}\chi_{\rm m}^{-1}(\mathbf{q})|M(\mathbf{q})|^2 + O(M^4)\right] \tag{25.7}$$

where $M(\mathbf{q})$ is the Fourier transform of $\mathbf{M}(\mathbf{r})$ and $\chi_{\rm m}(\mathbf{q})$ is given by

$$\chi_{\rm m}(q) = \frac{\chi_0(T)}{1 - J_{\rm eff}(q)n\chi_0(T)}$$

where $n = N/V$ (N is the number of the magnetic ions, V the volume), $J_{\rm eff} = I^2\chi(\mathbf{q})$, I being the electron-spin exchanges and $\chi_0(T)$ is the paramagnetic susceptibility $\chi_0(T) = S(S+1)/3T$. From (25.6c), the coupling term $F_{\rm sm}$ becomes

$$F_{\rm s,m} = -V\left\{n^2\frac{I^2}{2}\sum_{\mathbf{q}}[\chi_{\rm s}^{-1}(\mathbf{q},\Delta) - \chi_{\rm n}^{-1}(\mathbf{q})]|M(\mathbf{q})|^2\right\} \tag{25.8}$$

which depends on the difference of the conduction electrons susceptibility in the superconducting and normal state.

The magnetic structure can be obtained if the polarization of the magnetization is known. The helical structure

$$\mathbf{M}(\mathbf{r}) = (M\cos Qz, M\sin Qz, 0) \tag{25.9}$$

in (25.6b) has been considered as a possible structure. Until this point of the model, the order parameter was considered arbitrary large. Expanding $F_{s,m}(\Delta, M)$ in powers of Δ, one finds

$$F_{s,m}(\Delta, M) = V\Delta^2 a(2\pi)^3 \sum_{\mathbf{q}} g(\mathbf{q})|M(\mathbf{q})|^2 \tag{25.10}$$

where

$$a = \frac{N^2(0)I^2n^2\pi^2}{T}\,, \quad g(q) = \frac{1}{8\pi^2 p_0^2 q}$$

and we see that (25.10) has a q-dependence due to the non-local character of the interaction. The influence of the spin fluctuations on the critical temperature can be evaluated using these results. Indeed, taking $\Delta(\mathbf{r}) = \Delta$ and replacing $|M(\mathbf{q})|^2$ by its thermal average $\langle|\mathbf{M}(\mathbf{q})|\rangle$, we get

$$F(\Delta) = V\left[\frac{\alpha_0(T)}{2}|\Delta|^2 + \frac{\beta_0}{2}|\Delta|^4 + \frac{a}{2}|\Delta|^2 \int d^3\mathbf{q}\, g(\mathbf{q})\langle|\mathbf{M}(\mathbf{q})|^2\rangle\right] \tag{25.11}$$

and using now

$$\langle|M(\mathbf{q})|^2\rangle = 3T\chi_m/N\,, \tag{25.12}$$

we get the equation for T_c as

$$T_c = T_{c0} - nI^2N(0)\frac{3\pi^2}{2}\int d^3\mathbf{q}\, g(\mathbf{q})\chi_m(\mathbf{q}, T_c) \tag{25.13}$$

a result which coincides with the decreasing of the critical temperature in the weak pair breaking. The phase diagram of a superconductor with helical structure of the spin can be obtained from the free energy $F(\Delta, M, Q)$ which describes the electrons in the superconducting state interacting with the magnetic spins ordered in a helical structure.

We have to mention that the expansion of $F(\Delta, M, Q)$ in powers of Δ is not relevant because at the onset of magnetic order Δ is not small. Minimization of this free energy with respect to Δ and Q leads to an effective free-energy functional $F_s(m, T)$ for the superconducting helical state. Generally, it can be written as

$$F_s(M, T) = \frac{A(T)}{2}M^2 + \frac{\beta}{4}M^4 + F_0(T) \tag{25.14}$$

and the equation $A(T) = 0$ determines the transition temperature T_s from the paramagnetic to the helical state. If the helical magnetic order is described by (25.9) and the superconducting order parameter is considered as describing an homogeneous state, i.e., Δ = constant, the exchange interaction between electrons and spins can be treated as an external field $\mathbf{h}(\mathbf{r})$ acting on $\mathbf{M}(\mathbf{r})$. Thus we will take

$$\begin{aligned} \mathbf{h}(\mathbf{r}) &= h(\cos \mathbf{Q}\cdot\mathbf{r}, \sin \mathbf{Q}\cdot\mathbf{r}, 0) \\ h &= nMI \end{aligned} \tag{25.15}$$

neglecting the scattering effects. The microscopic theory of the helical order has been developed by Bulaevski *et al.*[27] using the Gor'kov equations. The free energy calculated from these results is complicated and for $h < qv_0$ and $T < T_{c0}$, the free energy $F(\Delta, m, Q)$ was obtained as

$$\begin{aligned} F(M,\Delta,0) = VN^2(0)n^2I^2\Big\{&\Big[1 - \frac{T}{T_{M0}} + \frac{Q^2}{p_0^2} + \frac{\pi^2\Delta}{2v_0 Q}\Big]M^2 \\ &+ \Big[f(S)\frac{T}{T_{M0}} - \frac{\pi^2\Delta}{2Qv_0}\frac{1}{16}\Big(\frac{nI}{\Delta}\Big)^2\Big]M^4\Big\} \\ &- VN(0)\Delta^2(1 + 2\ln\Delta(0)/\Delta) \end{aligned} \tag{25.16}$$

where

$$f(S) = \frac{3}{10}\frac{S^2 + S + \frac{1}{2}}{S^2(S+1)^2}\,.$$

We can see that the last term is the free energy of a BCS superconductor and the first two terms can be compared with Eq. (25.7). If we calculate $Q(M)$ and $\Delta(m)$ by minimizing (25.16) to Q and Δ, one obtains

$$Q(M) = Q_0\Big[1 - \frac{n^2}{48}\frac{I^2}{\Delta^2(0)}M^2\Big] \tag{25.17}$$

$$\Delta(M) = \Delta(0) - \frac{(nIM)^2 Q_0}{2\sqrt{3}p_0\Delta_0} \tag{25.18}$$

where Q_0 is given by (25.3).

With these results, (25.8) becomes

$$\begin{aligned} F_s(M,T) = VN(0)I^2[&M^2(T - T_1)/T_1 + M^4 f(S)(1 - T/T_1^*)] \\ &- \frac{1}{2}N(0)V\Delta^2(0) \end{aligned} \tag{25.19}$$

where

$$T_1 = T_{M0}\Big[1 - \frac{1}{4}\frac{Q_0^2}{p_0^2}\Big]\,.$$

and T^* is a characteristic temperature of the order

$$T^* \sim \Delta_0(\Delta_0/\varepsilon_0)^{\frac{1}{3}} . \tag{25.20}$$

For a strong magnetic coupling, the phase transition for the paramagnetic (superconducting) state to the helical state becomes of the first order.

The phase diagram T as function of T_{M0} is given in Fig. 27 where T_2 is the critical temperature of the first order phase transition which is connected with the changes of the polarization of $M(r)$. Another mechanism proposed for the explanation of the occurrence of the helical order is known as the electromagnetic interaction. Blound and Varma,[28] Matsumoto *et al.*[29] considered that the long wave component of the exchange interaction with wave length larger than the London penetration depth is screened by the persistent currents from the superconducting state.

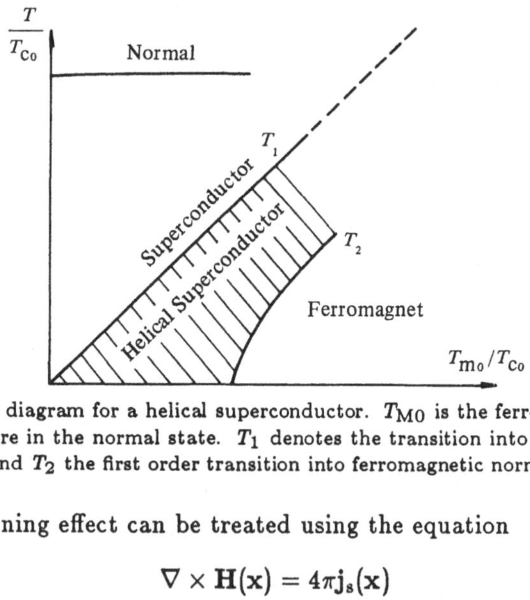

Fig. 27. Phase diagram for a helical superconductor. T_{M0} is the ferromagnetic transition temperature in the normal state. T_1 denotes the transition into helical superconducting state and T_2 the first order transition into ferromagnetic normal state.

The screening effect can be treated using the equation

$$\nabla \times \mathbf{H}(\mathbf{x}) = 4\pi \mathbf{j}_s(\mathbf{x}) \tag{25.21}$$

where the magnetic field is given by

$$\mathbf{B}(\mathbf{x}) = \mathbf{H}(\mathbf{x}) + 4\pi \mathbf{M}(\mathbf{x}) . \tag{25.22}$$

$\mathbf{B}(\mathbf{x})$ being the magnetic induction and $\mathbf{M}(\mathbf{x})$ the magnetization. In the linear response approximation, the current $\mathbf{j}_s$ will be given by

$$\mathbf{j}_s(\mathbf{x}) = -\frac{1}{4\pi\lambda^2} K(\mathbf{y}-\mathbf{x})\mathbf{A}(y) \tag{25.23}$$

where $K(\mathbf{z})$ is the kernel defined by (10.10). Now Eq. (25.21) becomes

$$\nabla^2\mathbf{B}(\mathbf{x}) = \frac{1}{\lambda^2}K(\mathbf{y}-\mathbf{x})\mathbf{B}(\mathbf{x}) + 4\pi\nabla^2\mathbf{M}(\mathbf{x}) \tag{25.24}$$

with the gauge $\nabla\cdot\mathbf{A} = 0$. From (25.23) and (25.24), the magnetic field $\mathbf{H}(\mathbf{x})$ is

$$\mathbf{H}(\mathbf{x}) = -4\pi\hat{F}(\mathbf{y}-\mathbf{x})\mathbf{M}(\mathbf{y}) \tag{25.25}$$

where

$$\hat{F}(\mathbf{y}-\mathbf{x}) = K(\mathbf{y}-\mathbf{x})[-\lambda^2\nabla^2 + K(\mathbf{y}-\mathbf{x})]\,. \tag{25.26}$$

Now, we can estimate the screening effect in a simple way. Let us introduce a molecular field $H_{\mathrm{m}}(\mathbf{x})$ acting on the local spin as a sum of the exchange field and the magnetic field generated by the persistent current

$$\mathbf{H}_{\mathrm{m}}(\mathbf{x}) = J(\mathbf{x}-\mathbf{y})\mathbf{M}(\mathbf{y}) + \mathbf{H}(\mathbf{x}) \tag{25.27}$$

where $J(\mathbf{x}-\mathbf{y})$ is the exchange interaction between spins. Equation (25.27) can now be written as

$$\mathbf{H}_{\mathrm{m}}(\mathbf{x}) = [J(\mathbf{x}-\mathbf{y}) - 4\pi F(\mathbf{x}-\mathbf{y})]\mathbf{M}(\mathbf{x})\,. \tag{25.28}$$

If we compare (25.27) with (25.28), we see that the persistent current gives rise to a screening effect which changes the exchange interaction as

$$J_{\mathrm{eff}}(\mathbf{q}) = J(\mathbf{q}) - 4\pi F(\mathbf{q}) \tag{25.29}$$

Kotani *et al.*[30] calculated the spin wave of the helical spin system. The method improved by Aldea and Turcu[31] considers the effect of the electromagnetic interaction of the spin-spin exchange interaction. Due to the transverse and longitudinal components of the Green function of the electromagnetic field, the exchange integral $J_{i,j}(\mathbf{q})$ has been obtained as

$$J_{i,j} = T_{ij}(\mathbf{q})J_{\perp}(\mathbf{q}) + L_{ij}(\mathbf{q})J_{\parallel}(\mathbf{q}) \tag{25.30}$$

where

$$T_{ij} = \delta_{ij} - q_iq_j/q^2\,, \qquad L_{i,j} = q_iq_j/q^2$$

and

$$J_{\perp}(\mathbf{q}) = J_0(\mathbf{q}) + \frac{2\pi N}{3V}(\mu_0 g)^2 - \frac{2N}{V}(\mu_0 g)^2\frac{K(\mathbf{q})}{\lambda^2q^2 + K(\mathbf{q})} \tag{25.31a}$$

$$J_{\parallel} = J_0(\mathbf{q}) - \frac{4\pi N}{3V}(\mu_0 g)^2 \tag{25.31b}$$

where $J_0(\mathbf{q})$ is the spin-spin exchange interaction in the absence of the electromagnetic effects, and has the form

$$J_0(q) = J_0(0) - D_0 q^2 \ . \tag{25.32}$$

The maximum of $J_1(\mathbf{q})$ has been calculated taking $K(\mathbf{q}) \simeq 1$ and the helical order has the wave vector

$$Q_{sw} = \frac{1}{\lambda}\left[\left(\frac{\bar{c}}{d}\right)^2 - 1\right]^{\frac{1}{2}} \tag{25.33}$$

where $\bar{c} = 2\pi(\mu_0 g)^2 n$ and $d = D_0/J(0)\lambda^2$.
Using the idea due to Maekawa *et al.*,[32] Tachiki[33] showed that the phase diagram of such a superconductor with helical ordered spins may present a region with a self-induced vortex state. The occurrence of this state is a function of the coupling constants and of the concentration of the magnetic ions. An interesting result obtained in Refs. 32, 33 is the occurrence of the first order phase transition from the type I superconductivity (near the magnetic critical temperature) to a type II superconductivity above the magnetic critical temperature.

Bulaevski *et al.*[34] considered the magnetic structures which may appear in the re-entrant superconductors due to the exchange and the electromagnetic dipolar interactions of the localized moments and electrons and the anisotropy. Such a realistic model was necessary because in the real compounds, it is impossible to explain all kinds of transitions only by the electromagnetic interaction. The role of the electromagnetic interaction is, in fact, to give rise to a nonuniform magnetic structure in the superconducting state but the anisotropy could modify the helical structure. In the presence of the magnetic anisotropy, the magnetic structure is dominated by a special phase which has a transverse character but the spins are arranged in the magnetic domains. The domain structure is one dimensional because the energy of the superconducting state depends weakly on the type of the magnetic structure. The transition between the superconducting state to the domain structure may be of the second order or of the first order. In the first case, the domain phase appears by the modification of the helical structures and in the second case, this phase appears by creation of series of nuclei with alternating opposite direction of magnetization.

A remarkable result concerning the coexistence between superconductivity, spatial modulated order and ferromagnetism has been obtained by Machida and Nakanishi.[35] The main idea proposed in Ref. 35, and supported by the experimental data, is that superconductivity helical order and ferromagnetism really coexist in many ternary compounds. The system of conduction electrons has been considered as one-dimensional interacting with a localized spin system. The calculations have been performed in the mean-field approximation and the self-consistent equations have been solved in terms of the Weierstrass functions. The sinusoidal modulated magnetic phase (observed experimentally) is more stable than the helical structure which appears as a metastable one. The superconducting state is also spatially modulated, and the solutions of the self-consistent equations showed the existence of a soliton lattice which has a two-energy gap structure and a spin-density polarization of the conduction electrons. Except the fact that the electronic system is considered as one-dimensional, this model seems to be reasonable to explain the coexistence between the three phases.

References

1. O. Fischer, M. Decroux, R. Chevrel and M. Sergent in *Proceedings of the Second Conference on Superconductivity in d - and - f - Bands Metals*, ed. D. H. Doughlass (Plenum Press, New York, 1976) pp. 175–1987.
2. S. K. Sinha, G. W. Crabtree, D. G. Hinks and H. Mook, *Phys. Rev. Lett.* **48** (1982) 950.
3. S. Maekawa and M. Tachiki, *Phys. Rev.* **B18** (1978) 4688.
4. K. Machida and D. Younger, *J. Low Temp. Phys.* **35** (1979) 449.
5. V. Jaccarino and M. Peter, *Phys. Rev. Lett.* **9** (1962) 290.
6. K. Machida and D. Younger, *J. Low Temp. Phys.* **35** (1979) 561.
7. T. Matsuura, S. Ichinose and Y. Nagaoka, *Prog. Theor. Phys.* **57** (1977) 713.
8. G. S. Grest, L. Koffey and K. Levin, *J. Magn. Magr. Mat.* **31–34** (1983) 501.
9. U. Poppe, *Physica* **108B** (1981) 805.
10. C. G. Kuper, M. Revzen and A. Ron, *Phys. Rev.* **B29** (1984) 466.
11. K. Maki and T. Tsuneto, *Prog. Theor. Phys.* (1964) 945.
12. M. Fujita, K. Machida and T. Masubara, *J. Low Temp. Phys.* **54** (1984) 535.
13. I. A. Privorotsky, *Zh. Exp. Teor. Phys.* **43** (1962) 2225 (*Soviet Phys. JETP* **16** (1963)).
14. W. Baltensperger and S. Sträsler, *Phys. Kond. Matterie* **1** (1963) 20.
15. P. Fulde and J. Keller in *Superconductivity in Ternary Compounds*, eds. O. Fischer and M. B. Maple (Springer, Heidelberg, 1982) p. 249.
16. K. Machida, N. Nomura and T. Matsubara, *Phys. Rev.* **822** (1980) 2307.
17. M. J. Nass, K. Levin and G. S. Grest, *Phys. Rev. Lett.* **46** (1981) 614.
18. G. Zwicknagl and P. Fulde, *Z. Phys.* **43** (1981) 23.
19. K. Machida, *J. Low Temp. Phys.* **37** (1979) 583.
20. Y. Suzumura, Y. Okabe and K. Yshio, *Prog. Theor. Phys.* **74** (1985) 211.

21. C. Ro and K. Levin, *Phys. Rev.* **B29** (1984) 6155.
22. T. V. Ramakrishnan and C. M. Varma, *Phys. Rev.* **B24** (1981) 137.
23. G. Zwicknagl and P. Fulde, *Z. Phys.* **B43** (1981) 23.
24. Y. Okabe and A. D. S. Nagi, *Phys. Rev.* **B28** (1983) 6290.
25. M. Crisan and T. Veres, *J. Low Temp. Phys.* **65** (1986) 425.
26. P. W. Anderson and H. Suhl, *Phys. Rev.* **116** (1959) 898.
27. L. N. Bulaevski, A. I. Rusinov and M. L. Kulic, *Solid State Commun.* **30** (1979) 59, *J. Low Temp. Phys.* **39** (1980) 256, *Solid State Commun.* **37** (1981) 671, **40** (1981) 683.
28. E. I. Blound and C. M. Varma, *Phys. Rev. Lett.* **42** (1979) 1079.
29. H. Matsumoto, H. Umezawa and M. Tachiki, *Solid State Commun.* **31** (1979) 157.
30. A. Kotani, S. Takahasi, M. Tachiki, T. Koyama, H. Matsumoto and H. Umezawa, *J. Phys. Soc. Jpn.* **49** (1981) 2144.
31. A. Aldea and I. Turcu, *Phys. Rev.* **B29** (1984) 6213.
32. S. Maekawa, M. Tachiki and S. Takahasi, *J. Mag. Mat.* **13** (1979) 324.
33. M. Tachiki, *Physica* **109–110** (1982) 1699.
34. L. Bulaevski, A. I. Buzdin, S. V. Panjurov and M. L. Kulic, *Phys. Rev.* **B28** (1983) 1370.
35. K. Machida and H. Nakanishi, *Phys. Rev.* **B30** (1984) 122.

VI
SUPERCONDUCTIVITY IN QUASI-ONE-DIMENSIONAL SYSTEMS

The problem of high temperature superconductivity became of interest since the discovery of the first mechanism which gives rise to the electron-electron attraction. However, in 1964, Little proposed the first high temperature mechanism which was based on a one-dimensional electronic system. The electron-electron attraction was supposed to be given by an excitonic mechanism. In the same year, Ginzburg proposed as a good candidate for the high temperature superconductivity; the systems composed of layers of metals separated by insulators or semiconductors. The organic compounds from the category TTF-TCNQ have been intensively studied after 1970 in connection with the high temperature superconductivity. At the present time, we know that in these systems which are quasi-one-dimensional, Charge-Density-Waves (CDW), Spin-Density-Waves (SDW) and superconductivity may occur. The critical temperature of the materials which present superconductivity and CDW or SDW is low but the transition from the normal state to one of these states presents a great interest. This interest is due to the fact that it is known that in the systems with low dimensions ($d = 1, d = 2$), the fluctuations of the order parameter suppress the ordered phases. However, these condensed phases exist and this can be explained only by the fact that a real system does not have a low dimensionality but in fact three dimensional and presents a high anisotropy. In this way, we can treat the system as a quasi-one-dimensional or quasi-bidimensional system.

Another important feature presented by these systems is that the superconductivity and Charge-Density-Waves (or Spin-Density-Waves) may coexist. This is not a trivial problem because the superconductivity and the other two condensed states appear in the systems of electrons and the SDW state breaks time-reversal symmetry, but the CDW state does not.

In the following, we will present a mean-field theory of the superconducting state in materials which present particular features of the Fermi surface so that the CDW and SDW states may occur.

26. Superconductivity and Charge-Density-Waves

The occurrence of the superconducting state in the highly anisotropic materials which behave as quasi-one-dimensional systems has been studied first by Levin *et al.*[1] for a one-dimensional electronic system. Later, using the analogy to the martensitic transition from A-15 materials discussed by Bilbro and MacMillan,[2] Balsiero and Falicov[3] considered the occurrence of the superconducting state in systems which present Charge-Density-Waves, both transitions being mediated by the electron-phonon interaction.

In a two-dimensional structure, the phonons can give rise to a CDW metallic state but for same cases the electronic systems present a behaviour similar to a semiconductor. In this way, it is not difficult to predict that the CDW state tends to suppress the superconductivity.

The CDW state was considered by Machida *et al.*[4] as a perturbation for the superconducting state, and the Hamiltonian which describes such a system is

$$\mathcal{H} = \sum_{\mathbf{p},\alpha} \varepsilon(\mathbf{p}) c^{\dagger}_{\mathbf{p}\alpha} c_{\mathbf{p}\alpha} - V \sum_{\mathbf{p},\alpha} (c^{\dagger}_{\mathbf{p}\alpha} c_{\mathbf{p}+\mathbf{Q},\alpha} + h.c.) \tag{26.1}$$

where V is a given parameter proportional to the order parameter of CDW. The critical temperature T_c for the superconducting state will be calculated from the equation

$$\Delta(\mathbf{r}) = g\pi T \sum_{\omega} \int d^3\mathbf{r}' K(\mathbf{r},\mathbf{r}';\omega) \tag{26.2}$$

where the kernel $K(\mathbf{r},\mathbf{r}';\omega)$ has been defined as

$$K(\mathbf{r},\mathbf{r}';\omega) = G^{\uparrow}(\mathbf{r},\mathbf{r}';\omega) G^{\downarrow}(\mathbf{r},\mathbf{r}';-\omega) \ . \tag{26.3}$$

The Fourier transform of (26.2) is

$$\Delta(\mathbf{q}) = g\pi T \sum_{\omega'} \int \frac{d^3\mathbf{q}'}{(2\pi)^3} K(\mathbf{q}, \mathbf{q}'; \omega)\Delta(\mathbf{q}') \tag{26.4}$$

where $K(\mathbf{q}, \mathbf{q}'; \omega)$ is defined as

$$\begin{aligned} K(\mathbf{q}, \mathbf{q}'; \omega) = \sum_{\mathbf{p}_1 \cdots \mathbf{p}_4} & G(\mathbf{p}_1, \mathbf{p}_3; \omega) G(\mathbf{p}_2, \mathbf{p}_4; -\omega)\delta(\mathbf{p}_1 + \mathbf{p}_3 - q) \\ & \times \delta(\mathbf{p}_2 + \mathbf{p}_4 - \mathbf{q}) \,. \end{aligned} \tag{26.5}$$

The relevant wave numbers are $\mathbf{q} = 0$ and $\mathbf{q} = \mathbf{Q}$ and for these two cases (26.4) become

$$\begin{aligned} & 1 - gK(0,0)\Delta(0) + gK(0,Q)\Delta(Q) = 0 \,, \\ & gK(Q,0) + (1 - K(Q,Q)) = 0 \end{aligned} \tag{26.6}$$

and the critical temperature is obtained from the equation

$$\begin{vmatrix} 1 - gK(0,0)\,; & gK(0,Q) \\ gK(Q,0)\,; & 1 - gK(Q,Q) \end{vmatrix} = 0 \tag{26.7}$$

which determines a new critical temperature T_c for the superconducting state which has to be greater than the BCS critical temperature $T_c^{\rm BCS}$. The kernel $K(\mathbf{q}, \mathbf{q}'; \omega)$ can be evaluated from (26.1) taking

$$\begin{aligned} G(\mathbf{p}, \mathbf{p}'; \omega) &= \frac{i\omega - \varepsilon(\mathbf{p} + Q)}{(i\omega - \varepsilon(\mathbf{p}))(i\omega - \varepsilon(\mathbf{p} + \mathbf{Q})) - |V|^2} \\ G(\mathbf{p}, \mathbf{p}'; \omega) &= \frac{V}{(i\omega - \varepsilon(\mathbf{p}))(i\omega - \varepsilon(\mathbf{p} + \mathbf{Q})) - |V|^2} \end{aligned} \tag{26.8}$$

where $\mathbf{Q} \equiv (0, 0, 2p_0)$ in cylindrical coordinate.
Using (26.8) the diagonal part of the kernel is

$$K(0, Q; \omega) = \frac{\pi N(0) V}{8\varepsilon_0 |\omega|} \ln \frac{|\omega|}{2\varepsilon_0} \,, \tag{26.9}$$

the other contributions being

$$K(Q, 0; \omega) = \pi N(0) O\left(\frac{V^2}{\varepsilon_0^2}\right) \tag{26.10a}$$

$$K(0, 0; \omega) = N(0)\left(1 - \frac{1}{8}\frac{V^2}{\varepsilon_0^2}\right) \,, \tag{26.10b}$$

with these results we get the critical temperature

$$\frac{T_c - T_c^{\rm BCS}}{T_c^{\rm BCS}} = -\frac{V^2}{8\varepsilon_0}\frac{1}{N(0)g}(1-\beta) \tag{26.11}$$

where

$$\beta = \frac{1}{8}[N(0)g]^2\left\{\pi\sum_{\omega}^{\omega_D}\frac{1}{\omega}\ln\frac{\omega}{\varepsilon_0}\right\}. \tag{26.12}$$

From this formula, we see that the superconducting state with the two order parameters $\Delta(0)$ and $\Delta(Q)$ do not differ very much from the ordinary BCS state because $\Delta(Q)/\Delta(0)$ is of the order of V/ε_0, which is assumed to be small.

Equation (26.2) has to be linearized for the calculation of the upper critical field H_{c2}. Taking in (26.2) $K(\mathbf{r},\mathbf{r}+\mathbf{R})$, we can expand $\Delta(\mathbf{r}+\mathbf{R})$ around $\mathbf{R}$, neglecting the higher-order term in $(V/\varepsilon_0)^2$ and we get

$$\begin{aligned}&\frac{1}{6}\sum_{i=x,y,z}(a_i + 2b_i\cos\mathbf{Q}\cdot\mathbf{r})\frac{\partial^2\Delta(\mathbf{r})}{\partial r^2} + [gK(0,0)-1]\Delta(\mathbf{r})\\ &\quad - 2gK(0,\mathbf{Q})\cos\mathbf{Q}\mathbf{r}\cdot\Delta(\mathbf{r}) = 0\end{aligned} \tag{26.13}$$

where

$$a_i = \sum_i \int d^3R_i e^{-i\mathbf{q}\cdot\mathbf{R}_i}K(\mathbf{q},\mathbf{q}) = \frac{7}{8}\varsigma(3)gN(0)\left(\frac{v_0}{\pi T_c}\right)^2, \tag{26.14}$$

$$b_i = \sum_i \int d^3R_i e^{-i\mathbf{q}\cdot\mathbf{R}_i}K(\mathbf{q}+\mathbf{Q},\mathbf{q}) \sim O\left(\frac{v_0}{\varepsilon_0}\right), \tag{26.15}$$

$$gK(0,0) - 1 = gN(0)\ln\frac{T_c}{T}. \tag{26.16}$$

Equation (26.13) is reduced to the Mathieu-type differential equation which differs from the Ginzburg-Landau equation by an additional potential term $\cos(\mathbf{Q}\cdot\mathbf{r})$ given by the off-diagonal elements $K(\mathbf{q},\mathbf{q}+\mathbf{Q})$ in the kernel.

Equation (26.13) can be written as

$$\begin{aligned}&-(\nabla - 2ie\mathbf{A})(2\widehat{m})^{-1}\cdot(\nabla - 2ie\mathbf{A})\Delta(\mathbf{r}) + k'\cos\mathbf{Q}\cdot(\mathbf{r}-\mathbf{r}_0)\\ &= 1/(2m\xi^2(T)\Delta(\mathbf{r}))\end{aligned} \tag{26.17}$$

where k' is a constant, and $(2\widehat{m})^{-1}$ is the effective-mass tensor

$$(2\widehat{m})^{-1} = \begin{bmatrix} M^{-1} & 0 & 0\\ 0 & m^{-1} & 0\\ 0 & 0 & m^{-1}\end{bmatrix} \tag{26.18}$$

which describes the electrons from a layered-type crystal with uniaxial symmetry.
If H is in the x-y plane, the potential vector is $\mathbf{A} = (Hz\sin\Phi, 0, H\cos\Phi)$, where Φ is the angle between H and the x-axis, the equation for H_{c2} is

$$\frac{1}{\xi_\perp^2(T)} = \left(\frac{M}{m}\right)^{1/2} f(\Phi) - |k| \exp\left[-\frac{(M/m)^{1/2}Q^2}{4h_{c2}(\Phi)f(\Phi)}\right] \tag{26.19}$$

where $h_{c2} = 2\pi H_{c2}/\Phi_0$ and

$$f(\Phi) = \left(\frac{M}{m}\cos^2\Phi + \sin^2\Phi\right)^{1/2}.$$

The coherence length $\xi_\perp$ is the coherence perpendicular to the layered plane. If H is parallel to the Oz axis, $\mathbf{A} = (0, Hx, 0)$, the critical field is

$$H_{c2}^{z}(T) = \frac{\Phi_0}{2m\xi_\perp^2(0)}\left(\frac{m}{M}\right)^{1/2}\left|\frac{T-T_c}{T_c}\right| \tag{26.20}$$

which presents a typical Ginzburg-Landau behaviour near T_c. The critical field $h_{c2}(T)$ gives a positive curvature near T_c but $H_{c2}^{z}(T)$ does not present this behaviour. The influence of the paramagnetism on the superconducting state has also been studied by Gabovich *et al.*,[5] and the main conclusion is that the Pauli limiting field H_p is enhanced by the CDW state.

A correct treatment like that given by Machida[6] has to consider the influence of the CDW state on the superconducting state, as well as the influence of superconductivity on the CDW state. The Hamiltonian which describes such a model is

$$\begin{aligned} \mathcal{H} = \sum_{\mathbf{p},\alpha} \varepsilon(\mathbf{p}) c_{\mathbf{p}\alpha}^{\dagger} c_{\mathbf{p}\alpha} - \sum_{\mathbf{p}\in D_1} (W c_{\mathbf{p}\alpha}^{\dagger} c_{\mathbf{p}+\mathbf{Q},\alpha} + h.c.) \\ - \sum_{\mathbf{p}} (\Delta C_{\mathbf{p}\uparrow}^{\dagger} c_{-\mathbf{p}\downarrow}^{\dagger} + h.c.) \end{aligned} \tag{26.21}$$

where W and Δ are the order parameters for the Charge-Density-Waves and the superconducting state is determined by the self-consistent equations

$$W = \frac{V}{2} \sum_{\mathbf{p}\in D_1} \sum_{\alpha} \langle c_{\mathbf{p}\alpha}^{\dagger} c_{\mathbf{p}+\mathbf{Q},\alpha} \rangle \tag{26.22}$$

$$\Delta = g \sum_{\mathbf{p}\in D_2} \langle c_{-\mathbf{p}\downarrow} c_{\mathbf{p}\uparrow} \rangle \tag{26.23}$$

D_1 being region of the Fermi surface which satisfies the nesting condition and where the CDW appears and D_2 the region of the superconducting electrons. Using the strong-coupling formalism, the Green function is

$$G^{-1}(\mathbf{p},\omega) = i\omega\hat{1} - \varepsilon(\mathbf{p})\hat{\tau}_3\hat{\sigma}_z + W\hat{\sigma}_x\hat{\tau}_3 + \Delta\hat{\sigma}_x\hat{\tau}_x \tag{26.24}$$

and with these results, we obtain the gap equations as

$$\ln\frac{T}{T_{w0}} = 2\pi T\sum_{\omega\geq 0}\left[\frac{1}{\sqrt{\omega^2+W^2+\Delta^2}} - \frac{1}{\omega}\right] \tag{26.25}$$

$$\ln\frac{T}{T_{c0}} = 2\pi T\frac{N_1(0)}{N(0)}\sum_{\omega\geq 0}\left[\frac{1}{\sqrt{\omega^2+W^2+\Delta^2}} - \frac{1}{\omega}\right] + \frac{N_2(0)}{N(0)}2\pi T\sum_{\omega>0}\left[\frac{1}{\sqrt{\omega^2+\Delta^2}} - \frac{1}{\omega}\right] \tag{26.26}$$

where

$$2\pi T_{w0}\sum_{0\leq\omega\leq E_B}\frac{1}{\omega} = \ln\frac{2\gamma E_B}{\pi T_{w0}} = \frac{1}{VN(0)} \tag{26.27a}$$

$$2\pi T_{c0}\sum_{0\leq\omega\leq\omega_D}\frac{1}{\omega} = \ln\frac{2\gamma\omega_D}{\pi T_{c0}} = \frac{1}{gN(0)}, \tag{26.27b}$$

E_B being the cut-off energy for CDW state, and $N(0) = N_1(0) + N_2(0)$ the total density of states from the region 1 and from the superconducting region 2. The temperature dependence of the two order parameters can be obtained from (26.25)–(26.26). Near $T \cong T_c$ and $T = 0$ we can perform analytical calculations. In the first case, we get

$$\Delta^2(T) = \frac{(8\pi T_c)^2}{7\varsigma(3)}\left[\frac{N(0)}{N_2(0)} - \frac{N_1(0)}{N(0)}\frac{T_{c0}}{T_{w0}}\right]\left|\frac{T_c - T}{T_{c0}}\right| \tag{26.28}$$

and if we take $T = 0$, the superconducting gap is

$$\frac{\Delta(0)}{T} = \frac{\pi}{\gamma} \tag{26.29}$$

and $W(0)$ is

$$\left[\frac{W(0)}{W_0}\right]^{1/2} = 1 - \left(\frac{T_c}{T_{w0}}\right)^2; \quad W_0 = 2E_B\exp\left[-\frac{1}{VN(0)}\right] \tag{26.30}$$

where $W(0)$ is the CDW order parameter in the absence of superconductivity.

Equation (26.30) shows that the BCS relation does not hold for the CDW state in this case.

Numerical calculations performed using (26.25)–(26.26) showed that there is a temperature domain in which superconductivity and the CDW state coexist. However, a correct calculation implies the calculation of the free energy of the two phases $\delta F = F(\Delta, W) - F(0,0)$. The calculation performed in Ref. 6 showed the stability of the coexisting phase.

27. Coexistence Between Spin-Density-Waves and Superconductivity

The coexistence between Spin-Density-Waves and superconductivity has been considered first by Machida[3] in the ternary rare-earth compounds in which the rare-earth atoms are arranged in a regular lattice. The simplest model for this problem proposed in Ref. 8 consists of dividing the momentum space into two regions (1) and (2). One assumes that the Fermi surface satisfies a nesting condition in region (1), which allows the SDW formation. The nesting wave vector $\mathbf{Q}$, which characterizes the SDW state is assumed to be $2\mathbf{Q} = \mathbf{G}$ ($\mathbf{G}$ is the reciprocal lattice vector of the lattice). The remaining region of the Fermi Surface is denoted as region (2) where one gets the superconducting gap.

The Hamiltonian which describes this model is

$$\mathcal{H} = \mathcal{H}_0 + \mathcal{H}_{\mathrm{SDW}} + \mathcal{H}_{\mathrm{BCS}} \tag{27.1}$$

where

$$\mathcal{H}_0 = \sum_{\mathbf{p},\alpha} \varepsilon(\mathbf{p}) c^\dagger_{\mathbf{p}\alpha} c_{\mathbf{p}\alpha} \tag{27.1a}$$

$$\mathcal{H}_{\mathrm{SDW}} = -\sum_{\mathbf{p}\in 1} \sum_{\alpha} [\alpha M c^\dagger_{\mathbf{p}\alpha} c_{\mathbf{p}+\mathbf{Q},\alpha} + h.c.] \tag{27.1b}$$

$$\mathcal{H}_{\mathrm{BCS}} = -\sum_{\mathbf{p}} [\Delta c^\dagger_{\mathbf{p}\uparrow} c^\dagger_{-\mathbf{p}\downarrow} + h.c.] \tag{27.1c}$$

where the magnetization M has been defined as

$$M = \frac{I}{2} \sum_{\mathbf{p}\in 1} \alpha \langle c^\dagger_{\mathbf{p},\alpha} c_{\mathbf{p}+\mathbf{Q},\alpha} \rangle \tag{27.2}$$

and the superconducting order parameter Δ is

$$\Delta = g \sum_{\mathbf{p}} \langle c_{-\mathbf{p}\downarrow} c_{\mathbf{p}\uparrow} \rangle \ . \tag{27.3}$$

The Green functions which are necessary to calculate the two other parameters in region (1) are

$$G_{\alpha\alpha}(\mathbf{p}+\mathbf{Q},\mathbf{p};\omega) = \frac{\alpha M(\omega^2 + \varepsilon^2(\mathbf{p}) + M^2 - \Delta^2)}{D} \tag{27.4}$$

$$F_{\uparrow\downarrow}(-\mathbf{p},\mathbf{p};\omega) = \frac{\Delta(\omega^2 + \varepsilon^2(\mathbf{p}) - M^2 + \Delta^2)}{D} \tag{27.5}$$

where

$$D(\omega) = [\omega^2 + \varepsilon^2(\mathbf{p}) + (M+\Delta)^2][\omega^2 + \varepsilon^2(\mathbf{p}) + (M-\Delta)^2] \ . \tag{27.6}$$

The Green functions in the region (2) are identical with the BCS Green functions because in this region the exchange integral I vanishes.

In terms of Green functions (27.4) and (27.5), the magnetization M and the order parameter Δ are

$$M = \frac{I}{2} \sum_{\mathbf{p}\in 1} \pi T \sum_{\omega,\alpha} \alpha G_{\alpha,\alpha}(\mathbf{p}+\mathbf{Q},\mathbf{p};\omega) \tag{27.7}$$

$$\Delta = g \sum_{\mathbf{p}} \pi T \sum_{\omega} F_{\downarrow\uparrow}(-\mathbf{p},\mathbf{p};\omega) \ . \tag{27.8}$$

From (27.5) and (27.6), we get

$$M = IN_1(0) \int_{-E_\mathrm{B}}^{E_\mathrm{B}} d\varepsilon \pi T \sum_{\omega} \frac{M(\omega^2 + \varepsilon^2 + M^2 - \Delta^2)}{D(\omega)} \tag{27.9}$$

$$\Delta = gN_1(0) \int_{-\omega_\mathrm{D}}^{\omega_\mathrm{D}} d\varepsilon \pi T \sum_{\omega} \frac{\Delta(\omega^2 + \varepsilon^2 - M^2)}{D(\omega)} + gN_2(0) \int_{-\omega_\mathrm{D}}^{\omega_\mathrm{D}} d\varepsilon \pi T \sum_{\omega} \frac{\Delta}{\omega^2 + \varepsilon^2 + \Delta^2} \tag{27.10}$$

where $N_1(0)$ is the density of states for the region 1 at the Fermi level and E_B is the cut-off energy of the SDW which is of the order of the conduction band width. Using now relations similar to (26.27a)–(26.27b), the temperature dependence of $M(T)$ and $\Delta(T)$ are

$$\ln\frac{T}{T_{s0}} = 2\pi T\sum_{\omega\geq 0}\Big[\frac{1}{2M}\Big(\frac{M+\Delta}{\sqrt{\omega^2+(M+\Delta)^2}}+\frac{\Delta-M}{\sqrt{\omega^2+(\Delta-M)^2}}\Big)-\frac{1}{\omega}\Big] \tag{27.11}$$

$$\begin{aligned}\ln\frac{T}{T_{c0}} &= \frac{N_2(0)}{N(0)}2\pi T\sum_{\omega\geq 0}\Big[\frac{1}{\sqrt{\omega^2+\Delta^2}}-\frac{1}{\omega}\Big]\\ &+\frac{N_1(0)}{N(0)}2\pi T\sum_{\omega\geq 0}\frac{1}{2\Delta}\Big[\Big(\frac{\Delta+M}{\sqrt{\omega^2+(\Delta+M)^2}}\\ &+\frac{\Delta-M}{\sqrt{\omega^2+(\Delta-M)^2}}\Big)-\frac{1}{\omega}\Big]\end{aligned} \tag{27.12}$$

where T_{s0} is the critical temperature for the SDW state without superconducting pairing and T_{c0} is the critical temperature of the superconducting state in the absence of the SDW state, and

$$N(0) = N_1(0) + N_2(0)\ .$$

If $T_{c0} < T_{s0}$, the critical temperature of the SDW state is higher than that of the superconducting state. The superconducting transition temperature T_c in the presence of the SDW state is generally lower than T_{c0} and can be obtained from (27.9)–(27.10) in the limit $\Delta \to 0$. From these two equations we get

$$\ln\frac{T_c}{T_{c0}} = \frac{N_1(0)}{N(0)}\pi T_c\sum_{\omega\geq 0}\Big[\frac{\omega^2}{\omega^2+[M(T_c)/2\pi T_c]^2}-\frac{1}{\omega}\Big] \tag{27.13}$$

which in the approximation $M(T_c)/T_c \ll 1$ gives

$$\frac{T_c}{T_{c0}} = 1-\frac{21\varsigma(3)}{2}\frac{N_1(0)}{N(0)}\Big[\frac{M(T_c)}{2\pi T_c}\Big]^2\ . \tag{27.14}$$

In the same Eqs. (27.9)–(27.10) if $M \ll E_B$ in the limit $\Delta \to 0$, we obtain

$$\begin{aligned}1 &= IN_1(0)2\pi T_c\sum_{0\leq\omega\leq E_B}\frac{1}{\sqrt{\omega^2+M^2}}\ ,\\ 1 &= gN_1(0)2\pi T_c\sum_{0\leq\omega<\omega_D}\omega^2(\omega^2+M^2)^{-3/2}+gN_2(0)\ln\frac{2\gamma\omega_D}{\pi T_c}\end{aligned} \tag{27.15}$$

which reduces to

$$\frac{T_c}{T_{c0}} \cong \left(\frac{T_{c0}}{eT_{s0}}\right)^{N_1(0)/N_2(0)} . \tag{27.16}$$

The temperature dependence of the order parameters has been performed numerically[8] and the coexistence region $M \neq 0$, $\Delta \neq 0$ has been obtained for $N_1(0)/N(0) = 0.05$ and $T_{c0}/T_{s0} = 0.5$. In order to analyse the coexistence between these two states we have to calculate the free energy $F(g, I)$ as

$$F(g,I) = F(0,0) + \int_0^g dg' \frac{\partial F(g',I)}{\partial g'} + \int_0^I \alpha I' \frac{\partial F(g,I')}{\partial I'} \tag{27.17}$$

which can be written as

$$\delta F = F(\Delta, M) - F(0,0) = \int_0^\Delta \alpha\Delta'\Delta'^2 \frac{d}{d\Delta'}\left(\frac{1}{g}\right) - \int_0^M dM'M'^2 d\left(\frac{1}{I'}\right)$$

and from (27.9)–(27.10), we get

$$\begin{aligned}\delta F(\Delta, M) = {} & N(0)\Delta^2 \ln\frac{T}{T_c} + N_1(0)M^2 \ln\frac{T}{T_{s0}} \\ & - 2N_1(0)\pi T \sum_{\omega \geq 0} \Big\{ \sqrt{\omega^2 + (\Delta + M)^2} \\ & + \sqrt{\omega^2 + (\Delta - M)^2} - 2\omega - \frac{\Delta^2 + M^2}{\omega} \Big\} \\ & - 4\pi T N_2(0) \sum_{\omega \geq 0} \left[\sqrt{\omega^2 + \Delta^2} - \omega^2 - \frac{\Delta^2}{2\omega} \right] .\end{aligned} \tag{27.18}$$

Near T_c, (27.18) becomes

$$\delta F(\Delta, M) = N(0)\Delta^2 \ln\frac{T}{T_c} + F(0, M) + O(\Delta^4) \tag{27.19}$$

where $F(0, M)$ depends only on the order parameter $M(T)$ considered near the superconducting critical temperature.
From this result and taking the order parameter as

$$\Delta^2(T) = \Delta(0)[1 - T/T_c]$$

the free energy becomes

$$\delta F[\Delta, M(T \approx T_c)] = -\delta F(0)[1 - T/T_{c0}] . \tag{27.20}$$

In a similar way we can evaluate the free energy at the lower critical temperature T_s at which the SDW state disappears. The free energy $\delta F[M, \Delta(T \cong T_{s0})]$ is

$$\delta F[M, \Delta(T \cong T_{s0})] = \delta F(0)[1 - T/T_{s0}]^2 \qquad (27.21)$$

and the entropy discontinuity ΔS is given by

$$\Delta S = N(0)\frac{\Delta^2(T_c)}{T_{c0}} + N_1(0)\frac{M^2(T_c)}{T_{s0}} . \qquad (27.22)$$

From this equation, we can see that at $T = T_c$, a first order phase transition takes place, a result which is different from the case of coexistence between CDW and the superconducting state.[6] A more realistic model for the coexistence between Spin-Density-Waves and Superconductivity has been proposed by Psaltakis and Fenton[7] on the physical idea that in this problem the superconductivity has to be treated as isotropic, in contrast to the earlier theory given by Machida and Matsubara[9] and the simple one presented by Machida.[8] In the model considered in Ref. 7, the SDW occur in the region of the Fermi surface (denoted by (1)) where the nesting condition is satisfied. In region (2), the nesting condition is not satisfied, but the band structure calculations of the organic superconductors showed that region (2) is very small. In such a model, the superconductivity coexists with the Spin-Density-Waves state if the coupling constant $g^{s,s}$ for singlet Cooper pairing is sufficiently strong within region (2) to drive the superconductivity over the entire Fermi surface, against the SDW effect in region (1) of the Fermi surface. In this case, we show that the superconducting pairing involves singlet amplitude constant in space and a comparably large to sinusoidal triplet amplitude. The mixed Cooper pairing is due to the sinusoidal SDW breaking time-reversal symmetry.

The Hamiltonian which describes this model has the form

$$\mathcal{H} = \mathcal{H}_0 + \mathcal{H}_{ss} + \mathcal{H}_{SDW} + \mathcal{H}_{TS} \qquad (27.23)$$

where

$$\mathcal{H}_0 = \sum_{\mathbf{p},\alpha} \varepsilon(\mathbf{p}) c^\dagger_{\mathbf{p}\alpha} c_{\mathbf{p}\alpha} , \qquad (27.23a)$$

$$\mathcal{H}_{ss} = \frac{1}{2} \sum_{\mathbf{p};\alpha,\beta} [(i\sigma_y)\Delta(\mathbf{p}) c^\dagger_{\mathbf{p},\alpha} c^\dagger_{-\mathbf{p},\beta} + h.c.] \tag{27.23b}$$

$$\mathcal{H}_{\rm SDW} = -\sum_{\mathbf{p}\in 1} \sum_{\alpha,\beta} [(\boldsymbol{\sigma}\cdot\mathbf{n}) M(\mathbf{p}^-,\mathbf{p}^+) c^\dagger_{\mathbf{p}^-,\alpha} c_{\mathbf{p}^+,\beta} + h.c.] \,, \tag{27.23c}$$

$$\mathcal{H}_{\rm TS} = -\frac{1}{2} \sum_{\mathbf{p}\in 1,\alpha,\beta} (\boldsymbol{\sigma}\cdot\mathbf{n}\, i\sigma_y)_{\alpha\beta} [\Delta_{-Q} c^\dagger_{-\mathbf{p}^-,\alpha} c^\dagger_{\mathbf{p}^-,\beta} + \Delta_Q c^\dagger_{-\mathbf{p}^+,\alpha} c^\dagger_{\mathbf{p}^+,\beta}] \,. \tag{27.23d}$$

The superconducting order parameters $\Delta(\mathbf{p})$ and $\Delta_{\pm Q}(\mathbf{p})$ are defined self-consistently by the equations

$$\Delta_i = \frac{1}{2} \sum_{j\in 1,2} g^{s,s}_{i,j} \sum_{\mathbf{p}\in j;\alpha,\beta} (i\sigma_y)_{\alpha,\beta} \langle c_{\mathbf{p}\alpha} c_{-\mathbf{p}\beta} \rangle \,, \tag{27.24}$$

$$M = \frac{1}{2} V^{\rm SDW} \sum_{\mathbf{p}\in 1,\alpha,\beta} (i\sigma_y \boldsymbol{\sigma}\cdot\mathbf{n})_{\alpha\beta} \langle c_{\mathbf{p}^-\alpha}, c_{-\mathbf{p}^-\beta} \rangle \,, \tag{27.25}$$

$$\Delta_{-Q} = \frac{1}{2} g^{\rm TS} \sum_{\mathbf{p}\in 1;\alpha,\beta} (i\hat{\sigma}_y \boldsymbol{\sigma}\cdot\mathbf{n})_{\alpha,\beta} \langle c_{\mathbf{p}^-\alpha} c_{-\mathbf{p}^-\beta} \rangle \tag{27.26}$$

where $\Delta_{-Q}(\mathbf{p}) = -\Delta_Q(\mathbf{p}) = \Delta_{-Q}\,\mathrm{sign}\,\mathbf{p}$.
In (27.23), we introduced the spinor $i\hat{\sigma}_y$ for the singlet pairs and $(\boldsymbol{\sigma}\cdot\mathbf{n})i\hat{\sigma}_y$ for the triplet pairs, and the notation $\mathbf{p}^\pm = \mathbf{p} \pm \mathbf{p}_0$ for the region (1). The direction of $\mathbf{p}$ is parallel to $\mathbf{a}\,\varepsilon(\mathbf{p})$ and is measured from the Fermi energy ε_0, and $\varepsilon(\mathbf{p}^\pm) = -\varepsilon(\mathbf{p}^\pm)$. The wave vector $\mathbf{Q}$ of the SDW is taken as $\mathbf{Q} = 2p_0 a$ and the polarization vector n is taken perpendicular to $\mathbf{a}$.

Using now the strong coupling formalism the Green functions have been calculated[7] and with these results Eqs. (27.24)–(27.26) become

$$\Delta_i = -\frac{1}{2} \sum_{j=1,2} g^{s,s}_{i,j} \sum_{\mathbf{p}\in j} \sum_{+,-} \frac{\Delta_j + M}{2E_{j\pm}(\mathbf{p})} \tanh \frac{E_{j\pm}(\mathbf{p})}{2T} \,, \tag{27.27}$$

$$M = \frac{1}{2} V^{\rm SDW} \sum_{\mathbf{p}\in 1} \sum_{+,-} \frac{M \pm \Delta_1}{2E_{1\pm}(\mathbf{p})} \tanh \frac{E_{1\pm}(\mathbf{p})}{2T} \,, \tag{27.28}$$

$$\Delta_{-Q} = \frac{1}{2} g^{\rm TS} \sum_{\mathbf{p}\in 1} \sum_{+,-} \frac{\Delta_{-Q} \pm |\varepsilon(\mathbf{p}^-)|}{2E_{1\pm}(\mathbf{p})} \tanh \frac{E_{1+}(\mathbf{p})}{2T} \tag{27.29}$$

where $E_{j\pm}(\mathbf{p})$ $(j = 1, 2)$ is defined by

$$E_\pm(\mathbf{p}) = \begin{cases} E_{1\pm}(\mathbf{p}) = [(\varepsilon(\mathbf{p}^-) \pm \Delta_{-Q}(\mathbf{p}))^2 + (\Delta_1 \pm M)^2]^{1/2} \\ E_{2\pm}(\mathbf{p}) = [\varepsilon^2(\mathbf{p}) + \Delta_2^2(\mathbf{p})]^{1/2} \,. \end{cases} \tag{27.30}$$

If $|g^{\mathrm{TS}}N(0)| \ll 1$, we can neglect Δ_{-Q} and equations for Δ (which is constant), and M are the same as those obtained in Ref. 8 and those presented as the first approach to this problem.

An interesting result can be obtained by linearizing Eqs. (27.27) and (27.29). If $T_{\mathrm{SDW}} \gg T_c$, we get

$$\Delta_{-Q} = \frac{g^{\mathrm{TS}} N_1(0)\Delta}{2 + g^{\mathrm{TS}} N_1(0)}$$

an equation which shows that if $T \to T_c$, $\Delta \to 0$ in the same time with Δ_Q. In a similar way for $T_{\mathrm{SDW}} \cong T_c$ (but with $M(T_c) \ll T_c$), we get

$$\Delta_{-Q} = g^{\mathrm{TS}} \frac{N(0)M(T_c)}{T_c} \Delta \, .$$

Now, we can estimate the influence of the anisotropic interaction on the superconducting state.

Let us introduce the notations

$$\lambda_{i,j} g_{i,j}^{s,s} N(0) \, , \tag{27.31}$$

and for the singlet Cooper gaps in region (1) respectively (2)

$$\Delta_1 = \Delta_e + \Delta_o \, , \quad \Delta_2 = \Delta_e - \Delta_o \tag{27.32}$$

where Δ_e and Δ_o are the even and odd parts with respect to regions (1) and (2) from the Fermi surface. The equation for the gaps will be linearized near T_c and we get (if $M = 0$) the critical temperature as:

a) for $\lambda_{22} = \lambda_{12} < 0; \lambda_{11} < \lambda_{21} < 0$ and $\lambda_{11} + \lambda_{22} = -2\lambda < 0$
$$T_{c0} = \frac{2\gamma E^{ss}}{\pi} \exp[-1/2\lambda] \quad \text{and } \Delta_1 = \Delta \, , \tag{27.33a}$$

b) for $-\lambda_{22} = \lambda_{12} = \lambda_{11} = \lambda_{21} = \lambda > 0$
$$T_{c0} = \frac{2\gamma E^{ss}}{\pi} \exp[-1/\lambda\sqrt{2}] \quad \text{and } \Delta_1 = -\Delta_2(\sqrt{2} - 1) \, , \tag{27.33b}$$

c) for $-\lambda_{22} = -\lambda_{12} = -\lambda_{21} = \lambda > 0$
$$T_{c0} = \frac{2\gamma E^{ss}}{\pi} \exp[-1/\lambda\sqrt{2}] \tag{27.33c}$$

where E^{ss} is the singlet cut-off.

The effect of the non-magnetic impurities can be studied using the standard methods but taking separately regions (1) and (2). As we expect, the critical temperature T_{SDW} in the presence of the impurities is strongly depressed according to the equation

$$\ln \frac{T_{\mathrm{SDW}}}{T_{\mathrm{SDW}}^0} = \Psi\left[\frac{1}{2}\right] - \Psi\left[\frac{1}{2} + \frac{1}{2\pi T_{\mathrm{SDW}}}\left(\frac{1}{\tau_{11}} + \frac{1}{\tau_{12}}\right)\right] \tag{27.34}$$

where $T^0_{\rm SDW}$ is the critical temperature in the absence of the impurities and the scattering times τ_{ij} are the electron scattering times defined in the usual way for impurity scattering from region j to region i.

The impurities influence on the superconductivity will be studied for $\Delta_{-Q} = 0$ and for $\tau_{11} = \tau_{12} = \tau_{22} = \tau_{21}$ and the critical temperature has been approximated[7] as

$$T_c = T_{c0}\left[1 - \frac{\pi}{8\tau_{12}T_{c0}}\left(1 \pm \frac{\lambda_{12}+\lambda_{21}}{[(\lambda_{11}-\lambda_{22})^2 + 4\lambda_{12}\lambda_{21}]^{1/2}}\right)\right] \tag{27.35}$$

for $M = 0$.

In (27.35), the upper or lower sign is required according to whether the upper or lower sign is required in the condition

$$\frac{\lambda_{11}+\lambda_{22} \pm [(\lambda_{11}-\lambda_{22})^2 + 4\lambda_{12}\lambda_{21}]^{1/2}}{(\lambda_{12}\lambda_{21} - \lambda_{22}\lambda_{11})} > 0\,.$$

The critical temperature is also affected by the anisotropy, and the combined effects are expressed by

a) $T_c = T_{c0} \quad \text{if} \quad \Delta_1 = \Delta_2\,,$

b) $T_c = T_{c0}\left[1 - \frac{\pi(1+1/\sqrt{2})}{8T_{c0}\tau_{12}}\right] \quad \text{for } \tau_{12}T_{c0} \gg 1\,,$

$$\Delta_2 = -\Delta_1\left[1 + \sqrt{2}\left(1 - \frac{\ln T_{c0}/T_c}{(\lambda\sqrt{2})^{-1} + \ln T_{c0}/T_0}\right)\right],$$

c) $T_c = T_{c0}\left[1 - \frac{\pi(1-1/\sqrt{2})}{8T_{c0}\tau_{12}}\right] \quad \tau_{12}T_{c0} \gg 1$

$$\Delta_1 = -\Delta_2\left[1 - \sqrt{2}\left(1 - \frac{\ln T_{c0}/T_c}{(\lambda\sqrt{2})^{-1} + \ln T_{c0}/T_c}\right)\right].$$

In the coexistence region, T_c and $T_{\rm SDW}$ have to be calculated numerically from the equations for the two gaps.

Performing the free energy calculations as in Ref. 7, one obtains an interesting result contained in the equation

$$\delta F = -\frac{1}{2}N(0)\Delta^2(m,\Delta,\Delta_Q) - \frac{1}{2}N_1(0)m^2(\Delta,\Delta_{-Q}) \tag{27.36}$$

which shows that at $T = 0$ the coexistence free energy is not an explicit function of the triplet gap Δ_Q.

The magnetic susceptibility at $T = 0$ has been calculated as

$$\chi_{\parallel}(T=0) = 0 \tag{27.37a}$$

$$\chi_{\perp}(T=0) = \mu_0^2 N_1(0)\left\{1 + \frac{M^2-\Delta_1^2}{2M\Delta_1}\ln\left|\frac{M+\Delta_1}{M-\Delta_1}\right|\right\} \tag{27.37b}$$

and because the larger value of Δ_{-Q} increases the triplet amplitude, $\chi_1(T = 0)$ has a non-zero value. This model seems to be appropriate for the explanation of the coexistence between the superconductivity and the SDW state in the $(TMTSF)_2X$ organic superconductors.

References

1. K. Levin, D. L. Mills and S. Cunninghan, *Phys. Rev.* **B10** (1974) 3821; **B10** (1974) 3832.
2. G. Bilbro and W. L. McMillan, *Phys. Rev.* **B14** (1976) 1887.
3. C. A. Balsiero and L. M. Falicov, *Phys. Rev.* **B20** (1979) 4457; *Phys. Rev. Lett.* **45** (1980) 662.
4. K. Machida, T. Koyama and T. Matsubara, *Phys. Rev.* **B23** (1981) 99.
5. A. M. Gabovich, E. A. Pashitskii and A. S. Shpigel, *Sov. Phys. JETP* **50** (1979), A. M. Gabovich and A. Shpigel, *J. Low Temp. Phys.* **51** (1983) 581.
6. K. Machida, *J. Phys. Soc. Jpn.* **53** (1984) 712.
7. G. C. Psaltakis and E. W. Fenton, *J. Phys. C. Solid State Phys.* **16** (1983) 3913.
8. K. Machida, *J. Phys. Soc. Jpn.* **50** (1981) 2195; **50** (1981) 3231; **51** (1982) 1420; **53** (1983) 1333.
9. K. Machida and F. Matsubara, *J. Phys. Soc. Jpn.* **50** (1981) 3231.

VII
UNCONVENTIONAL SUPERCONDUCTIVITY

The electron-electron attraction is essential for the occurrence of the superconducting pairing. In the majority of systems (metals, alloys, or compounds), this pairing is supposed to be given by the attraction between the electrons mediated by the electron-phonon coupling. However, the estimation of the critical temperature T_c in this model, taking the realistic parameters, pointed out the existence of an upper limit for this temperature determined by the phononic characteristic frequency. This limit for T_c is about 30 K and was not overcome for a long time.

As one of the most important purposes in superconductivity was the obtaining of the high temperature superconductivity, a lot of effort was done to discover new mechanisms which could give a higher critical temperature. These new mechanisms imply in fact some new models in which the electron-electron interaction is not mediated by phonons, and they are known as unconventional mechanisms for superconductivity.

In these new mechanisms, we consider instead of phonons, other bosons such as plasmons, excitons, spin fluctuations, in order to explain the occurrence of the superconductivity in different materials or to find out a new interaction which may give rise to a higher critical temperature. The discovery of the superconductivity in some materials which contain a high concentration of magnetic ions generated a special kind of electron pairing which was attributed to the quasiparticles called heavy fermions. The electron-electron attraction in these systems takes place between quasiparticles with heavy effective mass and is not mediated only by the phonons. Using a periodic Anderson model to describe the Kondo lattice, different models have been developed; some of them consider the electron-phonon interaction as well as hybridization and repulsion, others consider the charge and spin fluctuations as responsible for the pairing. For these systems, the pairing is anisotropic and the superconductivity may coexist with the other condensed phases in the quasiparticles which form a Fermi liquid.

In 1986, Müller and Bednorz discovered the high temperature superconductivity in A-B-Cu-O compounds with A = La, B = Ba, Sr, or Ca. All these high T_c superconductors present a two dimensional character of the band structure with an imperfect nesting. The first model proposed is due to Anderson who suggested the "resonanting valence bond".

Another direction which was developed by Alexandrov, Ranninger, Robaszkiewicz, Nasu, and Rice has, as the main idea, the occurrence of the bipolarons in these materials because of the large local lattice deformations which could give rise to an attractive one-site interaction. In the limit of large electron-phonon interaction, these bipolarons can be condensed (a Bose condensation) and this new phase contributes to the superconducting state. As the band structure is appropriate for the occurrence of the Spin-Density-Wave and Charge-Density-Waves, Machida, Kato and Ohkawa used many ideas from the theory of heavy fermions superconductivity to take into consideration the occurrence of the high temperature superconductors.

Hirch followed the similarity with the heavy fermion systems, and showed that in the systems with large one-site electron-electron repulsion, the effective interaction between nearest-neighbour electrons with antiparallel spins is an attractive interaction. Schüttler, Jarrel and Scalapino considered the model with negative U bipolaronic centers in perturbative methods which was combined with the Monte Carlo simulation to study the influence of the small excitonic impurities on the critical temperature.

The inter-layer electrons have been considered by Inoue, Takemori and Ohtaka in the formalism of the strong coupling theory of superconductivity. Abrahams, Varma and Schmitt-Rink showed the importance of the charge transfer which may mediate an electron-electron attraction. Morgenstern, Müller, Deutcher and Bednorz performed a numerical simulation on an xy-spin glass in two dimensions to explain the high temperature superconductivity in the new materials using the granular model of superconductivity.

28. Non-Phononic Mechanisms

The electron-phonon interaction, generally considered as responsible for the occurrence of electron pairs, has been considered a long time as giving an upper bound critical temperature. The raising of the critical temperature was predicted by considering different bosons as plasmons, excitons or acoustic plasmons[1,2] in order to replace the phonons as mediating bosons between two electrons which form the Cooper pairs.

The materials known as A-15 compounds have been considered a long time as the superconductors with the highest critical temperature. Many of these compounds exhibit also structural phase transitions at a critical temperature T_M above the critical temperature T_c. A simple non-phononic mechanism proposed by Geilikman[3] consists, from two electronic bands, of a "s-band" with a light effective mass and a "d-band" with a heavy effective mass. The indirect attractive interaction between s-electrons can become large to overcome the repulsive bare Coulomb interaction and the s-electrons system can exhibit superconductivity. For the A-15 compounds, the d-band is very narrow and the acoustic plasmons can mediate the pairing.

Another possibility to obtain the non-phononic pairing appears in the systems which might be called the superconducting Schottky barrier analysed by Allender, Bray and Bardeen[4] and Inkson and Anderson.[5] In this system, enhancement of the critical temperature of a metal is achieved by placing it in contact with a semiconductor.

The electrons in the metal interact by way of virtual excitations in semiconductor so that an extra attractive interaction is obtained.

All the non-phononic models (as well as the phononic model) can be analysed considering the electron-electron interaction $U(\varepsilon - \varepsilon', \mathbf{q})$ near the Fermi surface. The Cooper pairs appear if $U(\varepsilon - \varepsilon', \mathbf{q}) < 0$, and the general equation for the critical temperature is

$$T_c = 1.14\omega_D \exp\left[-\frac{1}{U(\varepsilon - \varepsilon', \mathbf{q})}\right] \tag{28.1}$$

where for the electron-phonon standard superconductor $U = g - \mu^*$ and $\omega_c = \omega_D$ is the phonon characteristic energy and the Fermi energy is $\varepsilon_0 > \omega_D$. The quantities g and μ^* are directly connected with the effective interaction between electrons by the dielectric function $\varepsilon(\omega, \mathbf{q})$ of the system

$$V_{\text{eff}} = \frac{4\pi e^2}{q^2 \varepsilon(\omega, \mathbf{q})} . \tag{28.2}$$

The connection is given by the equation

$$\mu^* - g = N(0)\langle V_{\text{eff}}(\mathbf{q}, \omega)\rangle \tag{28.3}$$

where the brackets imply averaging over the momentum transfer on the Fermi surface and $N(0)$ is the electron density of states on this surface. The existence of an effective value of the momentum $\mathbf{q}$ of the order of the Fermi momentum p_0 is essential for such an average. Equation (28.3) shows that if $\varepsilon(\mathbf{q}, 0) < 0$, the relation $g \leq \mu^*$ would hold obligatorily and the total interaction is repulsive. The superconducting state appears only when $g > \mu^*$ which shows that the critical temperature has an upper bound. At $g = \mu^*$, T_c as function of the characteristic frequency ω_c has a peak. The maximum of T_c occurs at $\omega_c = \varepsilon_0 \exp(-1/g)$ and is

$$T_c^{\max} = \varepsilon_0 \exp\left[-\frac{3}{g}\right] . \tag{28.4}$$

From (28.4), $T_c^{\max}$ was estimated[6] at $g = \mu^* = 0.5$ as 10 K but this estimation can be improved if we perform a more accurate analysis of the effective electron-electron interaction.

The simplest model is the system of interacting electrons in a medium consisting of independent oscillators. The dielectric function of these electrons is

$$\varepsilon(\omega, \mathbf{q}) = \varepsilon_r(\omega) = 1 + \frac{\omega_r^2(\varepsilon_0 - 1)}{\omega_r^2 - \omega^2} \tag{28.5}$$

and if

$$\omega_r < \omega \leq \sqrt{\varepsilon_0}\,\omega_r \, , \tag{28.6}$$

the effective interaction (28.2) is attractive. The frequency ω_r from (28.5) can be obtained from the equation $\omega_r^2(\varepsilon_0 - 1) = m^{-1}4\pi e^2 n_r$, where n_r is the concentration of the oscillators.

For the "jelium" model, the dielectric function is

$$\varepsilon(\omega, \mathbf{q}) = 1 + \frac{\chi^2}{q^2} - \frac{\omega_i^2}{\omega^2} \tag{28.7}$$

where the frequency ω_i is

$$\omega_i = \left[\frac{4\pi e^2 n_i Z^2}{M}\right]^{\frac{1}{2}} , \tag{28.8}$$

n_i being the concentration of the ions with the mass M and χ is the screening constant

$$\chi^2 = \frac{4e^2 m q_0^2}{\pi^2} \,. \tag{28.9}$$

In this model, the expression (28.2) becomes

$$V_{\text{eff}} = \frac{4\pi e^2}{q^2 + \chi^2}\left[1 + \frac{\omega^2(q)}{\omega^2 - \omega^2(q)}\right] \tag{28.10}$$

where $\omega(q)$ is the frequency of the longitudinal waves given by

$$\omega^2(q) = \frac{\omega_i^2 q^2}{q^2 + \chi^2} \tag{28.11}$$

and $0 < \omega < \omega(q)$.

These two models can be generalized taking for the dielectric function,

$$\varepsilon(q) = 1 + \frac{\chi^2}{q^2} - \frac{\omega^2}{\omega_i^2} + \frac{\Omega^2(\varepsilon_0 - 1)}{\Omega^2 - \omega^2} \tag{28.12}$$

and from the condition $\varepsilon(\omega, \mathbf{q}) = 0$, we get

$$\omega_-^2(q) = \omega_{\text{ph}}^2(q) = \frac{\omega_i^2 q}{\varepsilon_0 q^2 + \chi^2} \tag{28.13}$$

$$\omega_+^2(q) = \Omega_e^2(q) = \frac{(\varepsilon_0 q^2 + \chi^2)\Omega^2}{q^2 + \chi^2} \tag{28.14}$$

where we considered $\Omega \gg \omega_i$. From (28.13)–(28.14) we see that $\omega_{\text{ph}}(q)$ is associated with the phonons spectrum and $\Omega_e(q)$ with the electronic excitations. Such a model is often called excitonic model and from (28.14) we obtain

$$\Omega_e(0) = \Omega \,, \quad \Omega_e(2p_0) = \sqrt{\frac{\varepsilon_0 + \alpha}{1 + \alpha}}\,\Omega \tag{28.15}$$

where

$$\chi^2 = \alpha(2p_0)^2 \,, \quad \alpha = \frac{e^2 m}{\pi p_0}$$

and if $\varepsilon_0 \gg 1$ and $\alpha \ll 1$, we get the pole of the dielectric function $\varepsilon^{-1}(\omega, \mathbf{q})$ for $q = 2p_0$ as

$$\Omega_c(2p_0) = \sqrt{2\varepsilon_0}\,\Omega \,. \tag{28.16}$$

In this case, the effective electron-electron interaction is frequency dependent and from (28.2), we get

$$V_{\text{eff}} = \frac{4\pi e^2}{p_0^2}\left\{1 + \frac{\omega_i^2/\varepsilon_0^2}{\omega^2 - \omega_i^2/\varepsilon_0} + \frac{(\varepsilon_0 - 1)\Omega^2}{\omega^2 - \varepsilon_0\Omega^2}\right\} \tag{28.17}$$

where for $\alpha \ll 1$ we approximated $(2p_0)^2 + \chi^2 \cong 2p_0^2$ and the phonon frequency by $\omega_{\text{ph}} = \omega_c/\sqrt{\varepsilon_0}$.

A more accurate analysis can be performed if we consider the dielectric function as a response function. It is known that a response function $R(\omega, \mathbf{q})$ satisfies the Kramers-Kroning relation

$$R(\omega, \mathbf{q}) = R(\infty, \mathbf{q}) + \frac{1}{\pi}\int_0^\infty d\omega'^2 \frac{\operatorname{Im} R(\omega', \mathbf{q})}{\omega'^2 - \omega^2 - i\delta} \tag{28.18}$$

and in the static limit

$$R(0, \mathbf{q}) = R(\infty, \mathbf{q}) + \frac{1}{\pi}\int_0^\infty \frac{d\omega'^2}{\omega'^2} \operatorname{Im} R(\omega', \mathbf{q}) \,. \tag{28.19}$$

If we apply (28.19) for $1/\varepsilon(0, \mathbf{q})$ with the condition $\varepsilon(\infty, \mathbf{q}) = 1$, we get

$$\frac{1}{\varepsilon(0, \mathbf{q})} = 1 + \frac{1}{\pi}\int_0^\infty \frac{d\omega'^2}{\omega'^2} \frac{1}{\varepsilon(\omega', \mathbf{q})} \tag{28.20}$$

and from (28.18), we get for the dynamic dielectric function

$$\operatorname{Im} \varepsilon(\omega, \mathbf{q}) \geq 0, \quad \operatorname{Im} \varepsilon^{-1}(\omega, \mathbf{q}) < 0 \tag{28.21}$$

which in the particular case of the static limit gives equivalent to

$$\varepsilon(0, \mathbf{q}) \geq 1, \quad \varepsilon(0, \mathbf{q}) < 0 \,. \tag{28.22}$$

This result shows that the causality condition does not preclude negative values for $\varepsilon(0, \mathbf{q})$, only values between 0 and 1 are forbidden.

On the other hand, the dielectric function $\varepsilon(\omega, \mathbf{q})$ is also a response function and we can write

$$\varepsilon(0, \mathbf{q}) = 1 + \frac{1}{\pi}\int_0^\infty \frac{d\omega'^2}{\omega'^2} \operatorname{Im} \varepsilon(\omega', \mathbf{q}) \tag{28.23}$$

and using now (28.22) for this relation it would follow that a negative value for the static dielectric function is impossible and

$$\varepsilon(0, \mathbf{q}) \geq 1 \tag{28.24}$$

a relation which is valid even for $\mathbf{q} \to 0$.

The relations (28.22) and (28.24) are associated with the stability of the electronic system.

In order to discuss the stability of the electronic system in the formalism of the dielectric function, we write $\varepsilon^{-1}(\omega, \mathbf{q})$ as

$$\frac{1}{\varepsilon(\omega, \mathbf{q})} = 1 - \int_0^\infty dE^2 \frac{\sigma(E, \mathbf{p})}{E^2 - \omega^2 - i\delta} \tag{28.25}$$

where the spectral density $\sigma(E, \mathbf{p})$ is

$$\sigma(\omega, \mathbf{p}) = -\frac{1}{\pi}\mathrm{Im}\left[\frac{1}{\varepsilon(\omega, \mathbf{p})}\right] . \tag{28.26}$$

From (28.25), we get

$$\varepsilon(\omega, 0) = 1 + \frac{1}{\pi}\int_0^\infty dE^2 \frac{\mathrm{Im}\,\varepsilon(E, 0)}{E^2 - \omega^2 - i\delta} \tag{28.27}$$

where

$$\mathrm{Im}\,\varepsilon(\omega, 0) = \pi\sigma|\varepsilon|^2 \geq 0 . \tag{28.28}$$

Equation (28.25) can be written as

$$\frac{1}{\varepsilon(0, \mathbf{q})} = 1 - I \tag{28.29}$$

where

$$I = 2\int_0^\infty \frac{dE}{E}\sigma(E, \mathbf{q}) \geq 0 \tag{28.30}$$

and $\varepsilon(0, \mathbf{q})$ has to satisfy (28.22).

The first inequality implies $I < 0$ and the second $I > 0$. If we note that

$$\varepsilon(0, 0) = \lim_{q \to 0} \varepsilon(0, \mathbf{q}) \geq 1$$

the general behaviour of the dielectric function can be given by:

a. for all values of $\mathbf{q}$, the function $\varepsilon(0, \mathbf{q})$ satisfies

$$0 < \varepsilon^{-1}(0, \mathbf{q}) < 1 ,$$

b. the function $1/\varepsilon(0, \mathbf{q})$ has a zero for $q \neq 0$ and becomes negative.

If we analyse now the stability of the electronic system using thermodynamics, we can write the general condition

$$\frac{\delta^2 E[\rho]}{\delta\rho(0,\mathbf{p})\delta\rho(0,-\mathbf{p})} = -\frac{1}{F(0,\mathbf{p})} \, . \tag{28.31}$$

In (28.31), $E[\rho]$ is the energy of the system depending on the density ρ and the function F is

$$F(\omega,\mathbf{p}) = \frac{p^2}{4\pi}[\varepsilon^{-1}(\omega,\mathbf{p}) - 1] \, . \tag{28.32}$$

These results can be applied to a simple model in which $\delta\rho$ contains the charge variations due to the electrons and ions. The dynamic dielectric function $\varepsilon(\omega,\mathbf{p})$ is given by the polarization function

$$\varepsilon(\omega,\mathbf{p}) = 1 - \frac{4\pi}{p^2}\Pi(\omega,\mathbf{p}) \tag{28.33}$$

and

$$\Pi(\omega,\mathbf{p}) = \Pi_{\mathrm{i}}(\omega,\mathbf{p}) + \Pi_{\mathrm{e}}(\omega,\mathbf{p}) \, . \tag{28.34}$$

The ionic polarization function is

$$\Pi_{\mathrm{i}}(\omega,\mathbf{p}) = \frac{p^2\omega_{\mathrm{i}}^2}{4\pi[\omega^2 - \omega_{\mathrm{e}}^2(\mathbf{p}) + i\delta]} \tag{28.35}$$

and the electronic contribution is

$$\Pi_{\mathrm{e}}(\omega,\mathbf{p}) = \frac{p^2\omega_{\mathrm{e}}^2}{4\pi[\omega^2 - \omega_{\mathrm{c}}^2 - p^2/\chi^2 + i\delta]} \tag{28.36}$$

where

$$\omega_{\mathrm{e}} = [4\pi e\rho_{\mathrm{e}}/m]^{1/2} \tag{28.37}$$

is the plasmonic frequency for electrons and

$$\omega_{\mathrm{i}} = [4\pi Ze\rho_{\mathrm{i}}/M]^{1/2} \tag{28.38}$$

is the ionic plasmonic frequency.

The dielectric function $\varepsilon(\omega,\mathbf{p})$ can be calculated from (28.33) and (28.35)–(28.36), and we have

$$\varepsilon(\omega,\mathbf{p}) = 1 - \frac{\omega_{\mathrm{i}}^2}{\omega^2 - \omega_{\mathrm{c}}^2 p^2/\chi^2 + i\delta} - \frac{\omega_{\mathrm{i}}^2}{\omega^2 - \omega_0^2(p) + i\delta} \tag{28.39}$$

which gives

$$\varepsilon(0,\mathbf{p}) = 1 + \frac{\chi^2}{p^2} + \frac{\omega_{\rm i}^2}{\omega_0^2(p)} \,. \tag{28.40}$$

From (28.39), the spectral density $\sigma(\omega,\mathbf{p})$ was obtained as

$$\sigma(\omega,\mathbf{p}) = \omega_{\rm c}^2\delta\left[\omega^2 - \omega_{\rm c}^2\left(1+\frac{p^2}{\chi^2}\right)\right] + \omega_{\rm i}^2\left(\frac{p^2}{p^2+\chi^2}\right)$$
$$\times\, \delta\left[\omega^2 - \omega_{\rm i}^2\frac{p^2}{p^2+\chi^2} - \omega_0^2(\mathbf{p})\right] \tag{28.41}$$

and from this relation we get two kinds of elementary excitations: the plasmonic excitations with the energy

$$\omega_{\rm p}(\mathbf{p}) = \omega_{\rm c}\left[1+\frac{p^2}{\chi^2}\right]^{1/2} \tag{28.42}$$

and the phononic excitations

$$\omega_{\rm ph} = \left[\omega_0^2(\mathbf{p}) + \omega_{\rm i}\frac{p^2}{p^2+\chi^2}\right]^{1/2} \tag{28.43}$$

and this result shows that the electron-electron interaction has to be considered together with the electron-phonon interaction. The electron-phonon interaction $g(\mathbf{q})$ is enhanced and the new coupling constant is

$$\tilde{g}(q) = \frac{g(q)}{\varepsilon_{\rm e}(\omega,q)} \,. \tag{28.44}$$

In this approximation a standard perturbative calculation gives for the energy of phonons: the poles of the phononic Green function

$$\omega_{\rm ph}^2 = \frac{mZ}{3M}v_0^2q^2 \tag{28.45}$$

and $\varepsilon_{\rm e}(0,\mathbf{q})$ is given by (28.33) with $\Pi = \Pi_{\rm e}$.
The total effective interaction is

$$V_{\rm eff} = \frac{4\pi e^2}{q^2\varepsilon_{\rm e}(\omega,\mathbf{q})} + \frac{g^2(\mathbf{q})}{\varepsilon_{\rm e}^2(0,\mathbf{q})}\frac{\omega^2(\mathbf{q})}{\omega^2-\omega^2(\mathbf{q})} \tag{28.46}$$

where

$$\omega^2(\mathbf{q}) = \omega_{\rm i}^2\left[1 - \frac{g^2(q)}{V(q)}\left(1-\frac{1}{\varepsilon_{\rm e}(0,\mathbf{q})}\right)\right] .$$

If (28.46) is rewritten as

$$V_{\text{eff}} = \frac{4\pi e^2}{q^2 \varepsilon_{\text{eff}}(\omega, \mathbf{q})} \tag{28.47}$$

the effective dielectric function is

$$\varepsilon_{\text{eff}}(\omega, \mathbf{q}) = \varepsilon_{\text{e}}(\omega, \mathbf{q}) - \frac{\omega_{\text{i}}^2}{\omega^2} \tag{28.48}$$

which becomes

$$\varepsilon_{\text{eff}}(\omega, \mathbf{q}) = \varepsilon_{\text{e}}(\omega, \mathbf{q}) - \frac{g^2(q)}{V(q)} \frac{\omega_{\text{i}}^2(q)}{\omega^2 - \omega_{\text{i}}^2(1 - g^2(q)/V(q))} \,. \tag{28.49}$$

The condition of the lattice stability $\omega^2(\mathbf{q}) > 0$ gives

$$1 + \frac{g^2(q)}{V(q)} \frac{1 - \varepsilon_{\text{e}}(0, \mathbf{q})}{\varepsilon_{\text{e}}(0, \mathbf{q})} > 0 \tag{28.50}$$

and using for $\varepsilon_{\text{e}}(0, \mathbf{q})$ the Hubbard approximation $(\varepsilon_{\text{e}}^{-1}(\mathbf{q})) = 1 + \frac{U}{2}\Pi(0, \mathbf{q})$ where U is the exchange interaction, we have the condition

$$g^2(q) + \frac{U}{2} - V(q) < \frac{1}{|\Pi(0, \mathbf{q})|} \,. \tag{28.51}$$

With this result, we get, for the effective interaction (28.47), the properties:

$$\text{a)}\ \varepsilon_{\text{e}}(0, \mathbf{q}) > 0\,, \quad \varepsilon_{\text{eff}}\left[\frac{g^2}{V(q)} - 1\right] < 0 \tag{28.52a}$$

$$\text{b)}\ \varepsilon_{\text{e}}(0, \mathbf{q}) > 0\,, \quad \varepsilon_{\text{eff}}\left[\frac{g^2}{V(q)} - 1\right] > 0 \tag{28.52b}$$

which give the important result expressed by $\varepsilon_{\text{e}}(0, \mathbf{q}) > 0$ then $V_{\text{eff}}(0, \mathbf{q}) < 0$; namely, the effective electron-electron interaction is attractive if the static dielectric function of electrons is positive.

These general results give us the possibility to analyse the most important non-phononic mechanisms.

1) *The two-band mechanism*

This is a non-phononic three-dimensional mechanism proposed in Ref. 3 to explain the superconductivity in systems with d-electrons and s-electrons. The excitations which are supposed to mediate the attractive electron-electron interaction are the acoustic plasmons, and in the following we try to see if the critical temperature can be high on the hypothesis that between the d-electron and s-electron masses there exists the relation $m_d \gg m_s$.

The total dielectric function $\varepsilon(\omega, \mathbf{q})$ is

$$\varepsilon(\omega, \mathbf{q}) = 1 + 4\pi\alpha_s(\omega, \mathbf{q}) + 4\pi\alpha_d(\omega, \mathbf{q}) \tag{28.53}$$

where

$$\begin{aligned} \alpha_s(\omega, \mathbf{q}) &= -V(q)\Pi_{s,s}(\omega, \mathbf{q}) \\ \alpha_d(\omega, \mathbf{q}) &= -V(\mathbf{q})\Pi_{d,d}(\omega, \mathbf{q}) \end{aligned} \tag{28.54}$$

and

$$\Pi_{i,i}(\omega, \mathbf{q}) = \frac{p_{0i}^2 q}{2\pi^2} \int_{-1}^{1} x dx \frac{1}{\omega - q v_i x - i\delta} \tag{28.55}$$

with $v_i = p_{0i}/m_i$. If

$$q v_d/\omega < 1, \quad q v_s/\omega > 1 , \tag{28.56}$$

we have

$$\mathrm{Re}\,\Pi_{d,d} = \frac{p_{0d}^3 q^3}{3\pi^2 m_d \omega^2}; \quad \mathrm{Im}\,\Pi_{d,d}(\omega, \mathbf{q}) = 0 \tag{28.57}$$

and

$$\mathrm{Re}\,\Pi_{s,s} = -\frac{m_s p_{0s}}{\pi^2}; \quad \mathrm{Im}\,\Pi_{s,s}(\omega, \mathbf{q}) = \frac{-m_s^2 \omega}{2\pi q} . \tag{28.58}$$

From the condition $\varepsilon(\omega, \mathbf{q}) = 0$, we get from (28.53) the frequency

$$\omega^2 = \frac{\omega_d^2}{1 + (k_s/q)^2\left[1 + i\frac{\pi}{2}\frac{\omega}{q v_s}\right]} \tag{28.59}$$

where

$$k_s^2 = \frac{4\pi e^2 m_s p_{0s}}{\pi}; \quad \omega_d^2 = \frac{4\pi e^2 p_{0d}^3}{3\pi m_d} .$$

In the limit of small wave vector $\mathbf{q}$, the energy of the acoustic plasmons is

$$\omega(\mathbf{q}) = \frac{\pi}{4}\frac{\omega_d}{k_s^2} q = \frac{\pi}{12}\left(\frac{m_s n_d}{m_d n_s}\right) v_s q \tag{28.60}$$

and the amortization is

$$\gamma_q = \frac{\pi}{4}\frac{\omega_d}{k_s^2} q = \frac{\pi}{12}\left(\frac{m_s n_d}{m_d n_s}\right) v_s q \tag{28.61}$$

and the existence of these quasiparticles is assured by the condition

$$\omega_q > \gamma_q . \tag{28.62}$$

Tutto and Ruvalds[7] discussed the role of the acoustic plasmons in the electronic pairing and the plasmon coupling constant has been calculated as

$$\lambda_{\rm pl} = \frac{g_{\rm pl}^2 N(0)}{2\langle\omega_{\rm pl}\rangle}\left(\frac{q_{\rm M}}{p_0}\right)^2 \tag{28.63}$$

where $q_{\rm M}$ is the cut-off momentum and $\theta_{\rm pl} = q_{\rm M}s$ is the plasmonic characteristic energy.

For Nb_3Sn if $\lambda_{\rm ph} \simeq 0.5$, the value obtained from the normal data $T_c \simeq 18$ K is obtained only if $\lambda_{\rm pl} \simeq 0.2$ is considered. Direct observation of this effect can be seen by tunnelling and at the present state it is generally accepted that such an additional mechanism is effective in these materials.

2) *Tunnelling mechanisms*

According to the idea proposed in Ref. 4 and reconsidered in Ref. 5, a system consisting of a thin metal layer on a superconducting surface is a good candidate for a superconducting system. The metal and semiconductor film are in direct contact and there is no oxide layer or other barrier separating them. In this way, it is possible that some electronic state of the metal penetrates into the energy-gap region of the semiconductor. The pairing interaction is due to the exciton-electron and phonon-electron interactions and the excitonic contribution has been calculated using the Eliashberg equation with the extra-kernel:

$$K_{\rm ex} = \alpha_{\rm e}^2 \int_0^{\omega_{e0}} d\tilde{\omega} F_{\rm e}(\tilde{\omega})[(\omega' + \tilde{\omega} + \omega)^{-1} + (\omega' + \omega - \omega)^{-1}] \tag{28.64}$$

where

$$\alpha_{\rm e} = 0.5\lambda_{\rm ex}(\omega_{\rm g} + A)/\omega_{\rm g}\ ,$$

$\omega_{\rm g}$ is the semiconducting plasma energy and $\omega_{\rm e0}$ is the maximum exciton frequency. The function $F_{\rm e}(\tilde{\omega})$ is supposed to be a Lorentzian with width A and centred at the average gap width $\omega_{\rm g}$.

The dependence of the critical temperature T_c as function of different parameters: $A/\omega_{\rm g}$, ε_0 and $\lambda_{\rm ex}$ shows that high T_c are obtained only for unrealistic values of parameters. However, the important result obtained in this model is that the exciton-electron mechanism gives rise to a substantial enhancement in T_c compared with the phonon-electron mechanism.

The non-phononic mechanisms due to the electron-electron interaction via exchange bosons have been reconsidered by Grabowski and Sham[8] using the Eliashberg equations. These models give unrealistic high critical temperatures if the boson energy is greater than the Fermi energy. If we try to go beyond the Migdal approximation including self-energy, vertex corrections and spin fluctuations terms, we get a surprising result. These corrections are negligible for low boson energies and reduce T_c if the boson energy increases and becomes of the order $1/10\varepsilon_0$. The conclusion obtained in Ref. 8 is that the maximum T_c is 60 K if the Fermi energy is of the order of 10^5 K but an increasing of the bosons energy gives rise to a decreasing of T_c.

We have to mention that in the phononic mechanisms the maximum T_c has been estimated as 30 K for the Debye frequency $\theta_D \sim 100$ K.

29. Heavy-Fermions Superconductivity

Research on heavy fermions superconductivity is presently one of the most active feature in the theory of superconductivity.[8] The general properties of these new superconductors can be summarized as:

a) The systems $CeCu_2Si_2$, UBe_{13}, UPt_{13} become superconducting with a critical temperature $T_c \simeq 1$ K.

b) These systems have a large specific heat coefficient of magnitude $200(UPt_3)$ to $1000(CeCu_2Si_2)$ mJ/k^{2}mol, whereas that of the normal metals is of the order of a few mJ/k^{2}mol. Thus electrons with large effective mass, called "heavy fermions" must be present at the Fermi energy.

c) As the specific heat jump at T_c is of the same large magnitude as the specific heat in the normal state, the heavy fermions must be pairing.

d) The normal properties of these systems are in accordance with the general assumption that these systems are Kondo lattices.

A qualitative understanding of what may happen in these systems is connected with the existence of a kind of Kondo resonance peak at the Fermi energy, and the electrons in this resonance peak are "heavy fermions". This resonance peak is a feature of the f-electron spectrum which forms the "heavy fermions", and thus the f-electrons are paired up in the superconducting state. The pairing presents some particular aspects which show a pronounced difference to usual superconductivity. In the usual superconductors, the Fermi energy ε_0 (or the band width of the conduction electrons) is of the order 10^4 K, the critical temperature T_c is of the order of 10–20 K and the characteristic energy is the Debye temperature $\theta_D \cong 10^2$ K, so we have $T_c \ll \theta_D \ll W$. When the electrons in the Kondo resonance peak undergo the pairing, the situation is totally different; the width of the resonance peak is of the order of $T_K \simeq 10$ K (T_K is the Kondo temperature) which is only one magnitude order larger than T_c and the pairing mechanism provides for an energy scale θ as large as or even larger than the effective band width $W = T_K$; i.e., we have $\theta \geq T_K = W > T_c$.

From these considerations, it is easy to see that the standard methods for the theory of superconductivity have to be carefully applied. Usually, the heavy fermion systems are described by the Kondo lattice. The Hamiltonian for the Kondo lattice model is

$$\begin{aligned} \mathcal{H} = & \sum_{n,\mathbf{p},\alpha} \varepsilon_n(\mathbf{p}) c^\dagger_{n\mathbf{p}\alpha} c_{n\mathbf{p}\alpha} + \sum_{i,\alpha} E_f f^\dagger_{i\alpha} f_{i\alpha} + \sum_{i,j,\alpha} t_{ij} f^\dagger_{i\alpha} f_{i\alpha} \\ & + \frac{1}{\sqrt{N}} \sum_{n,\mathbf{p}} \sum_{\alpha,i} [V_n(\mathbf{p}) \exp(-i\,\mathbf{p}\cdot\mathbf{R}_i) c^\dagger_{n\mathbf{p}\alpha} f_{i\alpha} + h.c.] \\ & + \frac{1}{2} U \sum_i \sum_{\alpha,\beta} f^\dagger_{i\alpha} f_{i\beta} f^\dagger_{i\beta} f_{i\alpha} \end{aligned} \qquad (29.1)$$

where n is the band index for the itinerant-electrons with the energy $V_n(\mathbf{p})$, E_f is the energy of the localized electrons "f", t_{ij} is the transfer integral between nearest-neighbour f-electrons, $V_n(\mathbf{p})$ is the hybridization coupling constant and U is the Coulombic interaction.

A Fermi liquid theory of the heavy fermions assumes the f-electrons as itinerant, and in the model described by (29.1), this is possible by two mechanisms; the first one is due to the possibility of the transfer between adjacent f-electrons. The contribution of the transfer integral to the f band width W is of the order

$$W = 2Z|t| \qquad (29.2)$$

where Z is the coordination number of the lattice. The second mechanism is due to the hybridization between f orbitals and the conduction electrons, which is estimated as

$$\Gamma = \pi \sum_{\mathbf{n},\mathbf{p}} V_n(\mathbf{p}) V_n^*(\mathbf{p}) \delta(\mu - \varepsilon_n(\mathbf{p})) \sim \pi N(\varepsilon_0) \qquad (29.3)$$

where $N(\varepsilon_0)$ is the density of states of the conduction electrons at the Fermi level. In a good approximation, we can consider $N(\varepsilon_0) \sim 1/2W$ where W is the conduction band width. The coupling constant V is in fact a transfer integral between f orbitals and their adjacents s, p or d orbitals.

In the transition metals, the various transfer integrals between adjacent sites are of the same order, then we have $|t| \sim |V| \ll W$. In this case, hybridization between the f electrons and the conduction electrons can be neglected because $W \gg \Gamma$. For the transition metals if $|t| \ll V$, which is the case of the systems with strong atomic character of f electrons, the transfer integral "t" can be neglected and Eq. (29.1) reduces to an Anderson lattice. In the case $U \to \infty$ and $n_f \sim 1$, the f-electrons are almost localized, and in order to treat the strong correlation between these electrons we adopt the Luttinger method.[9-11] This method was first applied by Martin[12] to investigate if a system with the valence fluctuation phenomena is a metal or an insulator. The Luttinger sum rule, which states that total volume circled by the Fermi surface is given by the number of total electrons, has been demonstrated by Ohkawa[13] for the periodic Anderson lattice. The existence of the Fermi surface gives the possibility to describe the heavy fermions by the usual many-body methods as the Green function theory. For finite U, the single particle of the f-electrons can be written as

$$G_{f,f}^{-1}(\omega, \mathbf{p}) = i\omega - E_f(\mathbf{p}) - M(\omega, \mathbf{p}) - \sum(\omega, \mathbf{p}) \qquad (29.4)$$

where

$$E_f(\mathbf{p}) = E_f + \sum_f \exp[-i\,\mathbf{p} \cdot \mathbf{R}_{ij}]t_{ij} \qquad (29.5)$$

$$M(\omega, \mathbf{p}) = \sum_n \frac{|V_n(\mathbf{p})|^2}{i\omega + \mu - \varepsilon_n(\mathbf{p})} \qquad (29.6)$$

and the self-energy Σ can be calculated perturbatively in terms of U. In this model, the electronic specific heat coefficient γ is given by

$$\gamma = \frac{2\pi^2}{3} N^*(\mu) \qquad (29.7)$$

where the density of the quasi-particles is

$$N^*(\mu) = \frac{1}{N} \sum_{\mathbf{p}} \delta(\mu - E^*(\mathbf{p}))\ . \qquad (29.8)$$

In order to calculate the density (29.8), we expand $\Sigma(\omega, \mathbf{p})$ around the Fermi surface as

$$\sum(i\omega, \mathbf{p}) = \sum(p_0, i0) + (1 - \chi)i\omega + \frac{1}{m} p_0(p - p_0) \qquad (29.9)$$

an expansion which is correct if the quasi-particles are near the Fermi surface. We then get

$$E^*(\mathbf{p}) = \mu + \frac{1}{m^*} p_0(p - p_0) \qquad (29.10)$$

where

$$\frac{1}{m^*} = \frac{1}{Z}\Big[\frac{1}{m} + \frac{1}{m_0} + \sum_n \frac{|c_n|^2}{m_n}\Big] \tag{29.11}$$

m_0 being the mass of the f-electrons and $m_n(n \neq 0)$ the conduction electron mass. The constant Z has been defined by

$$Z = \chi + \sum_n |c_n|^2 \tag{29.12}$$

where

$$c_n = \frac{V_n(p_0)}{\mu - \varepsilon_n(p_0)} \tag{29.13}$$

and

$$\chi = 1 + \frac{\partial \Sigma(p_0, \omega)}{\partial \omega}\Big|_{\omega \to 0}\,. \tag{29.14}$$

The single particle Green function (29.4) can be approximated as

$$G_{l,n}(i\omega, \mathbf{p}) = \frac{c_l^\dagger c_n}{Z}\frac{1}{i\omega + \mu - E^*(\mathbf{p})} \tag{29.15}$$

where l and n are band indices, the value $n = 0$ corresponding to the f-band and $n \neq 0$ to the conduction band.

At this point we stress an approximation due to the special feature of the heavy fermion systems. Concretely, it is reasonable to assume $|m_0| \gg m_n(n \neq 0)$ and $|m| \gg m_n$ if the dispersion of $\Sigma(p_0, i0)$ is not so large as that of the conduction electrons. Then we see from (29.11) that the effective mass m^* of the quasi-particles depends on $|c_n|^2$. When m^* is large, $|c_n|^2 \ll 1$ for $n \neq 0$, which shows that the quasi-particles are mainly composed of f-electrons. This Fermi liquid description is complete if we prove that the normal Fermi liquid is possible in the large U limit, and calculate the two parameters χ and m.

In the approximation $t_{ij} = 0$ and the f levels are quite shallow ($n_f \to 0$) the effect of the Coulomb correlation can be neglected, and the ground state of the system is a normal Fermi liquid. If the f levels are so deep that $n_f \sim 1$, the f electrons are almost localized because of large U. The conduction electrons can mediate an exchange interaction and the ground state is magnetic in the limit $n_f \to 1$. As n_f satisfies $0 < n_f < 1$, it is sure that there exists a critical n_c where a first-order transition occurs from a non-magnetic to a magnetic state. This critical value has been calculated[13] considering the competition between the Kondo effect and the exchange interaction. The Kondo temperature has been calculated[13–15] as

$$T_k = \frac{2\Gamma}{\pi}\frac{1-n_f}{n_f} \tag{29.16}$$

and the exchange interaction[13] between the f electrons is of the form

$$J = \frac{Z\Gamma}{4\pi}\frac{F(2p_0R)}{\ln(D/\Gamma)} \tag{29.17}$$

where $F(x) = x^{-4}(x\sin x - \cos x)$ and R, the distance between spins, was taken of the order of the lattice constant a. The critical value n_c will be estimated from the condition

$$T_k \sim J \tag{29.18}$$

and using $D/\Gamma \gg 1, |F(2p_0R)| < 1$ we get $n_c \simeq 0.99$, a value very close to unity.

In the non-magnetic regime, the Fermi liquid parameters can be approximated by those of the dilute Kondo alloys because the spin fluctuations are similar in both systems.
Then

$$\chi = \frac{\pi\Gamma}{4T_k} \sim \frac{\pi^2}{8(1-h_f)} \tag{29.19}$$

and $m \to \infty$. In the Kondo lattice, $Z \gg 1$ and for $n_f = 0.95$ we get $x = 20$ and $Z = 20$; thus Z and x have the same order. In this model, the coefficient γ reaches 0.8 J/k^{2}mol for $T_K = 10$ K and $\varepsilon_0 = D/2$.

Another important aspect of the Fermi surface sum rule is concerning the density of states of the quasi-particles. If this density of states has a peak, the chemical potential is likely to be pinned around this peak. Koyama and Tachiki[16] analysed a class of materials as $CeCu_6$ which behave as heavy fermions system at very low temperatures. The numerical calculations of the density of states show the existence of a sharp peak near the Fermi level corresponding to the heavy fermions and a broad peak similar to the resonance peak in Kondo system. As the temperature increases, the heavy fermion system peak decreases and changes to the resonance peak. The electronic density of states for the heavy-fermion system described by the periodic Anderson model has also been calculated by Zlatic *et al.*[17] taking U as a perturbation. In this case, the self-energy has been calculated from

$$\sum(\omega, \mathbf{p}) = U^2\frac{\pi T}{N}\sum_{\mathbf{q},\omega} G(\omega, \mathbf{p}+\mathbf{q})\chi(\mathbf{q}, \omega+\omega') \tag{29.20}$$

where the susceptibility $\chi(\mathbf{q},\omega)$ has been considered in the lowest order approximation taking its expression for the free electrons.

The momentum dependence of the self-energy (29.20) has been calculated considering for the quasi-particles the energy $\varepsilon(p) \sim p$. Around the Fermi level, the self-energy $\Sigma(\omega \to 0, \mathbf{p})$ was obtained as

$$\Sigma(\omega \to 0, \mathbf{p}) = ap + bp^n \ln p \tag{29.21}$$

with positive constants a and b. If $\varepsilon(p) = \beta p$, one has $n = 1$. If $n > 1$, as it is expected from the Fermi liquid theory $n = 3$, then the non-logarithmic contribution dominates in $\Sigma(p)$. For large ω, the p-dependence in $\Sigma(\omega, \mathbf{p})$ can be neglected and

$$\Sigma(\omega, \mathbf{p}) = \frac{1}{4}\frac{U^2}{\omega + i\delta} \tag{29.22}$$

which gives rise to the symmetric peaks at the energies $\varepsilon = \pm U/2$. These peaks correspond to the single-particle excitations in the atomic limit. The calculation of $N(\varepsilon)$ using (29.21) shows the existence of a narrow peak centered at the Fermi level and two broad side peaks. This structure of $N(\varepsilon)$ arises from the interplay of correlations and interatomic hybridization. The difference between the results obtained in Refs. 16 and 17 can be done by numerical approximations.

A new method to describe the Kondo lattice was done by Tesanovic and Valls[18] using the so called "slave-boson-field theory". This method was introduced by Coleman,[19] Read and Newns[20–21] in the context of $1/N$-expansion (N is the degeneracy of the impurity f level) for the single impurity Kondo problem. If in the periodic Anderson Hamiltonian we take the limit of a very large $U(U \to \infty)$, we may replace this Hamiltonian by a slave-boson-field theory in which the additional variables $b_i(b_i^\dagger)$ are Bose operators and describe the availability of site i for hybridization.

Such a Hamiltonian has the form

$$\begin{aligned}\mathcal{H}_{\mathrm{SB}} &= \sum_{\mathbf{p},\alpha} \varepsilon(\mathbf{p}) c^\dagger_{\mathbf{p}\alpha} c_{\mathbf{p}\alpha} + \sum_{m,\mathbf{p}} (-E_{\mathrm{f}}) f^\dagger_{\mathbf{p}m} f_{\mathbf{p}m} \\ &+ V \sum_{i,m,\alpha} (c_{i\alpha} f^\dagger_{im} b_i + h.c.)\end{aligned} \tag{29.23}$$

with the constraint

$$n_{\mathrm{f}i} + b_i^\dagger b_i = 1\,. \tag{29.24}$$

The model is to be understood in the following way: a boson is created when an f electron which moves out of rare-earth site, and it is annihilated whenever an f electron moves in. The constraint (29.24) reflects the strong Coulombic repulsion and makes necessary the introduction of a new effective Hamiltonian,

$$\widetilde{\mathcal{H}}_{\rm SB} = \mathcal{H}_{\rm SB} + \sum_i \lambda_i (n_{fi} + b_i^\dagger b_i) \tag{29.25}$$

where λ_i is a chemical potential which for a lattice satisfies $\lambda_i = \lambda$. For the single impurity problem, such a Hamiltonian will give an exact solution for an f-level which has an infinite degeneracy and becomes a good approximation in the mean-field. Let us consider the single impurity case. The Bose field b_i has at low temperature a non-vanishing average value

$$\langle b_i \rangle = |b_0| \exp(i\Phi_i) = b_0 \tag{29.26}$$

in the framework of the mean-field approximation. For a dilute alloy, the phases Φ_i are uncorrelated but for a Kondo lattice it seems natural to assume the magnitude and the phase of the Bose field as fixed. Then the coherent macroscopic state will be generated by the conduction as well as the f electrons and we will approximate the Kondo lattice as a giant quasi-bound overall singlet state. The Hamiltonian which can describe such a state in the mean-field approximation is

$$\begin{aligned} \mathcal{H}_{\rm MF} = & \sum_{\mathbf{p},\alpha} \varepsilon(\mathbf{p}) c_{\mathbf{p}\alpha}^\dagger c_{\mathbf{p}\alpha} + \sum_{\mathbf{p},m} \widetilde{E}_{\rm f} f_{\mathbf{p}m}^\dagger f_{\mathbf{p}m} \\ & \times V \sum_{i,\alpha} [c_{i\alpha} f_{i\alpha}^\dagger |b_0| \exp(i\Phi_0) + h.c.] \end{aligned} \tag{29.27}$$

where $\widetilde{E}_{\rm f} = -E_{\rm f} + \lambda$. The new quantity b_0 will be determined from the self-consistent condition

$$\lambda b_0 = -V\pi T \sum_{\mathbf{p},\omega,\alpha} A_\alpha(\omega, \mathbf{p}) \tag{29.28}$$

where

$$A_\alpha(\omega, \mathbf{p}) = \langle f_{\mathbf{p}\alpha} c_{\mathbf{p}\alpha}^\dagger \rangle \tag{29.29}$$

is a new anomalous Green function. We get the expression for (29.29) if we use the motion equation method to calculate the Green function for the conduction electrons

$$G_{\rm c}(\omega, \mathbf{p}) = \langle T(c_{\mathbf{p}} c_{\mathbf{p}}^\dagger) \rangle \tag{29.30}$$

and for the f-electrons

$$G_f(\omega, \mathbf{p}) = \langle T(f_{\mathbf{p}} f_{\mathbf{p}}^\dagger)\rangle \ . \tag{29.31}$$

The Green functions (29.29)–(29.31) obtained from (29.26) are

$$G_c(\omega, \mathbf{p}) = \frac{i\omega - \tilde{E}_f}{(i\omega - \tilde{E}_f)(i\omega - \varepsilon(\mathbf{p})) - |\tilde{V}|^2} \tag{29.32a}$$

$$G_f(\omega, \mathbf{p}) = \frac{i\omega - \varepsilon(\mathbf{p})}{(i\omega - \tilde{E}_f)(i\omega - \varepsilon(\mathbf{p})) - |\tilde{V}|^2} \tag{29.32b}$$

$$A(\mathbf{p}, \omega) = \frac{\tilde{V}}{(i\omega - \tilde{E}_f)(i\omega - \varepsilon(\mathbf{p})) - |\tilde{V}|^2} \ . \tag{29.32c}$$

From (29.28), we can determine the temperature T_{MF}, below which the coherent state exists. If $T < T_{MF}$ from (29.24), we get

$$\langle n_f \rangle = 1 - |b_0|^2 \ll 1 \tag{29.33}$$

and $\lambda(T_{MF})$ is expected to be of the order of E_f. Equation (29.28) can be approximated by

$$E_f = -V^2 \pi T \sum_{\omega, \mathbf{p}} \frac{1}{(i\omega)[i\omega - \varepsilon(\mathbf{p})] - |\tilde{V}|^2} \tag{29.34}$$

and neglecting $|\tilde{V}|^2$, we get

$$T_{MF} = 1.14\, \omega_D \exp[-E_f/\Gamma] \tag{29.35}$$

where $\Gamma = N^0(0)|V|^2$ and $N^0(0)$ is the density of the conduction electrons. The expression (29.35) looks like a critical temperature and in fact below this temperature a coherent state consisting of heavy fermions appears. Similarly to T_{MF} we can obtain $\tilde{V}$ from the equation

$$\lambda(0) = -V^2 \pi T \sum_{\mathbf{p}, \omega} [(i\omega - \tilde{E}_f)(i\omega - \varepsilon(\mathbf{p})) - |\tilde{V}|^2]^{-1} \tag{29.36}$$

and we get

$$|\tilde{V}| = \omega_D \exp(-E_f/\Gamma) \ . \tag{29.37}$$

Compared this result with (29.35) gives

$$T_{MF} \cong T_K = \frac{|\tilde{V}|^2}{\omega_D} \ . \tag{29.38}$$

The poles of the Green functions (29.32) will give the energy spectrum which corresponds to two hybridized bands. The states in this spectrum are linear combinations of f-electrons and conduction electrons wave functions. From the poles of the Green functions, we obtain the energies of the two bands

$$E_{\pm}(\mathbf{p}) = \frac{1}{2}(\varepsilon(\mathbf{p}) + \tilde{E}_f) \pm \frac{1}{2}\{(\varepsilon(\mathbf{p}) - \tilde{E}_f)^2 + 4|\tilde{V}|^2\}^{1/2} . \qquad (29.39)$$

In the realistic systems there is hybridization with more than one conduction band, and in order to ensure the metallic behaviour of the system we take the Fermi level near the top of $E_-(\mathbf{p})$ band. An eigenstate of (29.2) with the particle-hole symmetry is

$$|q\rangle = a(\varepsilon)|f\rangle + b(\varepsilon)|c\rangle \qquad (29.40)$$

where $|q\rangle$ describes the quasi-fermion wave function, and

$$a^2(\varepsilon) = \frac{|\tilde{V}|^2}{[(|\varepsilon| + |\tilde{V}|^2/D)^2 + |\tilde{V}|^2]} \qquad (29.41a)$$

$$b^2(\varepsilon) = \frac{(|\varepsilon| + |\tilde{V}|^2/D)^2}{[(|\varepsilon| + |\tilde{V}|^2/D)^2 + |\tilde{V}|^2]} . \qquad (29.41b)$$

The density of states near the Fermi level has been obtained as

$$N(\varepsilon) = N^0(\varepsilon)\left[1 + \frac{|\tilde{V}|^2}{(|\varepsilon| + |\tilde{V}|^2/D)^2}\right] \qquad (29.42)$$

which at the Fermi level can be approximated as

$$N(0) = N^0(0)\frac{D^2}{|\tilde{V}|^2} \qquad (29.43)$$

a relation which shows that there is a large enhancement in the density of states $N^0(0)$ and the new density of states is

$$N(0) \gg N^0(0) . \qquad (29.44)$$

The Fermi velocity is

$$v_0 = \left.\frac{dE_-(p)}{dp}\right|_{\varepsilon=0} \cong v_0^0 \frac{|\tilde{V}|^2}{D^2} \ll v_0^0$$

which shows that the quasi-particles described by (29.40) are induced heavy fermions. The fluctuations of the mean-field solution for b will give rise to an effective interaction between the heavy fermions. If we introduce the fluctuations of b_i by the relation

$$b_i = b_0 + B_i \ , \tag{29.45}$$

the fluctuations contribution is given by

$$\mathcal{H}_{\mathrm{B}} = \lambda \sum_{\mathbf{q}} B^{\dagger}_{\mathbf{q}} B_{\mathbf{q}} + \frac{V}{\sqrt{N}} \sum_{\mathbf{p},\mathbf{q}} (c_{\mathbf{p}} f^{\dagger}_{\mathbf{p}+\mathbf{q}} B_{\mathbf{q}} + h.c.) \ . \tag{29.46}$$

The fluctuations of the Bose field are similar with the phonons in the electron-phonon interaction and the total Hamiltonian can be written as

$$\tilde{\mathcal{H}} = \mathcal{H}_{\mathrm{MF}} + V^2 \tilde{\lambda}^2 \sum_{\mathbf{p},\mathbf{p}',\mathbf{q}} \frac{1}{\omega^2 - \omega^2(\mathbf{q})} c^{\dagger}_{\mathbf{p}'-\mathbf{q}} f^{\dagger}_{\mathbf{p}+\mathbf{q}} c_{\mathbf{p}} f_{\mathbf{p}'} \tag{29.47}$$

where the first term is given by (29.2) and $\omega(q) = cq$ is the energy of fluctuations, c being the speed of the sound which corresponds to the propagation of long wavelength fluctuations in the slave-boson fields. The speed of propagation of slave bosons is v_p^0 and satisfies

$$v_p^0 \gg c > v_0$$

where $c = (E_0 \Gamma)^{1/2}/2p_0$.

Using this result, the last term from (29.47) can be replaced by the static limit with the scale of the interaction which is smaller than the kinetic energy of the heavy fermions.

The effective interaction has been recalculated in the Random Phase Approximation and for Γ/E_0 this is singular near $2p_0$. In this model, the effective hybridization gives rise to a coherent state characterized by the occurrence of the heavy non-interacting fermionic excitations.

The occurrence of superconductivity in these systems implies the explanation of the origin of the attractive interaction between the f electrons. The effective repulsive interaction between the quasi-particles is due to the one-site repulsive Coulomb interaction modified by the hybridization and has the same form as that obtained by Yoshida and Yamada,[22] namely,

$$\Gamma_{\uparrow\downarrow} = \frac{\pi^2 \Gamma^2}{4 T_{\mathrm{K}}} \tag{29.48}$$

an expression which shows that this form (in which the retardation effects have been considered) is of the short range. The first mechanisms proposed for the explanation of the attraction between fermionic quasi-particles considered the electron-phonon interaction as responsible for the effective attraction. Even the role of the phonons can be considered in different ways. Razafimandimby *et al.*[23] pointed out that the Kondo-volume collapse is important for the electron-phonon coupling. The same idea has been used by Mayake *et al.*[24] who showed the existence of a weak coupling regime for the d-pairing and a strong coupling regime for the s-pairing.

Ohkawa[25] proposed an attractive interaction between adjacent f electrons due to one-phonon exchange. The main results of this theory is an attractive interaction which is sensitively dependent on the angle between f-site-ligands. If this angle is smaller than $\pi/2$ (for a simple *fcc* structure), the one-phonon exchange results give rise to an attractive interaction between f-electrons. The model has been reconsidered by the author[26] taking into consideration the Kondo effect which reduces the charge susceptibility. The screening due to the Kondo effect maintains the attractive electron-electron interaction due to the electron-phonon coupling. An important problem which has to be discussed in connection with the electron-phonon mechanism is the validity of the Migdal theorem (See Appendix from Chap. 2). Because the phonon frequency is much larger than the band width of the quasi-particles ($\omega_D \gg T_K$) this theorem cannot be applied to the heavy-fermion superconductors. In the long-wave length and static limit, the electron-phonon interaction becomes small and the standard methods can be applied. Eliashberg[27] analysed the validity of the Migdal theorem and pointed out the existence of a strong dynamic renormalization of the energy of the conduction electrons. The renormalization factor $\lambda_e = \partial t/\partial \varepsilon$ has been calculated as function of the Ginzburg-Landau parameter and the result $\chi \sim \lambda_e^{3/2}$ shows the possibility of the occurrence of high values for the upper critical field.

Another mechanism considered by different authors as responsible for the attraction between the heavy fermions is due to the spin-fluctuations excitations. Even in the simple Kondo volume collapse mechanism, the modulation of the Kondo temperature is in fact equivalent to the consideration of the spin fluctuations. The importance of the spin-fluctuation has been stressed first by Varma[28] and reconsidered by Miyake *et al.*[30,31] in connection with the form of the pair wave function. Due to the strong on-site repulsion in the singlet channel, the pair wave function has a node in the origin and the singlet state is anisotropic. Varma[28,29] and Anderson[32] analysed the analogy of the superconductivity in the heavy fermion systems with the superfluidity of the ^{3}He where the ferromagnetic paramagnons lead to a triplet pairing. Using this idea Miyake *et al.*[30,31] suggest that the antiferromagnetic spin fluctuations suppress conventional singlet pairing. Hirsch[33] suggested that the triplet pairing is also suppressed but the anisotropic singlet pairing is favored. Another important feature of this problem is the existence of the lines of zeros of the gaps $\Delta(\mathbf{k})$ on the Fermi surface. If the pair interaction has the form

$$V_l = \left\{ \begin{matrix} 3 \\ -1 \end{matrix} \right\} 2 \int_0^1 dx \cdot x P_l(1 - 2x^2)[-J(2p_0 x)] \tag{29.49}$$

where $J(q)$ is the exchange interaction, it is easy to see that the sign of (29.49) is dependent on the variation of $J(q)$ for ferromagnetic or antiferromagnetic spin fluctuations respectively. In the first case $V_0 > 0, V_1 < 0$, as for ^{3}He, whereas $V_0 > 0$ and $V_1 > 0$ for the antiferromagnetic fluctuations. The realistic models considered the crystal symmetry and a complete discussion was given by Volovik and Gor'kov.[34]

An interesting result was obtained by Jichu *et al.*[35] taking into consideration the attractive interaction between the heavy fermions mediated by the spin-density and charge-density fluctuations. This effective pairing is positive for the singlet pairs and depends on the sign of the polarization operator $\Pi_0(\mathbf{q})$ for the triplet pairs.

The electron-phonon interaction, strongly enhanced by the Kondo-volume collapse, is large and it is important to note that without this mechanism the interaction is weak.

On the other hand, the electron-electron interaction due to the phonons appears to be caused by the dynamical effect and the heavy fermions cannot follow completely the phonons.

We may conclude this discussion, concerning the effective attraction between the heavy-fermions, with the simple result that the enhanced electron-phonon interaction and the other non-phononic mechanisms are responsible for the occurrence of the superconductivity in these materials.

The Kondo lattice has been described by Yoshimori and Kasai[36] by a periodic Anderson model with a small dispersion in the f-band written as

$$E_f(p) = \alpha\varepsilon(p) \tag{29.50}$$

where $\alpha \ll 1$ is a parameter. Using this model, Tachiki and Maekawa[37] calculated the critical temperatures for the conduction electrons and for the f-electrons. In fact, this is a two-fluid model and the superconductivity is done by the conduction electrons on the f-electrons. Using the Gor'kov equations for these two kinds of quasi-particles they obtained the critical temperatures for the conduction electrons:

$$T_c^c = 1.14 T_K \exp\left[-\frac{1}{N(0)g_c}\left(\frac{\alpha+\tilde{\gamma}}{\alpha}\right)\right] \tag{29.51}$$

and for the f-electrons:

$$T_c^f = 1.14 T_K \exp\left[-\frac{\alpha(\tilde{\gamma}+\alpha)}{N(0)g_f}\right] \tag{29.52}$$

$\tilde{\gamma}$ being a parameter given by

$$\tilde{\gamma} = 1 - \left(\frac{\partial\tilde{\Sigma}(\omega)}{\partial(i\omega)}\right)_{i\omega=0} \tag{29.53}$$

where the self-energy $\tilde{\Sigma}$ describes the interaction between the conduction electrons and f-electrons and can be calculated using instead of Γ from the Kondo problem the expression $\tilde{\Gamma} = \alpha/\pi N(0)$. In (29.51)–(29.52), $\tilde{\alpha} = \alpha/\gamma$, $N(0)$ is the density of states of the quasi-particles at the Fermi level and $\tilde{\gamma} \gg 1$. These two critical temperatures have been calculated in the approximation

$$T_K < |\varepsilon| < \omega_D \tag{29.54}$$

which shows that the cut-off energy from BCS has to be replaced by T_K. The critical temperature T_c^c is very small because $\alpha \ll 1$ and the factor $(\alpha+\tilde{\gamma})/\alpha$ is very large. The superconducting temperature which is considered to be relevant is given by (29.52) which is greater than T_c^c. However, T_c^f contains $g_f = g_f^0 - \Gamma_{\uparrow\downarrow}$, where the attraction between f-electrons g_f^0 is reduced by the repulsion $\Gamma_{\uparrow\downarrow}$. Using (29.48) the critical temperature (29.52) becomes

$$T_c = 1.14\, T_K \exp\left[-\frac{\alpha(\tilde{\gamma}+\alpha)}{N(0)g_f - \alpha^2/4T_N N(0)}\right] \tag{29.55}$$

which shows that the contribution of the repulsive interaction is strongly reduced by the small factor α^2.

A model with anisotropic pairing has been analysed[38–40] in the mean field approximation starting with the Hamiltonian

$$\mathcal{H}_{\text{eff}} = \sum_{\mathbf{p},\alpha} E(\mathbf{p}) c_{\mathbf{p}\alpha}^\dagger c_{\mathbf{p}\alpha} - \frac{1}{2}\sum_{\mathbf{p},\alpha,\beta}\left[\Delta_{\alpha\beta}(\mathbf{p}) f_{\mathbf{p}\alpha}^\dagger f_{\mathbf{p}\alpha} + h.c.\right] \tag{29.56}$$

where

$$\Delta_{\alpha,\beta}(\mathbf{p}) = \Delta^{(1)}_{\alpha,\beta} + \Delta^{(2)}_{\alpha,\beta} \tag{29.57}$$

$$\Delta^{(1)}_{\alpha,\beta} = -\Gamma_{\uparrow\downarrow}\frac{1}{NZ^2}\sum_{\mathbf{q}}\langle f_{-\mathbf{q}\alpha} f_{\mathbf{q}\beta}\rangle \tag{29.58}$$

and

$$\Delta^{(2)}_{\alpha,\beta} = g\frac{1}{Z^2}\sum_{\mathbf{q}}\sqrt{6}\langle f_{-\mathbf{q}\alpha} f_{\mathbf{q}\beta}\rangle\ . \tag{29.59}$$

The first component of the gap given by (29.58) is due to the repulsive interaction. The second component given by (29.59) is due to an attractive interaction which acts between the nearest neighbours f-electrons from a cubic lattice, and $f_s(\mathbf{q})$ is given by

$$f_s(\mathbf{q}) = \sqrt{\frac{2}{3}}[\cos q_x a + \cos q_y a + \cos q_z a]\ . \tag{29.60}$$

This problem has six degrees of freedom which can be classified according to the symmetry as follows:

1. even parity (s or $d\gamma$-symmetry)

$$\Delta^{(2)}_{\uparrow\downarrow}(\mathbf{p}) = \sum_{\mu} A_\mu \delta\cos(p_\mu a) \tag{29.61}$$

and $\Delta^{(2)}_{\alpha\alpha} = 0$. In this case $A_x = A_y = A_z$ for the s-symmetry and $A_x + A_y + A_z = 0$ for $d\gamma$-symmetry.

2. odd parity ($p\gamma$-symmetry)

$$\begin{aligned}\Delta^{(2)}_{\alpha\beta} &= \sum_{\gamma} B_{\alpha\beta}\delta\sin(p_\gamma a)\\ \Delta^{(1)}_{\alpha\beta} &= 0\ .\end{aligned} \tag{29.62}$$

The transition temperature has been calculated as

$$T_c = 1.14 T_K \exp[-1/F] \tag{29.63}$$

where

$$F = \frac{2a_0(a_2 - a_1)}{(a_1 - a_2) + [(a_2 - a_0)^2 + 4a_0(a_2 - a_1)^2]^{1/2}} \tag{29.64}$$

with

$$a_0 = \frac{\Gamma_{\uparrow\downarrow} N(0)}{Z^2}$$
$$a_1 = \frac{g\langle f_s(q)\rangle^2 N(0)}{Z^2} \tag{29.65}$$
$$a_2 = g\frac{\langle f_s(q)\rangle^2 N(0)}{Z^2} .$$

The effective coupling constant F given by (29.64) can be approximated as

$$F = \begin{cases} a_2 - a_0 & \text{for } a_2 > a_0 \\ a_2 - a_1 & \text{for } a_2 < a_0 \end{cases} \tag{29.66}$$

and we can see that T_c is a decreasing function of a_0, but is finite for very large values of a_0, which is proportional to the repulsion. In the heavy-fermion systems, $Z \gg 1$ and a_1 can compete with a_0 in some cases. In this model the energy gap vanishes along lines on the Fermi surface in the anisotropic s-symmetry. The energy gap $\Delta_{sp}(p, T)$ was calculated near the critical temperature as

$$\Delta_{sp}(p,T) = \Delta_0[f_s(p) - \langle f_s(p)\rangle R]\left|\frac{T_c - T}{T_c}\right|^{1/2} \tag{29.67}$$

where

$$R = \left[a_0 \ln \frac{1.14 T_K}{T_c}\right] \Big/ \left[1 + a_0 \ln \frac{1.14 T_K}{T_c}\right] . \tag{29.68}$$

The Fermi surface exists in the Brillouin zone, but the equation

$$\Delta_{sp}(p,T) = 0 \tag{29.69}$$

also defines a surface.

If the two surfaces intersect, the energy gap vanishes along lines. The average of Eq. (29.67) on the Fermi surface identically vanishes for $R = 1$. If there exists only a single Fermi surface, then the energy gap always vanishes along lines in the limit $R \to 1$, namely, in the limit of the strong correlations.

Matsuura *et al.*[38] proposed a model based on the periodic Anderson model for the s-wave electrons with a wave vector dependence of the interaction between electrons. This potential has the form

$$V(p) = \Gamma_{\uparrow\downarrow}\frac{\chi^2}{p^2 + \chi^2} - \tilde{g}\exp\left[-\frac{p}{\chi_D}\right] \tag{29.70}$$

where the first term is the screened Coulomb interaction and the second term is the attractive part given by the phonon-mediated interaction with a cut-off χ_D. In this model, above the critical value

$$\frac{\tilde{g}}{\Gamma_{\uparrow\downarrow}} = 0.53$$

the critical temperature T_c starts to have relevant values.

We have to mention that these models which imply essentially the existence of a electron-electron pairing (isotropic or anisotropic) cannot explain the behaviour of the materials such as UBe_{13} and the influence of doping on these materials. There is a large class of metallic compounds such as $U_{1-x}Th_xBe_{13}$, $U_{1-x}Th_xPt_3$ which seem to present a more complicated behaviour, the main point being the existence of the second phase transition which indicates the existence of the Spin-Density-Waves below the critical temperature T_c.

The new state implies a delicate balance among various ground states superconducting state, magnetic state and a normal Fermi liquid state in these heavy-fermion systems.

On the other hand, we have to stress that the anisotropic superconducting state with pointlike or linelike gapless region on the Fermi surface is given by the presence in the electronic system of a strong one-site repulsion which presents the isotropic singlet pairing. Then, the new model should contain the repulsive interaction in the $l = 0$ channel of the partial wave decomposition as an essential ingredient, in addition to the attractive interaction in the $l = 1$ channel (p-pairing case) or $l = 2$ channel (d-pairing case). The microscopic origin of these interactions was discussed by Miyake *et al.*[39] and Ohkawa and Fukuyama.[40]

Machida and Kato[41] showed that the superconducting states with odd-parity or even-parity have an inherent instability towards the Spin-Density-Wave state. The two states have a common characteristic in this case: they appear in strongly correlated electronic systems.

In the d-channel the electron-electron interaction is attractive and we can use the Hamiltonian

$$\begin{aligned}\mathcal{H} = \sum_{\mathbf{p},\alpha} \varepsilon(\mathbf{p}) c^{\dagger}_{\mathbf{p}\alpha} c_{\mathbf{p}\alpha} - M \sum_{\mathbf{p}} (c^{\dagger}_{\mathbf{p}+\mathbf{Q}\uparrow} c_{\mathbf{p}\downarrow} + h.c.) \\ - \sum_{\mathbf{p},\alpha} (\Delta_d(\mathbf{p}) c^{\dagger}_{\mathbf{p},\alpha} c^{\dagger}_{-\mathbf{p},-\alpha} + h.c.) \qquad (29.71)\end{aligned}$$

where the first term is the kinetic energy described by a tight binding model on a two-dimensional square lattice (the lattice constant $a = 1$), namely

$$\varepsilon(\mathbf{p}) = -t(\cos p_x + \cos p_y) \ . \tag{29.72}$$

This band is assumed to be half-filled to ensure the SDW nesting with the wave vector $\mathbf{Q} \equiv (\pi, \pi)$. The order parameter of this state is

$$M = U \sum_{\mathbf{p}} \langle c^{\dagger}_{\mathbf{p}+\mathbf{Q}\uparrow} c_{\mathbf{p}\downarrow} \rangle \tag{29.73}$$

where U is the repulsive one-site Coulomb interaction considered on the s-channel of the electron-electron interaction. The d-pairing state is described by the order parameter:

$$\Delta_{\rm d}(p) = \Delta_{\rm d} \tau_{\rm d}(p) = \frac{\Delta_{\rm d}}{\sqrt{2}} (\cos k_x - \cos k_y) \tag{29.74}$$

where we assume the attractive electron-electron interaction of the form $g_{\rm d}(\mathbf{p}, \mathbf{p}') = g_{\rm d} \tau_{\rm d}(\mathbf{p}) \tau_{\rm d}(\mathbf{p}')$. The self-consistent equation for (29.74) is

$$\Delta_{\rm d}(\mathbf{p}) = \sum_{\mathbf{p}', \alpha} g_{\rm d}(\mathbf{p}, \mathbf{p}') \langle c_{\mathbf{p}'\alpha} c_{-\mathbf{p}', -\alpha} \rangle \tag{29.75}$$

and the energy $\varepsilon(\mathbf{p})$ and the order parameter $\Delta_{\rm d}$ satisfy

$$\begin{aligned} \Delta_{\rm d}(\mathbf{p}) &= \Delta_{\rm d}(-\mathbf{p}) = -\Delta_{\rm d}(\mathbf{p} + \mathbf{Q}) \\ \varepsilon(\mathbf{p}) &= -\varepsilon(\mathbf{p} + \mathbf{Q}) = \varepsilon(\mathbf{p} + 2\mathbf{Q}) \ . \end{aligned} \tag{29.76}$$

With these considerations we can use the Gor'kov equations to calculate the order parameters (29.73) and (29.75). The relevant Green functions are

$$G_{21} = \frac{M}{(i\omega)^2 + \varepsilon^2(\mathbf{p}) + \Delta_{\rm d}^2(\mathbf{p}) + M^2} \tag{29.77}$$

and

$$F^{\dagger}_{11} = \frac{\Delta_{\rm d}(\mathbf{p})}{(i\omega)^2 + \varepsilon^2(\mathbf{p}) + \Delta_{\rm d}^2(\mathbf{p}) + M^2} \ . \tag{29.78}$$

Using these Green functions, we get the equations for M and $\Delta_{\rm d}(\mathbf{p})$ as

$$1 = U\pi T \sum_{\omega} \int \frac{d^2\mathbf{p}}{(2\pi)^2} \frac{1}{(i\omega)^2 + \varepsilon^2(\mathbf{p}) + \Delta_{\rm d}^2(\mathbf{p}) + M^2} \tag{29.79}$$

$$1 = g_{\rm d}\pi T \sum_{\omega} \int \frac{d^2\mathbf{p}}{(2\pi)^2} \frac{\tau_{\rm d}^2(\mathbf{p})}{(i\omega)^2 + \varepsilon^2(\mathbf{p}) + \Delta_{\rm d}^2(\mathbf{p}) + M^2} \ . \tag{29.80}$$

Linearizing (29.72) near the Fermi surface in the approximation $t \gg \omega, \Delta$ Eqs. (29.79)–(29.80) become

$$1 = \frac{U}{\pi t}\pi T \sum_{\omega} [(i\omega)^2 + \Delta_{\rm d}^2 + M^2]^{-1/2} \ln \frac{2\pi t}{\sqrt{(i\omega)^2 + M^2}} \tag{29.81}$$

$$1 = \frac{g_{\rm d}}{\pi t}\pi T \sum_{\omega} \left\{ [(i\omega)^2 + \Delta_{\rm d}^2 + M^2]^{-1/2} \ln \frac{2\pi t}{\sqrt{(i\omega)^2 + M^2}} - \frac{1}{2\Delta_{\rm d}} \ln \frac{\sqrt{(i\omega)^2 + \Delta_{\rm d}^2 + M^2} + \Delta_{\rm d}}{\sqrt{(i\omega)^2 + \Delta_{\rm d}^2 + M^2} - \Delta_{\rm d}} \right\} . \tag{29.82}$$

The critical temperature for the Spin-Density-Waves, $T_{\rm N}$, can be calculated from (29.81)–(29.82) if $M = 0$ and $T \to T_{\rm N}$. An analytical solution can be obtained only near the superconducting critical temperature T_c as

$$\frac{T_{\rm N0} - T_{\rm N}}{T_{\rm N0}} \cong \left(\frac{\Delta_{\rm d}}{2\pi T_{\rm N0}} \right)^2 \frac{7}{2} \varsigma(3) . \tag{29.83}$$

In the adopted mean field approximation the magnetic state appears via a second order transition and $T_{\rm N}$ is depressed by the presence of the d-pairing from the superconducting state. The odd parity states can appear in these systems and we consider the p-pairing. In this case, the last term of the Hamiltonian (29.71) is

$$\mathcal{H}_s = - \sum_{\mathbf{p},\alpha} (\Delta_{\rm p}(\mathbf{p}) c^{\dagger}_{\mathbf{p},\alpha} c^{\dagger}_{-\mathbf{p},-\alpha} + h.c.) \tag{29.84}$$

where

$$\Delta_{\rm p}(\mathbf{p}) = \Delta_{\rm p} \tau_{\rm p}(\mathbf{p}) = \frac{\Delta_{\rm p}}{\sqrt{2}} (\sin p_x + \sin p_y) \tag{29.85}$$

and $g_p(\mathbf{p}, \mathbf{p}') = g_{\rm p} \tau_{\rm p}(\mathbf{p}) \tau_{\rm p}(\mathbf{p}')$.

Using the same method we get the self-consistent equations for M and $\Delta_{\rm p}$ (if $t \gg \omega, \Delta_{\rm p}$):

$$1 = \frac{U}{\pi t}\pi T \sum_{\omega} \left\{ \frac{\ln 2\pi t / \sqrt{(i\omega)^2 + \Delta_{\rm p}^2 + M^2}}{[(i\omega)^2 + M^2]^{1/2}} - \frac{[2((i\omega)^2 + \Delta_{\rm p}^2)((i\omega)^2 + M^2) - M^2((i\omega)^2 + \Delta_{\rm p}^2 + M^2)]}{2((i\omega)^2 + M^2)^{3/2}((i\omega)^2 + \Delta_{\rm p}^2 + M^2)^2} \right\} \tag{29.86}$$

$$1 = \frac{g_{\rm p}}{\pi t}\pi T \sum_{\omega} \left\{ \frac{1}{\Delta_{\rm p}} \left(\frac{\pi}{2} - \tan^{-1} \frac{[(i\omega)^2 + M^2]^{1/2}}{\Delta_{\rm p}} \right) - \frac{M^2[(i\omega)^2 + M^2]^{1/2}}{((i\omega)^2 + \Delta_{\rm p}^2 + M^2)^2} \right\} . \tag{29.87}$$

In this case taking $M \to 0, T \to T_{\rm N}$ we get near T_c,

$$\frac{T_{\rm N0} - T_{\rm N}}{T_{\rm N0}} = \left(\frac{\Delta_{\rm p}}{2\pi T_{\rm N0}} \right)^2 \frac{21}{2} \varsigma(3) (\ln t / T_{\rm N0})^{-1} \tag{29.88}$$

the extra term $(\ln t/T_{\rm N0})^{-1}$ appearing due to the logarithmic divergence of the density of states. Taking $\Delta_{\rm d} \to 0, \Delta_{\rm p} \to 0$ we can calculate $T_c^{(\rm d)}$ and get the relation $T_c^{(\rm p)} < T_c^{(\rm d)}$ which shows that the magnetic state suppresses the p-pairing and not the d-pairing, if both gaps satisfy $\Delta(\mathbf{p}) = -\Delta(\mathbf{p}+\mathbf{Q})$. If these states satisfy the condition $\Delta(\mathbf{p}) = \Delta(\mathbf{p}+\mathbf{Q})$, then we can prove $T_c^{(\rm p)} > T_c^{(\rm d)}$ which demonstrates the importance of the combined symmetry relations, namely, the parity and the translation by Q to get a stable superconducting state in the presence of the magnetic state. Such a behaviour has also been suggested for high T_c oxide superconductors which is a system with strong electron correlations. However, it is until now an open question if this type of model can explain the high temperature superconductivity.

30. High-Temperature Superconductivity

The discovery of the high temperature superconductivity in Cu based oxides by Bednorz and Müller[43] has initiated a great number of theoretical investigations in order to understand the mechanism of the occurrence of this new feature of superconductivity. Before discussing the special characteristics of superconductivity in these kinds of materials we analyse the properties of two types of these materials in the normal state.

The first material is La_2CuO_4 pure and doped $La_{1.85}Sr_{0.15}$ CuO_4, and the second type is $YBa_2Cu_2O_{7-\delta}$. These materials contain two important structure elements: a) CuO_6 octahedrons and b) CuO_4 squares. Due to the first elements by sharing corners, they form a two-dimensional layered structure, the octahedrons being Jahn-Teller distorted, which splits the e_g degeneracy of the Cu(3d) states, and the antibonding $3d_{x^2-y^2}$ state becomes the highest in energy. The parameters of hybridization for Cu $3d_{x^2-y^2}$ and O($2p_x$, $2p_y$) states are $V_{\rm pd} = 1.8$ eV. The Coulomb integral $U_{\rm dd} = 32$ eV for Cu($3d_{x^2-y^2}$) and $U_{\rm pp} = 20$ eV for O(2p) and $U_{\rm ap} = 7$ eV are very large compared to the hybridization integrals.

In the materials of the second type, the structure elements b) CuO_4 form a one-dimensional chain by sharing corners. The Fermi surface of La_2CuO_4 is a half filled square lattice and at the points $(k_x, k_y) = (0, \pm\frac{\pi}{a})$ and $(\pm\frac{\pi}{a}, 0)$ the group velocity $\mathbf{v}(\mathbf{k}) = 0$. The density of states has a logarithmic singularity at the Fermi energy ε_0. The Fermi surface obtained by the band calculation[44,45] looks somewhat different from the idealized square lattice due to the dispersion of the energy in the K_z direction. The calculated Fermi surface predicts hole conductivity for doped systems which seems to be in agreement with the experiments.

The calculated Fermi surface has the nesting properties and the system is susceptible to the Charge-Density-Wave or Spin-Density-Wave transitions.

On the other hand, the correlation between the electrons is strong and this suggested the model based on the Hubbard Hamiltonian. A general conjecture is that the Coulomb interactions between electrons on different sites, which have been left out in the other model, may become important for this superconducting state.

The superconducting state presents the following characteristics:
— the gap ratio $2\Delta(0)/T_c$ which is 3.53 in the BCS superconductors is found to be much larger.
— a Josephson current has been detected.
— the flux quantum is $\Phi_0 = \pi/2e$.
— the phonon must be important in the mechanism because T_c shows an isotropic effect when ^{16}O is replaced by ^{18}O in $La_{1.85}Sr_{0.15}CuO_4$ but not in $YBa_2Cu_3O_7$.

The first model was proposed by Anderson[47] based on an older model proposed in 1973 to explain the insulating and magnetic properties of semiconductors. The insulating phase is supposed to be done by the "resonating-valence bond" state, and this phase as well as the magnetic one is favoured by low spin, low dimensionality and magnetic fluctuations. In such pure materials as La_2CuO_4 there exist singlet pairs in the insulating phase and these pairs become charged superconducting pairs when the insulator is sufficiently doped. The problem was considered by Anderson *et al.*[47-49] and it points out the importance of the Coulomb correlations which give rise to the structure distortions.

The two-dimensional character of the system explains the nesting of the different portions of the Fermi surface which give rise to the antiferromagnetism. The mechanism for superconductivity seems to be predominantly electronic and magnetic, although weak phonon interaction may favour the state.

Starting from this simple model we will present different models. We start with the models which are based on the electron-phonon interaction. Rice[50] initiated a model based on the occurrence of the Charge-Density-Waves which is the cause of the semiconducting properties. Using the adiabatic approximation for the phonons and the Hartree approximation for the Hubbard term, the gap from the electronic spectrum is

$$\Delta \sim t \exp(-\pi/\sqrt{4\lambda}) \tag{30.1}$$

where t is the hopping integral for the electrons, $\lambda = V^2/Kt, V$ being the electron-displacement coupling constant and K the coupling constant of the elastic energy of the oxygen atoms.

If the lattice deforms locally in the system there appear polarons, and for a strong lattice deformation the one-site interaction between two electrons with opposite spins is strong and this system is called a bipolaron. The occurrence of the bipolarons can be considered as due to a very strong electron-phonon interaction, the bipolarons being locally trapped, and the hole system can be considered as nearest-neighbouring electron pairs with opposite spins coupled to the lattice deformations. Prelovsek *et al.*[51] considered that the superconducting state appears by a Bose condensation of the polarons and estimated the Bose condensation temperature as

$$T_c = 1.75 \times 10^5 \left[\frac{\delta^2}{a^4 c^2 m_B m_Z}\right]^{1/3} \tag{30.2}$$

where δ is the dopant concentration, m_B and m_Z are the bipolaron effective masses in the plane and in the Z-direction respectively, and a, c are constants of the lattice. The specific feature of this model is the coexistence between the bipolarons and the Charge-Density-Waves.

At the present time there are many models which have common points as well as essential differences. So we will try to present the most important of these models and at the end we will consider their common features.

a) *Bipolaronic model*

This model has as the basic idea the reconsideration of the electron-phonon interaction in the narrow-band crystals. Anderson[32] and Chakraverty *et al.*[53,54] showed the occurrence of the bipolarons in the limit of a strong electron-phonon coupling ($g \geq 2.5$). The formation of the bipolarons in the non-metallic transition-metals compounds was shown[55–57] and this is in fact equivalent to the occurrence of the Cooper pairs. Alexandrov and Ranninger[57] showed that in a narrow-band crystal the strong electron-phonon coupling gives rise to the bipolarons, which are equivalent under certain conditions to a superfluid charged Bose system. The model has been reconsidered by Alexandrov *et al.*[58] using the observation that the strong-coupling condition

$$gN(0) > 1$$

is similar to that for the small polaron formation

$$\frac{2zg^2\omega}{D} > 1 \tag{30.3}$$

where $2zg^2\omega$ is the effective phonon-mediated attraction and $N(0) \sim D^{-1}$ is the density of states at the Fermi level, D is the bandwidth, ω is the characteristic phonon frequency in the system, and z is the number of nearest neighbours. Due to the condition (30.3), the electronic bandwidth D is drastically reduced and the narrow polaronic bandwidth is

$$W = D\exp(-g^2)\ . \tag{30.4}$$

If the attraction interaction between two polarons is stronger than the Coulomb repulsion, then a condensed state appears, the properties of this state being determined by the value of the bipolaron binding energy

$$A \cong 2\varepsilon_p - U \tag{30.5}$$

where $\varepsilon_p = g^2\omega$ and U is the Coulomb repulsion.

If the polaron-polaron interaction is attractive but small ($A \ll W$), the small polarons form spatially overlapping Cooper pairs with superconducting properties similar to the usual BCS superconductivity. The critical temperature has not exactly the same expression because

$$\omega \gtrsim \varepsilon_0 - W \tag{30.6}$$

which is in fact a violation of the adiabatic limit.

Another superconducting state called "*bipolaronic superconductor*" has been considered by Alexandrov *et al.*[59] and is supposed to appear due to the existence of the local pairs of fermionic carriers which are bound together by an attractive mechanism. In the site representation, a narrow-band superconductor can be described by the Hamiltonian

$$\mathcal{H} = \sum_{i,i'} T_{i,i'} c_i^\dagger c_{i'} + \sum_{i,\mathbf{q}} [g_i(\mathbf{q})(c_i^\dagger c_i b_\mathbf{q} + h.c.)] + U \sum_{i,i'} c_i^\dagger c_{i'}^\dagger c_{i'} c_i + \sum_\mathbf{q} \omega(\mathbf{q}) b_\mathbf{q}^\dagger b_\mathbf{q} \tag{30.7}$$

where $T_{i,i'}$ denotes the hopping integral, $g_i(\mathbf{q})$ the electron-phonon coupling, $U_{i,i'}$ the Coulomb interaction, and $\omega(\mathbf{q})$ the phonon energy. Using the Lang-Firsov[60] transform S_1, we get

$$\mathcal{H}_\mathrm{p} = \hat{S}_1 \mathcal{H} \hat{S}_1^{-1} = \mathcal{H}_0 + \mathcal{H}_1 \tag{30.8}$$

where

$$\hat{S}_1 = \exp\left[\sum_{i\mathbf{q}} (\omega^{-1}(\mathbf{q}) c_i^\dagger c_i b_\mathbf{q} g(\mathbf{q}) + h.c.)\right] . \tag{30.9}$$

In (30.8), $\mathcal{H}_0$ and $\mathcal{H}_1$ are given by

$$\mathcal{H}_0 = \sum_i (T_i - \varepsilon(\mathrm{p})) c_i^\dagger c_i + \sum_{i,i'} v_{i,i'} c_i^\dagger c_{i'}^\dagger c_{i'} c_i + \mathcal{H}_\mathrm{ph} \tag{30.10}$$

and

$$\mathcal{H}_1 = \sum_{i,i'} \hat{\sigma}_{i,i'} c_i^\dagger c_{i'} \tag{30.11}$$

where $\mathcal{H}_\mathrm{ph}$ is the last term from (30.7) and

$$\varepsilon_\mathrm{p} = \sum_\mathbf{q} \omega^{-1}(\mathbf{q}) |g(\mathbf{q})|^2 \tag{30.12}$$

$$v_{i,i'} = U_{i,i'} - \sum_\mathbf{q} \frac{1}{\omega(\mathbf{q})} |g(\mathbf{q})|^2 \exp(i\, \mathbf{q} \cdot (\mathbf{i} - \mathbf{i}')) . \tag{30.13}$$

The energy ε_p is known as the polaronic level shift, $v_{i,i'}$ is the polaron-polaron interaction (which has to be attractive for the superconducting state) and $\hat{\sigma}_{i,i'}$ is the kinetic energy of the small polaron. If we use instead of (30.12)–(30.13) the thermal average values (with the phonon density matrix) the energy $\hat{\sigma}_{i,i'}$ becomes

$$\sigma_{i,i'} = T_{i,i'} \exp(-g^2) \tag{30.14}$$

$$g^2 = \sum_\mathbf{q} \omega^{-2}(\mathbf{q}) \cot g \frac{\omega(\mathbf{q})}{2T} |g(\mathbf{q})|^2 [1 - \cos \mathbf{q} \cdot (\mathbf{i} - \mathbf{i}')] . \tag{30.15}$$

In order to describe the superconducting state we have to consider the interaction $v_{i,i'}$. Alexandrov and Ranninger[57] considered the case $|v_{i,i'}| \gg \sigma_{i,i'}$ in which the bipolarons, in the low density limit, have an excitation spectrum as a quantum superfluid.

Another possibility to obtain a superconducting state is to consider the properties of strongly coupled polarons. Alexandrov *et al.*[59] considered $\sigma/|v|$ as a small perturbation and eliminated the single polaronic state from (30.8). A new transform S_2 defined by

$$\langle \mathrm{f}|S_2|\mathrm{p}\rangle = \sum_{i,i'} \langle \mathrm{f}|\widehat{\sigma}_{i,i'} c_i^\dagger c_{i'}|\mathrm{p}\rangle (E_\mathrm{f} - E_\mathrm{p})^{-1} \tag{30.16}$$

and applied to the Hamiltonian (30.8)

$$\widetilde{\mathcal{H}} = \exp(S_2)\mathcal{H}_\mathrm{p} \exp(-S_2) \tag{30.17}$$

gives

$$\widetilde{\mathcal{H}} = \sum_{m\neq m'} [\widetilde{v}(\mathbf{m}-\mathbf{m}')b_m^\dagger b_m b_{m'}^\dagger b_{m'} - t(\mathbf{m}-\mathbf{m}')b_\mathbf{m}^\dagger b_{\mathbf{m}'}] \,. \tag{30.18}$$

In the limit $A \gg \omega, t(\mathbf{m}-\mathbf{m}')$ and $\widetilde{v}(\mathbf{m}-\mathbf{m}')$ have been calculated as

$$t(\mathbf{m}-\mathbf{m}') = (2T_{m,m'}^2/A)\exp(-g^2) \tag{30.19}$$

$$\widetilde{v}(\mathbf{m}-\mathbf{m}') = 4v(\mathbf{m}-\mathbf{m}') + 2T_{m,m'}^2/\Delta \tag{30.20}$$

and in the opposite limit $T \lesssim \Delta \lesssim \omega$, these parameters become

$$t(\mathbf{m}-\mathbf{m}') = (2T_{m,m'}^2/A)\exp(-2g^2) \tag{30.21}$$

$$\widetilde{v}(\mathbf{m}-\mathbf{m}') = 4v(\mathbf{m}-\mathbf{m}') + (2T_{m,m'}^2/A)\exp(-g^2) \,. \tag{30.22}$$

These results show that the effective bipolaronic mass and the effective interaction are strongly dependent on the ratio A/ω and that both of them increase rapidly with increasing A/ω. The Hamiltonian (30.18) can be transformed using the pseudospins

$$S_m^z = \frac{1}{2} - b_m^\dagger b_m \,, \quad S_m^x = \frac{1}{2}(b_m + b_m^\dagger) \,, \quad S_m^y = \frac{i}{2}(b_m - b_m^\dagger) \tag{30.23}$$

as

$$\widetilde{\mathcal{H}} = \sum_m S_m^z \Big[\mu + \sum_{m'\neq m} \widetilde{v}(m-m')S_{m'}^z\Big] + \sum_{m\neq m'} t(\mathbf{m}-\mathbf{m}')[S_m^x S_{m'}^x + S_m^y S_{m'}^y] \tag{30.24}$$

where μ is the chemical potential of bipolarons, and will be determined from the condition of conservation of the number of bipolarons

$$\frac{1}{N}\sum_m \langle S_m^z \rangle = \frac{1}{2} - n\,. \tag{30.25}$$

The mean field approximation has been used[57] to analyse this model and one obtains in the energy of the excitations, a gap proportional to t which gives an incorrect result leading to an exponential dependence of the specific heat. The true excitations spectrum consists of pseudomagnons with a gapless dispersion. Taking into consideration the quantum fluctuations, we can perform the thermodynamics of this system in the Random Phase Approximation.[59]

If we define the order parameter for the superconducting state by

$$S^x = \langle S_m^x \rangle = \frac{1}{2}[R^2 - (2n-1)^2]^{1/2}\,, \tag{30.26}$$

R is given by the self-consistent equation

$$\frac{1}{R} = \frac{1}{N}\sum_{\mathbf{k}} \frac{1}{\omega(\mathbf{k})} \coth\frac{\omega(\mathbf{k})}{2T} \tag{30.27}$$

where the excitation spectrum is given by

$$\omega(\mathbf{k}) = R[(t - t_k\cos^2\theta + \tilde{v}_k \sin^2\theta)(t - t_k)]^{1/2} \tag{30.28}$$

with

$$\cos\theta = (2n-1)/R \tag{30.29}$$

t_k and $\tilde{v}_k$ being the Fourier transforms of $t(\mathbf{m})$ and $v(\mathbf{m})$. Using now the condition $S^x = 0$ we can obtain T_c the critical temperature, from (30.27) by the equation

$$\frac{1}{2n-1} = \frac{1}{N}\sum_{\mathbf{k}} \coth[(2n-1)(t-t_k)/2T_c]\,. \tag{30.30}$$

In the case of a dilute system $|n| \ll 1, T_c$ is given by

$$T_c = \frac{3.31(na^{-3})^{3/2}}{m^*}(1 - 0.54n^{2/3}) \tag{30.31}$$

where $m = 1/ta^2$ is the effective bipolaronic mass in a lattice with the constant "a". In the high density limit, $|2n-1| \ll 1$, (30.30) gives

$$T_c \cong \frac{t}{2}\left[C^{-1} - \frac{(2n-1)^2}{3}\right] \tag{30.32}$$

where C is a constant which depends on the lattice type. This model is equivalent to the negative U Hubbard Hamiltonian but with a temperature-dependent narrow band W. Then the phenomenological negative U Hubbard Hamiltonian is applicable to polaronic system only in the limit $\omega \gg U$ and with D being replaced by a temperature-dependent W. In the limit $\omega \ll \Delta$, the bipolaronic Hamiltonian can be parametrized by an extended negative $-U$ Hubbard model in the strong-coupling limit.

A model of coexisting bipolarons, free electrons, and BCS pairs was proposed[62] for the intermediate electron-phonon coupling. The polaronic model has been reconsidered by Nasu[63] on the basis of an attractive analysis of the electron-phonon coupling. The effects of the electron-phonon coupling are in fact determined by the magnitude of the phonon energy $\omega(\mathbf{q})$, the electron-phonon coupling g, and the transfer-energy t. In the case of $g \ll t$ in the system appear excitations called large polarons. These are in fact electrons interacting with the phonon cloud, which is very thin but extended. If $g \gg t$, the excitations become small polarons excitations consisting of electrons with the phononic cloud in the same site, and with a considerable mass enhancement. As the energy $\omega(\mathbf{q})$ increases, the difference between the large and small polarons disappears because of the quantum phononic effects. If $\omega \gg g, t$ the phonon can follow the motion of the electron and this situation is in fact the inverse-adiabatic limit in which some molecular crystals are expected to appear. A phase diagram given in Ref. 63 as function of t, g and $\omega(\mathbf{q})$ shows the existence of the superconducting region and a bipolaronic insulator region. In the mean-field approximation, the superconducting gap is maximum for $\omega \sim t$. In the inverse-adiabatic limit there is a region where the small polarons dominate and the superconducting state becomes more stable than the Charge-Density state because the retardation effect is absent. The collective excitations of the superconducting state change their nature continuously from the BCS-pair breaking excitations to the superfluid type, with the increasing of g/t.

b) *Strong electron correlation model*

The special feature of the electron-phonon interaction can change the effective interaction between the localized electrons and it can become attractive if the local phonon frequency is high. This model, called the "negative-U centers", has been proposed by Ting *et al.*[64] considering each one-site bipolaron as a negative U center. Using the strong coupling formalism, it was shown that the critical temperature of s-electrons is much smaller than the critical temperature T_b below which the bipolarons appear. If the superconducting state is considered to be given by the bipolarons, the critical temperature T_c satisfies $T_{c0} < T_c < T_b$.

Hirch[66] used the Monte Carlo simulation to show that an effective attraction between nearest-neighbour antiparallel spins arises in the Hubbard model even in the limit of a large site electron-electron repulsion.

Schüttler *et al.*[67] also showed, using a Monte-Carlo method combined with many-body methods, that the critical temperature of a superconductor is strongly enhanced by a small concentration of excitonic negative-U centers. The influence of the electron-phonon interaction on the strong correlation electrons has been studied by Kuramoto and Watanabe[68] using a model described by the Hamiltonian

$$\mathcal{H} = -t\sum_{i,j,\alpha} c_{i\alpha}^{\dagger} c_{j\alpha} + \frac{U}{2}\sum_{i,\alpha} n_{i,\alpha} n_{i,-\alpha} + \sum_{\mathbf{q}} \omega(\mathbf{q}) b_{\mathbf{q}}^{\dagger} b_{\mathbf{q}} + \frac{1}{N}\sum_{i,\mathbf{q}} \exp(i\,\mathbf{q}\cdot\mathbf{R}_i) g(\mathbf{q}) n_i (b_{\mathbf{q}} + b_{-\mathbf{q}}^{\dagger}) \tag{30.33}$$

where the first two terms are the Hubbard contribution and the last two terms describe the influence of the electron-phonon interaction. The model of two-dimensional square lattice has been considered where the electron-phonon interaction can be eliminated by a canonical transformation and (30.33) written as

$$\mathcal{H} = \mathcal{H}_{\rm BP} + \sum_{i,j,\alpha} [-\tilde{t} c_{i\alpha}^{\dagger} c_{i\alpha} + 2J(\mathbf{S}_i \cdot \mathbf{S}_j - \frac{1}{4} n_i n_j)] \tag{30.34}$$

with

$$\mathcal{H}_{\rm BP} = -\sum_{i,j;\mathbf{q}} \omega(\mathbf{q}) \left[\frac{g(\mathbf{q})}{\omega(\mathbf{q})}\right]^2 \exp(i\,\mathbf{q}\cdot(\mathbf{R}_i - \mathbf{R}_j)) n_i n_j \tag{30.35}$$

where $J = 2t^2/U$, and $\mathbf{S}_i$ is the spin operator.
The Hamiltonian (30.35) represents an attractive interaction induced by phonons which is essential in the bipolaron theory. The second term of

(30.34) represents the superexchange between the neighbouring spins and this interaction is not reduced by the polaron effect because virtual hopping does not accompany the phonon cloud. If $t \gg J$, a bound state of two electrons can appear, but this state will be stabilized only if $J = 2t^2/U$ overcomes the kinetic energy. If t is much larger than J, the bound state can be formed without the occurrence of the polarons. On the other hand, the polaronic effect may be strong enough to make $t \ll J$ and the bound state can be stabilized by this mechanism. If the d-pairing is considered, the critical temperature can be calculated using the mean-field approximation from

$$1 = \frac{2J}{N} \int d\varepsilon \frac{1}{4\varepsilon} N(\mathbf{k}, \varepsilon) \tanh \frac{\varepsilon}{2T_c} (\cos k_x - \cos k_y) \tag{30.36}$$

an equation which shows that the critical temperature has a strong dependence on the band filling. Using for $\varepsilon(k)$, the expression $\varepsilon = -2\tilde{t}(\cos k_x + \cos k_y)$, we get from (30.36)

$$T_c = \tilde{A} \exp[-16\tilde{t}/n_e] \tag{30.37}$$

where n_e is the number of electrons from the band, and A is a constant. We may conclude that in this case the superexchange and the polaronic effects can give rise to the pairing effect.

Emery[69] showed that it is possible to obtain superconductivity from repulsive interaction, the binding of a pair being a consequence of their coupling to other degrees of freedom such as spin-fluctuations. In the strong-coupling limit, it is possible to get an attractive interaction of the order t^2U but the occurrence of the superconducting state is not sure.

c) *The inter-layer model*

The crystal structure of the high temperature compounds suggests a two-dimensional character of the conduction electrons. The density of states is large near the band center which increases the electron-phonon coupling constant g in favour of the critical temperature T_c. However, the Coulomb pseudopotential μ^* is stronger in two dimensions as compared with the usual three-dimensional expression. A simple model adopted by Inoue *et al.*[70] supposes that the electrons propagate freely in each layer but the overlap integral between different layers is negligible. An electron from a layer interacts with another electron from the neighbouring layer through the exchange of three-dimensional phonons and they form the Cooper pairs.

The critical temperature has been calculated in the strong coupling theory by linearization of the equations:

$$\omega(Z(\omega)-1)=\pi T\sum_{\omega'}\frac{\omega'}{\sqrt{\omega'^2+\Delta^2}}\lambda_{\rm I}(\omega-\omega') \tag{30.38a}$$

$$Z(\omega)\Delta(\omega)=\pi T\sum_{\omega'}\frac{\Delta(\omega')}{\sqrt{\omega'^2+\Delta^2}}[\lambda_{\rm II}(\omega-\omega')-\mu^*] \tag{30.38b}$$

where μ^* is the Coulomb pseudopotential and can be neglected for the inter-layer pairs.

The coupling constant $\lambda_{\rm I}$ and $\lambda_{\rm II}$ are considered to be of the form

$$\lambda_{\rm I}=\lambda_{\rm intra}I(\omega)\,,\quad \lambda_{\rm II}(\omega)=\lambda_{\rm inter}I(\omega) \tag{30.39}$$

where

$$I(\omega)=\int_0^\infty d\Omega\alpha^2(\Omega)F(\Omega)\frac{2\Omega}{\omega^2+2\Omega} \tag{30.40}$$

and the critical temperature T_c can be calculated in the weak coupling approximation as

$$T_{\rm c}^{(1)}=1.14\omega_{\rm D}\exp\left[-\frac{1+\lambda_{\rm intra}}{\lambda_{\rm inter}}\right] \tag{30.41}$$

the parameter $\lambda_{\rm inter}-\lambda_{\rm intra}=\Lambda$ appearing as a measure of a pair-breaking effect in (30.41). The critical temperature due to the intra-layer coupling is

$$T_{\rm c}^{(2)}=1.14\omega_{\rm D}\exp\left[-\frac{1+\lambda_{\rm intra}}{\lambda_{\rm intra}-\mu^*}\right] \tag{30.42}$$

and if we compare (30.41) with (30.42), we see that Λ works as an effective Coulombic pseudopotential and decreases $T_{\rm c}^{(1)}$. Then, in fact, we have a competition between these two mechanisms, and if $\mu^*\gg\Lambda$ the inter-layer will appear, while if $\mu^*\ll\Lambda$ the intra-layer pairing is favourable. The validity of such a mechanism can be proved if we consider the influence of the non-magnetic impurities. The intra-layer coupling is not affected but the inter-layer pairing is destroyed. A high concentration of non-magnetic impurities has to be used because the coherence length of these materials is very small.

d) *Spin-fluctuations models*

The absence of the isotropic effect in some high temperature superconductors suggested the possibility of a non-phononic mechanism, from which we will present one which takes into consideration the antiferromagnetic spin-fluctuations. This mechanism is in direct connection with the two-dimensional character of the Fermi surface which also presents a perfect nesting. Using an electronic system with the energy

$$\varepsilon(\mathbf{p}) = -2t(\cos p_x + \cos p_y) - \mu \tag{30.43}$$

(where μ is the chemical potential which determines the Fermi energy position) we calculate the spin susceptibility

$$\chi(\mathbf{q},\omega) = \frac{\chi_0(\mathbf{q},\omega)}{1 - U\chi_0(\mathbf{q},\omega)} \tag{30.44}$$

where $U > 0$ is the one-site Coulomb interaction. In the almost half-filled band $\mu \simeq 0$ and $\chi_0(\mathbf{q},\omega)$ becomes very large near $Q = (\pm 1, \pm 1)$.

For two-dimensional electronic system $\chi_0(\mathbf{q}) = \chi_0'(\mathbf{q}) + i\chi_0''(\mathbf{q})$ where the imaginary part $\chi_0''(\mathbf{q})$ is

$$\chi''(\mathbf{q},\omega) \propto -\ln[|\omega| + v_0|\mathbf{q} - \mathbf{Q}|/t] \tag{30.45}$$

and we see that $\chi''(\mathbf{q},\omega)$ approaches a constant if $\omega \to 0$ near the wave vector $\mathbf{Q}$. The inelastic-scattering rate can be calculated taking the processes where an electron is scattered to an unoccupied state across the Fermi surface while exciting a particle-hole pair. This rate is given by the integral over ω and $\mathbf{q}$ in (30.45) taking $\omega \sim T$ and if $\chi'' \sim \omega$ one obtains the standard Fermi liquid result $\tau_s^{-1} \propto T^2/\varepsilon_0$. In the case of χ'' given by (30.45) (for $|\omega| < |\mathbf{q} - \mathbf{Q}|$), we get

$$\tau_s^{-1} \propto T \tag{30.46}$$

and if we analyse the inelastic scattering-time for these materials we see a good agreement with the resistivity behaviour in the normal state. Using these observations Lee and Read[71] suggested the existence of a large energy scale as well as a d-pairing. The standard relation $2\Delta_0 = 3.5T_c$ has to be replaced by $\Delta \simeq \tau_s^{-1}(T_c)$, which seems to be reasonable if we consider the tunnelling measurements which give $2\Delta_0 = 13T_c$. In the superconducting state, the imaginary part of χ is

$$\chi_0''(Q,\omega) = \begin{cases} 0; & \omega < 2\Delta_0 \\ \frac{1}{4\pi t}\frac{\omega}{\sqrt{\omega^2 - 4\Delta^2}} \ln \frac{8t}{\sqrt{\omega^2 - \Delta^2}}; & \omega > 2\Delta_0 \end{cases} \tag{30.47}$$

and the real part is

$$\chi_0' \propto \ln^2|2\Delta_0 - \omega| \tag{30.48}$$

and the singularity from $\omega = 2\Delta_0$ is given by the density of states. Using (30.47)–(30.48) in (30.44) we see that $\chi(\mathbf{Q},\omega)$ has a pole at $\omega \le 2\Delta_0$ which indicates that the spin fluctuations give rise to pairing.

These considerations are in agreement with the models given by Machida and Kato,[72] Kopaev and Rusinow,[74] in which the superconducting transition temperature can be enhanced by the density wave instabilities. An expression of tight binding energy including the nearest ($t_0 > 0$) and next nearest ($t_1 > 0$) neighbour transfer integrals which has the form

$$\varepsilon(\mathbf{p}) = -t_0(\cos p_x + \cos p_y) - t_1 \cos p_x \cos p_y \tag{30.49}$$

has been used in the Hamiltonian

$$\mathcal{H} = \mathcal{H}_0 + \mathcal{H}_{sc} + \mathcal{H}_{SDW} \tag{30.50}$$

which describes a superconductor with anisotropic pairing and Spin-Density-Waves (SDW). In (30.50)

$$\mathcal{H}_0 = \sum_{\mathbf{p},\alpha} \varepsilon(\mathbf{p}) c^\dagger_{\mathbf{p}\alpha} c_{\mathbf{p}\alpha} \tag{30.51}$$

$$\mathcal{H}_{sc} = \sum_{\mathbf{p}} [\Delta(\mathbf{p}) c^\dagger_{\mathbf{p}\uparrow} c^\dagger_{-\mathbf{p}\downarrow} + h.c.] \tag{30.52}$$

$$\mathcal{H}_{SDW} = \sum_{\mathbf{p}} M[c^\dagger_{\mathbf{p}+\mathbf{Q}\uparrow} c_{\mathbf{p}\downarrow} + h.c.] \tag{30.53}$$

where

$$M = -U \sum_{\mathbf{p}} \langle c^\dagger_{\mathbf{p}+\mathbf{Q}\uparrow} c_{\mathbf{p}\downarrow}\rangle \tag{30.54}$$

and the superconducting gap is

$$\Delta(\mathbf{p}) = -\sum_{\mathbf{p}'} g(\mathbf{p},\mathbf{p}')\langle c^\dagger_{\mathbf{p}'\uparrow} c^\dagger_{-\mathbf{p}'\downarrow}\rangle\ . \tag{30.55}$$

The attractive interaction $g(\mathbf{p},\mathbf{p}')$ will be written as $g(\mathbf{p},\mathbf{p}') = g\tau(\mathbf{p})\tau(\mathbf{p}')$ and then $\Delta(\mathbf{p}) = \Delta\tau(\mathbf{p})$. The self-consistent equations for $\Delta(\mathbf{p})$ and M can be easily obtained, and $\Delta(\mathbf{p})$ can be classified according to the combined symmetry: parity $\Delta(-\mathbf{p}) = \pm\Delta(\mathbf{p})$ (Even or Odd) and translation symmetry by the nesting vector: $\Delta(\mathbf{k}+\mathbf{Q}) = \pm\Delta(\mathbf{k})$ (Even or Odd). In the first case, parity-even (E) and translation-odd (O), we have an anisotropic s-like pairing $\Delta_s(\mathbf{p}) = \Delta_s\tau_s = \frac{\Delta_s}{\sqrt{2}}(\cos p_x + \cos p_y)$ or d-like pairing with

$\Delta_d = \Delta_d \tau_d = \Delta_d/\sqrt{2}(\cos p_x - \cos p_y)$. In the second case, parity-odd (O) and translation-odd (O) and parity-even (E) and translation-even (E), we get the isotropic s-pairing $\Delta(\mathbf{p}) = \Delta_s$ and p-like pairing states.

In the first case, T_c calculated in the mean-field approximation can be enhanced but in the second case it is always depressed by the order parameter (30.54). We have to mention that the case of the Charge-Density-Wave (CDW) which may coexist with the isotropic singlet state is a particular case of the first case, and the mechanism proposed in Refs. 72–73 does not decrease the critical temperature. It is important to analyse the Fermi surface in the SDW state, as well as in the CDW state.

A simple calculation using the quasi-particles spectrum

$$E^{(2)}_{\rm SDW} = \left(\sqrt{\gamma_{\mathbf{p}}^2 + M^2} \pm \delta_{\mathbf{p}}\right)^2/(t_0^2 - t_1^2)$$

and

$$E^{(2)}_{\rm CDW} = \left(\delta_k \pm \sqrt{\gamma_{\mathbf{p}}^2 + W^2}\right)^2 ,$$

where $\gamma_{\mathbf{p}} = -t_0(\cos p_x + \cos p_y)$, $\delta_{\mathbf{p}} = t_1 \cos p_x \cos p_y$, ($W$ is the order-parameter for the Charge-Density state), shows that for $M > t_1$, the Fermi surface of the system disappears, the system becomes an insulator, and there is no possibility of the occurrence of the superconducting state. The condition $M < t_1$ will assure the existence of the Fermi surface. It was shown[72,73] that as M increases the density of states increases due to the occurrence of the Van Hove singularities in the spectrum.

The main conclusion reached from this model is that the existence of an anisotropic pairing due to the density fluctuations and the attractive interaction seems to be non-phononic. The importance of the charge transfer in the high-temperature superconductors was stressed by Varma *et al.*[75,76] who showed that the charge transfer between the nearest neighbour cations and anions is unscreened. In this way, an excitonic resonance can appear in the particle-hole spectrum, and the electron-electron interaction mediated by the exchange of this resonance can give rise to a high critical temperature. We will finish this paragraph without a definite conclusion concerning the microscopic description of the high temperature superconductivity. There are still some models which have not been presented because they are very close to the most important results discussed here.

At the present time, there are models which try to build up a theory similar to the heavy fermions superconductivity[77] or to find a connection with the spin-glass theory[78] in two dimensions. The RVB model has been improved considering that this state is characterized by two kind of excitations; neutral spin fermions called "spinons" which correspond to impaired spins and charged bosons called "holons" which correspond to an empty state. These new excitations introduced by Anderson *et al.*[79] have been used by Oguri and Maekawa[80] for the explanation of the normal state properties. The influence of the interlayer pairing has been considered by Wheatley *et al.*,[81] Schneider and Baeriswyl[82] and Birman *et al.*[83]

The anisotropic electron-phonon attraction has recently been described using the strong-coupling theory by Prohammer.[84] Using the same formalism Enz[85] proposed a mechanism based on the charge fluctuations in the Cu-O subsystem. Hirch[86] proposed a new mechanism based on the interaction of a hole with the outer electrons in atoms with nearly filled shells. Possibly one of these is in the right direction for a new model which can explain this new feature of superconductivity.

References

1. V. L. Ginzburg, *Usp. Fiz. Nauk* (1970) 185; *Sov. Phys.-Usp.* **13** (1970) 335.
2. J. Ruvalds, *Adv. Phys.* **30** (1981) 667.
3. B. T. Geilikman, *Usp. Fiz. Nauk Sov. Phys.-Usp.* **9** (1966) 142.
4. D. Allender, J. Bray and J. Bardeen, *Phys. Rev.* **B7** (1973) 1020.
5. J. C. Inkson and P. W. Anderson, *Phys. Rev.* **B8** (1973) 4429.
6. M. L. Cohen and P. W. Anderson in *"Superconductivity in d- and f-Band Metals"* ed. D. H. Doughlass (AIP, New York) p. 17 (1972).
7. I. Tutto and J. Ruvalds, *Phys. Rev.* **B19** (1979) 5641.
8. M. Grabowski and L. J. Sham, *Phys. Rev.* **B29** (1984) 6132.
9. G. R. Stewart, *Rev. Mod. Phys.* **56** (1984) 755; F. Steglich in *"Theory of Heavy Fermions and Valence Fluctuations"*, eds. T. Kasuya and T. Saso, *Springer Series in Solid State Sciences* **62** (1985) 23.
10. J. M. Luttinger and J. C. Ward, *Phys. Rev.* **118** (1960) 1417.
11. J. M. Luttinger, *Phys. Rev.* **119** (1960) 1153.
12. R. M. Martin, *Phys. Rev. Lett.* **48** (1982) 362.
13. F. J. Ohkawa, *J. Phys. Soc. Jpn.* **53** (1984) 1389.
14. F. J. Ohkawa, *J. Phys. Soc. Jpn.* **52** (1983) 3886.
15. F. J. Ohkawa, *J. Phys. Soc. Jpn.* **53** (1984) 2697.
16. T. Koyama and M. Tachiki, *Phys. Rev.* **B34** (1986) 3272.
17. V. Zlatić, S. K. Ghatak and K. H. Beneman, *Phys. Rev. Lett.* **57** (1986) 1263.
18. Z. Tesanovic and O. T. Valls, *Phys. Rev.* **B34** (1986) 1918.
19. P. Coleman, *Phys. Rev.* **B29** (1984) 3035.
20. N. Read and D. M. Newns, *J. Phys. C.* (1984) L1055.

21. N. Read, *J. Phys.* **C18** (1985) 2651.
22. K. Yoshida and K. Yamada, *Prog. Theor. Phys.* **53** (1975) 1286.
23. H. Razafimandimby, P. Fulde and J. Keller, *Z. Phys.* **B54** (1984).
24. H. Miyake, T. Matsuura and Y. Nagaoka, *Prog. Theor. Phys.* **72** (1984) 1063.
25. F. J. Ohkawa, *J. Phys. Soc. Jpn.* **53** (1984) 3568; **53** (1984) 3577.
26. F. J. Ohkawa, *J. Phys. Soc. Jpn.* **56** (1987) 713.
27. G. M. Eliashberg, *Pisma JETP* **45** (1987) 28.
28. C. M. Varma, *Comments Solid State Phys.* **11** (1985) 221.
29. C. M. Varma, *Bull. Am. Phys. Soc.* **29** (1984) 404; *Phys. Rev. Lett.* **55** (1985) 2723.
30. K. Miyake, S. Schmitt-Rink and C. M. Varma, *Phys. Rev.* **B34** (1986) 6554.
31. K. Miyake, *Magn. Mat.* **63-64** (1987) 411.
32. P. W. Anderson, *Phys. Rev.* **B30** (1984) 1549.
33. J. E. Hirsch, *Phys. Rev. Lett.* **54** (1985) 1317.
34. G. E. Volovik and L. P. Gor'kov, *JETP Lett.* **39** (1984) 674; *Sov. Phys. JETP* **61** (1985) 843.
35. H. Jichu, A. D. S. Nagi, B. Jin, T. Matsuura and Y. Kurada, *Phys. Rev.* **35** (1987) 1692.
36. A. Yoshimori and H. Kasai, *J. Magn. Mat.* **31-34** (1983) 475.
37. M. Tachiki and S. Maekawa, *Phys. Rev.* **B29** (1984) 2497.
38. T. Matsuura, K. Miyake, H. Jichu and Y. Kuroda, *Prog. Theor. Phys.* **72** (1984) 402.
39. K. Miyake, T. Matsuura, H. Jichu and Y. Nagaoka, *Prog. Theor. Phys.* **72** (1984) 1063.
40. F. J. Okhawa and H. Fukuyama, *J. Phys. Soc. Jpn.* **53** (1984) 4344.
41. K. Machida and M. Kato, *Phys. Rev. Lett.* **58** (1987) 1986.
42. M. Kato and K. Machida, *J. Phys. Soc. Jpn.* **56** (1987) 2136.
43. J. G. Bednorz and K. A. Müller, *Z. Phys.* **B64** (1986) 188.
44. J. E. Hirch and D. J. Scalapino, *Phys. Rev. Lett.* **56** (1986) 2732.
45. L. F. Mattheiss, *Phys. Rev. Lett.* **58** (1987) 1028.
46. J. Xu, A. J. Freeman and J. H. Xu, *Phys. Rev. Lett.* **58** (1987) 1035.
47. P. W. Anderson, *Science* **235** (1987) 1196.
48. P. W. Baskaran, Z. Zou and P. W. Anderson, *Solid State Commun.* **63** (1987) 857.
49. P. W. Anderson, G. Baskaran, Z. Zou and T. Hsu, *Phys. Rev. Lett.* **58** (1987) 2790.
50. T. M. Rice, *Z. Phys.* **B67** (1987) 141.
51. P. Prelovsek, T. M. Rice and F. C. Zhang, (preprint ETH-1986).
52. P. Anderson, *Phys. Rev. Lett.* **34** (1975) 953.
53. B. K. Chakraverty and C. Schlenker, *J. Phys. (Paris)* **37** (1976) C4–353.
54. B. K. Chakraverty, *J. Phys. Lett. (Paris)* **40** (1979) L99.
55. S. Lakkis, C. Schlenter, B. K. Chakraverty, R. Buder and R. Marezio, *Phys. Rev.* **B14** (1976) 1429.
56. B. Chakraverty, M. J. Sienko and J. Bonnerot, *Phys. Rev.* **17** (1978) 3781.
57. A. Alexandrov and J. Ranninger, *Phys. Rev.* **B23** (1981) 1796.
58. A. Alexandrov and J. Ranninger, *Phys. Rev.* **B24** (1981) 1164.

59. A. Alexandrov, J. Ranninger and S. Robaszkiewicz, *Phys. Rev.* **B33** (1986) 4526.
60. I. G. Lang and Yu. A. Firsov, *Sov. Phys. JETP* **60** (1984) 856.
61. S. Robaszkiewicz, R. Micnas and K. A. Chao, *Phys. Rev.* **B23** (1981) 1447; **B24** (1981) 1579.
62. J. Ranninger and S. Robaskiewicz, *Physica* **135B** (1985) 468.
63. K. Nasu, *Phys. Rev.* **B35** (1987) 1748.
64. C. S. Ting, D. N. Talwar and L. Nagi, *Phys. Rev. Lett.* **45** (1980) 1213.
65. C. S. Ting and D. Y. Xing, *Proc. Int. Workshop on "Novel Mechanisms of Superconductivity"* Berkeley (1987).
66. J. E. Hirch, *Phys. Rev. Lett.* **54** (1985) 1317.
67. H. B. Schüttler, M. Jarrell and D. J. Scalapino, *Phys. Rev. Lett.* **58** (1987) 1147.
68. Y. Kuramoto and T. Watanabe, *Solid State Commun.* **63** (1987) 821.
69. V. J. Emery, *Phys. Rev. Lett.* **26** (1987) 2794.
70. M. Inoue, T. Takemori and K. Ohtaka, *Solid State Commun.* **63** (1987) 201.
71. P. A. Lee and N. Read, *Phys. Rev. Lett.* **58** (1987) 2691.
72. K. Machida and M. Kato, *Jpn. J. Appl. Phys.* **26** (1987) L660.
73. K. Machida and Kato, *Phys. Rev. Lett.* **58** (1987) 1986.
74. Yu. V. Kopaev and A. I. Rusinov, *Phys. Lett.* **A121** (1987) 300.
75. C. M. Varma, S. Schmitt-Rink and E. Abrahams, *Solid State Commun.* **62** (1987) 681.
76. E. Abrahams, S. Schmitt-Rink and C. M. Varma *"Yamada Conference on Superconductivity in Highly Correlated Fermion Systems"* (Sendai, 1987).
77. F. J. Ohkawa, *J. Phys. Soc. Jpn.* **56** (1987) 2617.
78. I. Morgenstern, K. A. Müller and J. G. Bednorz (to be published in *Z. Phys.* **B**).
79. P. W. Anderson and Z. Zou, *Phys. Rev. Lett.* **60** (1988) 132.
80. A. Oguri and S. Maekawa, *Physica* **C156** (1988) 679.
81. J. M. Wheatley, T. C. Hsu and P. W. Anderson, *Phys. Rev.* **B37** (1988) 5897.
82. T. Schneider and D. Baeriswyl, *Z. Phys.* **B7** (1988) 5.
83. J. Birman and J. P. Lu, *Mod. Phys. Lett.* **B11-12** (1988) 1297.
84. M. Prohammer, *Physica* **C157** (1989) 4.
85. C. P. Enz, *Helv. Phys. Acta* **62** (1989) 122.
86. J. E. Hirch, *Phys. Lett.* **A451** (1989).

INDEX